Lecture Notes in Computer Science

Lecture Notes in Bioinformatics 14971

The series Lecture Notes in Bioinformatics (LNBI) was established in 2003 as a topical subseries of LNCS devoted to bioinformatics and computational biology.

The series publishes state-of-the-art research results at a high level. As with the LNCS mother series, the mission of the series is to serve the international R & D community by providing an invaluable service, mainly focused on the publication of conference and workshop proceedings and postproceedings.

Roberta Gori · Paolo Milazzo · Mirco Tribastone
Editors

Computational Methods in Systems Biology

22nd International Conference, CMSB 2024
Pisa, Italy, September 16–18, 2024
Proceedings

Springer

Editors
Roberta Gori
University of Pisa
Pisa, Italy

Paolo Milazzo
Università di Pisa
Pisa, Italy

Mirco Tribastone
IMT School for Advanced Studies Lucca
Lucca, Italy

ISSN 0302-9743 ISSN 1611-3349 (electronic)
Lecture Notes in Bioinformatics
ISBN 978-3-031-71670-6 ISBN 978-3-031-71671-3 (eBook)
https://doi.org/10.1007/978-3-031-71671-3

LNCS Sublibrary: SL8 – Bioinformatics

This Springer imprint is published by the registered company Springer Nature Switzerland AG
The registered company address is: Gewerbestrasse 11, 6330 Cham, Switzerland

Preface

This volume contains the papers presented at CMSB 2024: 22nd International Conference on Computational Methods in Systems Biology, held on September 16–18, 2024 in Pisa. The conference website can be found at the following address: https://biomedinfo.di.unipi.it/cmsb2024/.

The CMSB annual conference series was initiated in 2003. The previous editions of the conference were held in Rovereto, Italy (2003), Paris, France (2004, 2011), Edinburgh, UK (2005, 2007), Trento, Italy (2006, 2010), Rostock, Germany (2008), Bologna, Italy (2009), London, UK (2012), Klosterneuburg, Austria (2013), Manchester, UK (2014), Nantes, France (2015), Cambridge, UK (2016), Darmstadt, Germany, (2017), Brno, Czech Republic (2018), Trieste, Italy (2019), Konstanz, Germany (2020, online), Bordeaux, France (2021, hybrid), Bucharest, Romania (2022, hybrid), and Luxembourg City, Luxembourg (2023).

The conference brings together researchers from across the biological, mathematical, computational, and physical sciences who are interested in the modeling, simulation, analysis, inference, design, and control of biological systems. It covers the broad field of computational methods and tools in systems and synthetic biology and their applications. Topics of interest to the conference include, but are not limited to, methods and tools for biological systems analysis, modeling, and simulation; high-performance methods for computational systems biology; identification of biological systems; applications of machine learning; network modeling, analysis, and inference; automated parameter and model synthesis; model integration and biological databases; multiscale modeling and analysis methods; design, analysis, and verification methods for synthetic biology; methods for biomolecular computing and engineered molecular devices; data-based approaches for systems and synthetic biology; optimality and control of biological systems; modeling, analysis, and control of microbial communities. The conference welcomes new theoretical results with potential applications to systems and synthetic biology, as well as novel applications and case studies of existing methods, tools, or frameworks.

This year, there were 23 submissions of regular and tool papers. Each submission was reviewed by 3 program committee members. The committee decided to accept 16 papers. The program also includes 4 invited talks, by Juliana Bowles (University of St. Andrews, UK), Madalena Chaves (Inria, Université Côte d'Azur, France), Karoline Faust (KU Leuven, Belgium), and Corrado Priami (University of Pisa, Italy), and 1 invited tutorial by Irina Balaur (University of Luxembourg, Luxembourg). The abstracts of the invited talks and tutorial are included in the front matter of this volume. Finally, we received 9 extended abstracts for poster presentations and 10 extended abstracts for highlight talks not included in this volume. The conference followed a single-blind review process managed through the EasyChair platform.

The CMSB 2024 conference was co-located with the 4th International Workshop on Reaction Systems, providing participants with the opportunity to engage in both events and enrich their experience through interdisciplinary interactions. The workshop was organized in collaboration with Moreno Falaschi (University of Siena, Italy), Linda Brodo (University of Sassari, Italy), and Roberto Bruni (University of Pisa, Italy). The workshop program included a keynote tutorial by Ion Petre (University of Turku, Finland) and a keynote talk by Maciej Koutny (Newcastle University, UK), and 9 regular talks. The abstracts of the keynote talk and tutorial are included in the front matter of this volume.

We would like to thank all the members of the Program Committee and our reviewers for their thoughtful and diligent work. The organization of CMSB 2024 was supported by the Next Generation EU project MEDICA (PRIN 2022, CUP: I53D23003720006) and by the project "Hub multidisciplinare e interregionale di ricerca e sperimentazione clinica per il contrasto alle pandemie e all'antibioticoresistenza (PAN-HUB)" funded by the Italian Ministry of Health (POS 2014–2020, project ID: T4-AN-07, CUP: I53C22001300001). The Workshop on Reaction Systems was supported by the Master on Bioinformatics and Data Science of the University of Siena. Finally, the organization of CMSB 2024 was supported by the University of Pisa, and in particular by the Department of Computer Science.

July 2024

Roberta Gori
Paolo Milazzo
Mirco Tribastone

Organization

Program Committee

Alona Ben-Tal	University of Auckland, New Zealand
Daniela Besozzi	University of Milano-Bicocca, Italy
Luca Bortolussi	University of Trieste, Italy
Linda Brodo	University of Sassari, Italy
Roberto Bruni	University of Pisa, Italy
Luca Cardelli	University of Oxford, UK
Milan Ceska	Brno University of Technology, Czech Republic
Eugenio Cinquemani	Inria Grenoble-Rhône-Alpes, France
David Doty	University of California, Davis, USA
François Fages	Inria Saclay, France
Jérôme Feret	École normale supérieure, France
Roberta Gori (Co-chair)	University of Pisa, Italy
Ashutosh Gupta	Indian Institute of Technology Bombay, India
Monika Heiner	Brandenburg Technical University Cottbus-Senftenberg, Germany
Jane Hillston	University of Edinburgh, UK
Paolo Milazzo	University of Pisa, Italy
Pedro T. Monteiro	University of Lisbon, Portugal
Jun Pang	University of Luxembourg, Luxembourg
Loïc Paulevé	CNRS/LaBRI Bordeaux, France
Andrei Paun	University of Bucharest, Romania
Sabine Peres	University of Lyon, France
Ion Petre	University of Turku, Finland
Tatjana Petrov	University of Konstanz, Germany
Gleb Pogudin	École Polytechnique, France
David Šafránek	Masaryk University, Czech Republic
Albin Salazar	École normale supérieure, France
Carolyn Talcott	SRI International, USA
Mirco Tribastone (Co-chair)	IMT School for Advanced Studies Lucca, Italy
Max Tschaikowski	Aalborg University, Denmark
Adelinde Uhrmacher	University of Rostock, Germany
Andrea Vandin	Sant'Anna School of Advanced Studies Pisa, Italy
Verena Wolf	Saarland University, Germany
Christoph Zechner	Max Planck Institute of Molecular Cell Biology and Genetics, Germany

Additional Reviewers

Moreno Falaschi
Gerrit Grossmann
Fiete Haack
Felix Jahn
Justin N. Kreikemeyer
Julien Martinelli
Francesca Randone
Kerian Thuillier
Athénaïs Vaginay

Organizing Committee

Linda Brodo	University of Sassari, Italy
Roberto Bruni	University of Pisa, Italy
Martina Cirinciani	University of Pisa, Italy
Moreno Falaschi	University of Siena, Italy
Silvia Giulia Galfré	University of Pisa, Italy
Roberta Gori (Co-chair)	University of Pisa, Italy
Paolo Milazzo (Co-chair)	University of Pisa, Italy
Giang Pham	University of Pisa, Italy
Mirco Tribastone (Co-chair)	IMT School for Advanced Studies Lucca, Italy

Abstracts of Invited Talks and Tutorials

CMSB Invited Talks

A Reasoning-Based Framework for Safe Prescribing in the Elderly

Juliana Bowles

University of St. Andrews, UK

Abstract. Medication errors are an extremely common cause of patient harm worldwide and yet mostly preventable. In particular, the therapeutic effectiveness and safety of drugs over time is a significant concern, especially for patients over 65. It is the case that age-related physiology changes renal elimination and liver metabolism, and affects pharmacokinetics and pharmacodynamics. Moreover, factors such as physiologic changes, decline in functional abilities, reduced homoeostatic mechanisms, diseases associated with ageing and frailty, as well as polypharmacy (when 5 or more medications are taken) are risks associated with developing clinically significant drug-drug interactions. For these reasons, a group of medications known as Potentially Inappropriate Medications (PIMs) can be dangerous in the elderly and should be avoided. However, tackling PIMs effectively in practice is challenging, time-consuming and error-prone since the criteria underlying the definition of PIMs are complex and subject to frequent updates. This talk presents a comprehensive automated framework intended for a hospital setting which combines formal rules to represent PIMs criteria and inference engines to detect, solve or minimise undesirable side-effects of PIMs.

Circadian Clocks: Modeling and Characterization of Cycle Dynamics

Madalena Chaves

Inria, Université Côte d'Azur, France

Abstract. Many living organisms have evolved a circadian timing system to synchronize their physiology and behavior with the Earth's light and dark cycle. The core mechanism underlying circadian rhythms is a gene regulatory network consisting of about 5 genes whose expressions oscillate with a period around 24 hours. A large variety of mathematical models have been proposed to describe circadian rhythms: but how to compare their respective properties, cycle dynamics, or agreement with biological data? Focusing on the mammalian circadian clock, we developed methods to analyze and systematically characterize the oscillatory dynamics. We propose a segmentation of the circadian cycle into eight stages based on three of the core clock proteins, which provides a benchmark for comparison and validation of circadian cycles generated by a large range of models.

Tools for Microbial Network Analysis and Community Modelling

Karoline Faust

KU Leuven, Belgium

Abstract. Microbial communities play important roles in a number of applications, ranging from health and food production to waste water treatment and bioremediation. To engineer microbial communities, we need to know how community members interact with each other and their environment. However, it is hard to systematically elucidate these interactions experimentally. Methods to infer interactions from microbial sequencing data have received much attention, but evaluations have shown them to be of low accuracy. In the first part of my talk, I will focus on how to extract biological hypotheses from noisy microbial networks. In this context, I will present a number of tools we developed for this task, including manta, anuran, mako and microbetag. While microbial networks are useful for explorative studies, quantitative models are required to design consortia or to predict targeted interventions to shift communities towards desired compositions. I will present a few modelling approaches commonly applied to microbial communities, which we implemented in the miaSim R package, and illustrate their limitations on data we collected for synthetic gut communities grown in controlled conditions. Finally, I will discuss possible next steps.

Tools for Microbial Network Analysis and Community Modelling

Digital Health Opportunities

Corrado Priami

University of Pisa, Italy

Abstract. Digital health is a broad term that covers telemedicine (interpreted as distance monitoring of patients and digital interaction with healthcare providers); wearable technologies to collect physiological parameters in real time; electronic health records to have access everywhere to the clinical history of patients; digital therapies and computational development of therapies; health system engineering to optimize the operations and costs of healthcare systems to enhance the quality of the service; analytics to make medicine predictive and preventive by exploiting the data to develop biomarkers and research hypotheses. Digital health is becoming pervasive in the scientific community and in clinical practice. It spans different market sectors like pharmaceutical, medical, biotechnological and even telemedicine and patient-centric care. Digital health is characterized by highly multi- and interdisciplinary approaches and its scientific foundations lie in many different disciplines including bioinformatics, computation and systems biology, artificial intelligence, cybersecurity, nutrition, medicine and medical engineering, robotics, simulation, sensor networks, ethics and law, and economics. The talk will explore where the community of CMSB can have an impact and make a difference by also presenting some examples of how digital health can provide added value at the intersection of physicians and patients.

Graph Databases for Integration, Exploration and Analysis of Complex Data

Irina Balaur

University of Luxembourg, Luxembourg

Abstract. Graph databases (GDBs) are becoming increasingly popular across scientific disciplines (including finance, physics, biomedicine and social network analysis), being highly suitable to store, query and integrate complex heterogeneous data. In this tutorial, I will introduce GDB approaches for integration, exploration and analysis of data describing complex systems. In a GDB, data is represented using a graph: an object or a concept is represented as a node, and relationships between objects are represented as edges. To ensure a common foundation, I will start with a summary of relational databases, then I will detail two of the most frequently used GDB paradigms: i) the Resource Description Framework (RDF), which is a World Wide Web Consortium (W3C: w3.org/TR/rdf-concepts) standard vocabulary term used to describe resources and their relationships in the form of triples, and ii) the labelled property graph model (e.g., https://www.neo4j.com), where properties of both data and inter-relationships can be stored as attributes of the nodes/edges of the graph. I will outline relevant tools and analytical methods (including path finding, connectivity analysis and subgraph mining) and I will provide practical examples of the use of GDBs in systems biomedicine, such as in the study of neurodegenerative and cardiovascular diseases.

Graph Databases for Integration, Exploration and Analysis of Complex Data

Workshop on Reaction Systems Invited Talks

Mining Reactions in Reaction Systems with Discrete Concentrations

Maciej Koutny

Newcastle University, UK

Abstract. A fundamental feature of Reaction Systems is that all possible interactions of biochemical reactions are based on the mechanisms of facilitation and inhibition, i.e., the products (entities) of reactions may facilitate or inhibit each other. Another key feature of the Reaction System model is that it represents reactions, states, and dynamic processes using (tuples of) finite sets of entities (or molecules), and so it directly captures the qualitative aspects of systems, rather than their quantitative aspects. A situation often encountered in practical applications is that a reaction system with discrete concentrations is only partially specified, e.g., it is not known what the actual concentration of a reactant or inhibitor should be. In such cases, the possibility of an effective automated derivation of the missing details would provide an attractive design approach. To show how such an idea might work, this talk will discuss parametric reaction systems with parameters representing under-specified components of hypothetical reactions. The key development to be covered by the talk is a method for replacing the parameters in such a way that the resulting Reaction System operating in a specified external environment satisfies a given temporal logic formula.

Tutorial on Reaction Systems. A Set-Theoretical Modeling Framework Based on Facilitation, Inhibition, and Interaction with the Environment

Ion Petre

University of Turku, Finland

Abstract. Reaction systems (RS) is a model of computation inspired by the functioning of the living cell. Its main focus is on reactions and facilitation/inhibition dynamical inter-play between reactions. A reaction is described through its reactants, its inhibitors, and its products. The products of a reaction may introduce inhibitors to another reaction, temporarily blocking its triggering. This leads to a novel way of describing dynamical systems that allows an explicit trace of WHY a certain property emerges in the system. It is a flexible model of computation that allows both a qualitative, as well as a quantitative view on a system, a set-theoretical framework for computation, and the flexibility to model intricate biological behavior.

This tutorial offers a basic introduction to reaction systems, with no prerequisites needed except computational maturity. It introduces the basic notions of reaction system and reviews a number of research directions motivated by biological considerations. We discuss some of the unique features of a reaction systems: its explicit description of inhibition, its approach to competing on resources, its non-permanency philosophy, and the continuous interplay with the environment. We demonstrate the descriptive power of reaction systems through a number of examples, culminating in an RS-based model of the receptor tyrosine kinase signaling network in breast cancer.

Contents

Process Calculi and Rewriting Techniques for Analyzing Reaction Systems

Demis Ballis[1], Linda Brodo[2], Moreno Falaschi[3], and Carlos Olarte[4(✉)]

[1] DMIF, University of Udine, Udine, Italy
[2] Dipartimento di Scienze economiche e aziendali, Università di Sassari, Sassari, Italy
[3] Dipartimento di Ingegneria dell'Informazione e Scienze Matematiche, Università di Siena, Siena, Italy
[4] LIPN, CNRS UMR 7030, Université Sorbonne Paris Nord, Villetaneuse, France
`olarte@lipn.univ-paris13.fr`

Abstract. Reaction Systems (RSs) are a successful computational framework inspired by biological systems, where entities are produced by reactions and can be used to enable or inhibit other reactions. In RSs interaction with the environment is modeled by a linear sequence of sets of entities called context. In this paper, we present a new interaction language, called **ccReact** for implementing and specifying RSs. **ccReact** extends the classic RS computational model by allowing the specification of possibly recursive, nondeterministic conditional context sequences, enhancing the interactive capabilities of the models. We provide a rewriting logic semantics for **ccReact**, making it executable in the Maude system. This enables the simulation and the verification of (reachability) properties, and model checking temporal (LTL and CTL) formulas for our extended RSs. We analyze a breast cancer case study and show that our model checking analyses can in some cases improve the administration of monoclonal antibodies therapeutic treatments.

Keywords: reaction systems · model checking · SOS · rewriting logic · Maude

1 Introduction

Reaction Systems (RSs) [18] are a computational framework inspired by systems of living cells. Since their introduction, RSs have shown to be a general computational model with several applications ranging from the modeling of biological phenomena [7,8,12,23], and molecular chemistry [33] to theoretical foundations of computing [25,26]. A RS is built from a finite set of entities and a finite set

Research supported by the Next Generation EU project PNRR ECS00000017 "THE - Tuscany Health Ecosystem" (Spoke 3 - CUP B63C22000680007), Next-GenerationEU projects MEDICA (PRIN 2022, CUP_B53D23013170006), and DELICE (PRIN PNRR 2022, P20223T2MF).

R. Gori et al. (Eds.): CMSB 2024, LNBI 14971, pp. 1–18, 2024.
https://doi.org/10.1007/978-3-031-71671-3_1

of reactions acting on entities. A reaction is a triple (R, I, P) where R is the set of *reactants* (entities whose presence is needed to enable the reaction), I is the set of *inhibitors* (entities whose absence is needed to enable the reaction) and P is the set of *products* (entities that are produced if the reaction takes place).

The behavior of a RS is typically specified as a rewrite system that models *processes interacting* with an external *context*. The external context specifies the entities provided by the environment at each step of the process. The current system state is determined by the union of the entities coming from the environment with those produced in the previous state. A state transition is then determined by applying all and only the enabled reactions in the current state.

The specification of RSs may be an error-prone activity, since system states typically contain a large number of entities which might be deeply interconnected due to the form of the reactions in play. In this scenario, the verification of RSs is a fundamental albeit nontrivial task which requires a proper automated support.

Rewriting Logic (RL) [31] is a very general *logical* and *semantic framework* that is particularly suitable for modeling and analyzing complex, nondeterministic systems. Rewriting Logic is efficiently implemented in the high-performance language Maude [24], which seamlessly integrates functional, logic, concurrent, and object-oriented computations. The specification unit in RL is a *rewrite theory* $\mathcal{R}$, that combines a set of rewrite rules R modeling the concurrent transitions of a system, and an equational theory $\mathcal{E}$, representing system states as terms of an algebraic datatype.

Contributions. This paper presents a new interaction language, called **ccReact**, for specifying RSs. The language is equipped with an *executable* rewriting logic semantics, suitable for both simulation and verification. **ccReact** extends the classic RS computational model by allowing the specification of context sequences that model external stimuli coming from the environment. It exploits process algebraic operators (such as action prefix, sum and recursion) and offers the possibility to express conditional statements. The advantage of having a language for specifying contexts is twofold. First, we can recursively define contexts that possibly exhibit complex nondeterministic behaviors in a precise way, as sometimes have already appeared in the literature [20,21,28]; second, we are able to represent a collection of experiments within a single semantic object, so that the consequences of some variation in the context sequence can then be more easily compared and analyzed. Our approach thus overcomes the usual limitation of *in silico* experiments, where a single context sequence is synthesized and one can only observe the outcomes for that specific input without the possibility of analyzing context changes in real time.

The semantics of **ccReact** is formalized within the RL framework, thus making available for it the rich set of Maude analysis tools. Our rewrite theory can be directly used to: (1) simulate biological experiments; (ii) perform reachability analysis; and (iii) model-check CTL and LTL temporal properties. The usefulness of our approach is evaluated on a biological case study in [27] that analyses the behavior of the protein signaling networks for the HER2-positive breast cancer subtype in the presence of different combinations of monoclonal antibody drugs. The authors in [27] aim at achieving the best drug treatment for three different breast cancer representative cell lines: BT474, SKBR3 and HCC1954. Our analyses not only corroborate previous results in the literature concerning this system, but also allow us to validate new hypotheses.

Organization. Section 2 recalls the basics of RSs and introduces the case study used as a running example. Section 3 first introduces the syntax for **ccReact** and then its operational semantics, and we prove that it adequately captures the behavior of RS interactive processes (Theorems 1 and 2). After a brief overview on RL, Sect. 4 provides a RL characterization of the operational semantics of **ccReact**. In Sect. 5, we show how our analysis methods can be used to validate already known and new hypotheses about the case study. Section 6 reviews related work and Sect. 7 concludes the paper. The complete Maude specification and case studies are available in the companion tool [11].

2 Reaction Systems

In this section we recall some basic notions about RSs that are relevant to this work. For a full discussion refer to [18]. We also introduce our case study.

We use the term *entities* to denote generic molecular substances (e.g., atoms, ions, molecules) that may be present in the states of a biochemical system. The main mechanisms that regulate the functioning of a living cell are *facilitation* and *inhibition*. These mechanisms are based on the presence and absence of entities and are reflected in the basic definitions of RSs.

Definition 1 (Reaction Systems). *Let S be a (finite) set of entities. A reaction in S is a triple $a = (R, I, P)$, where $R, I, P \subseteq S$ are finite sets and $R \cap I = \emptyset$. We use $rac(S)$ to denote the set of all reactions on the set of entities S. A Reaction System (RS) is a pair $\mathcal{A} = (S, A)$ s.t. S is a finite set of entities, and $A \subseteq rac(S)$ is a finite set of reactions.*

The sets R, I, P are the sets of *reactants*, *inhibitors*, and *products*, respectively. Due to biological considerations, the sets R and P are usually nonempty: there is no creation from nothing ($R \neq \emptyset$), and if a reaction takes place then something is produced ($P \neq \emptyset$). A reaction can take place whenever all of its reactants are present in a given state while none of its inhibitors is present. If this happens, the reaction is enabled and creates its products. More formally,

Definition 2 (Reaction Result). *Given a (finite) set of entities S, and a subset $W \subseteq S$, we define the following:*

1. *Let $a = (R, I, P) \in rac(S)$ be a reaction in S. The result of reaction a on W, denoted by $res_a(W)$, is defined by: $res_a(W) \triangleq \begin{cases} P & \text{if } en_a(W) \\ \emptyset & \text{otherwise} \end{cases}$ where the* enabling *predicate is defined by $en_a(W) \triangleq (R \subseteq W) \wedge (I \cap W = \emptyset)$.*
2. *Let $A \subseteq rac(S)$ be a finite set of reactions. The result of A on W, denoted by $res_A(W)$, is defined by: $res_A(W) \triangleq \bigcup_{a \in A} res_a(W)$.*

RSs are based on three assumptions: **no permanency**, i.e., any entity vanishes unless it is sustained by a reaction; **no counting**, the basic model of RSs is very abstract and qualitative, i.e. the quantity of entities that are present in a cell is not considered; **threshold nature of resources**, it is assumed that either an entity is available for all reactions, or it is not available at all.

Since living cells are seen as open systems that interact with the external environment, the behavior of a RS is formalized through the notion of *interactive process*, that is, a process that may react to external stimuli.

Definition 3 (Interactive Process). *Let* $\mathcal{A} = (S, A)$ *be a RS and let* $n \geq 0$*. An n-step* interactive process *in* $\mathcal{A}$ *is a pair* $\pi = (\gamma, \delta)$ *s.t.* $\gamma = \{C_i\}_{i \in [0,n]}$ *is the* context sequence *and* $\delta = \{D_i\}_{i \in [0,n]}$ *is the* result sequence*, where* $C_i, D_i \subseteq S$ *for any* $i \in [0, n]$*,* $D_0 = \emptyset$*, and* $D_{i+1} = res_{\mathcal{A}}(D_i \cup C_i)$ *for any* $i \in [0, n-1]$*. We call* $\tau = W_0, \ldots, W_n$ *with* $W_i \triangleq C_i \cup D_i$*, for any* $i \in [0, n]$ *the* state sequence.

The context sequence γ represents the environment interacting with the RS. It specifies the *input* that comes from the environment. The result sequence δ is entirely determined by γ and the reactions in A. As expected, distinct context sequences may lead to distinct outcomes for the same reaction set A, but a context sequence γ determines a *unique* result sequence δ.

Case Study. We consider the case study in [27] concerning the protein signaling networks in breast cancer and focus on a particular subtype of breast cancer: the HER2-positive subtype. This cancer subtype is characterized by an over expression of the human epidermal growth factor receptor 2 (HER2, also named ErbB-2). This is a receptor tyrosine kinase (RTK), and it is part of the epidermal growth factor (EGF) receptor family also composed by ErbB-1, ErbB-3, and ErbB-4. These four receptors form different homo- or hetero-dimers via growth factor activation (for example EGF for ErbB-1 or heregulin, HRG, for ErbB-3), that activate a signaling pathway that can contribute to the breast cancer tumor genesis and progression. The therapeutic treatment is typically based on monoclonal antibody drugs. Unfortunately, the therapy is often weakened by drug resistance that can be caused by the pathway deregulatory activity or by the ErbB-family receptors that allow to bypass the drug activity.

This case study focuses on a treatment where different drug combinations are tested to improve drug efficiency. The experimental analyses in [27] consider: three different kinds of breast cancer, identified by the cell lines BT474, HCC1954, and SKBR3; three types of drugs: erlotinib (e), pertuzumab (p), trastuzumab (t), provided alone or in combination; and two types of input stimulus: EGF and HRG. Input stimuli enable the activation of the EGF receptors and start the signaling pathways to be analyzed. Experiments in [27] aims at finding specific *attractors* in steady states, that is, entities that are key in the tumor genesis and progression which are present in a stationary system configuration. The search is carried out w.r.t. a persistent stimulus that includes both EGF and HRG and different drug combinations. Both short- and long-term experiments have been conducted using two distinct classes of attractors: attractors

AKT, ERK1/2, p70S6K (short-term analysis), and attractors RPS6, RB (long-term analysis). Table 1 summarizes the results of [27] for the six considered systems (one per kind of cancer cell line, short and long term versions). Each row of the table presents the efficacy of a drug combination under the same external stimulus for the three kinds of cancer (x stands for no drug treatment). The treatment is effective when attractors are not present in a steady state (cell value is equal to 0). In [27], a qualitative model, based on boolean networks, is used to describe the systems under consideration. To generate our reactions for the six systems in [27], we follow the translation method in [13] which allow Boolean networks to be directly converted into reaction sets. The resulting RSs we obtain for the six systems in Table 1, from left to right, include 39, 39, 34, 47, 50 and 55 reaction rules respectively.

Table 1. Short-term and long-term experiments from [27].

	Short-term									Long-term					
	BT474			HCC1954			SKBR3			BT474		HCC1954		SKBR3	
	AKT	ERK1/2	p70S6K	AKT	ERK1/2	p70S6K	AKT	ERK1/2	p70S6K	RPS6	RB	RPS6	RB	RPS6	RB
x	1	1	1	1	1	1	1	1	1	1	0	1	0	1	0
e	0	1	1	1	0	1	0	0	0	0	0	0	0	0	0
p	1	1	1	0	0	0	0	0	0	0	0	0	0	0	0
t	1	1	1	1	1	1	1	1	1	1	0	1	0	1	0
e,p	1	1	1	1	0	1	0	0	0	0	0	0	0	0	0
e,t	0	1	1	1	0	1	0	0	0	0	0	0	0	0	0
p,t	1	1	1	0	0	0	0	0	0	0	0	0	0	0	0
e,p,t	1	1	1	1	0	1	0	0	0	0	0	0	0	0	0

Below we use some of the involved entities and reactions to illustrate the above definitions.

Example 1. Let $\mathcal{A} = (S, A)$ be the RS where $S = \{\mathsf{egf}, \mathsf{e}, \mathsf{p}, \mathsf{erbb1}, \mathsf{erk12}, \mathsf{p70s6k}\}$, and $A = \{a_1, a_2, a_3\}$ with $a_1 = (\{\mathsf{egf}\}, \{\mathsf{e}, \mathsf{p}\}, \{\mathsf{erbb1}\})$, $a_2 = (\{\mathsf{egf}\}, \emptyset, \{\mathsf{erk12}\})$, $a_3 = (\{\mathsf{erk12}\}, \emptyset, \{\mathsf{p70s6k}\})$. Consider the context sequence $\gamma = C_0, \cdots, C_3$, with $C_0 = \{\mathsf{egf}, \mathsf{e}\}$, $C_1 = \emptyset$, $C_2 = \{\mathsf{egf}\}$, and $C_3 = \emptyset$. Then, $\pi = (\gamma, \delta)$ in an interactive process where $\delta = D_0, \cdots, D_3$, with $D_0 = \emptyset$, $D_1 = \{\mathsf{erk12}\}$, $D_2 = \{\mathsf{p70s6k}\}$, and $D_3 = \{\mathsf{erbb1}, \mathsf{erk12}\}$. The resulting state sequence is $\tau = W_0, \cdots, W_3$, where $W_0 = \{\mathsf{egf}, \mathsf{e}\}$, $W_1 = \{\mathsf{erk12}\}$, $W_2 = \{\mathsf{egf}, \mathsf{p70s6k}\}$, and $W_3 = \{\mathsf{erbb1}, \mathsf{erk12}\}$.

3 ccReact: A Language for Executing and Analyzing RSs

Drawing inspiration from process algebras, this section introduces the syntax and the operational semantics for **ccReact**, a language of processes that allows for the definition of external contexts interacting with a RS. The purpose of **ccReact** processes is to provide stimuli to the RS in the form of a context C, i.e., a subset of the set of entities *controlled* by the environment. Our language includes constructors to model the situation in *in silico* experiments where different alternative inputs need to be tested and some of them may depend on the current state of the system. This opens up the possibility of designing new experiments as shown later in Sect. 5. Next we define the syntax of **ccReact**:

Definition 4 (ccReact Processes). *The language of processes is defined as:*

$$
\begin{array}{lll}
c & ::= e \mid \overline{e} \mid c \wedge c \mid c \vee c & \textit{(Conditions)} \\
p & ::= C \mid c \rightarrow C & \textit{(Prefixes)} \\
K, M, N & ::= \mathbf{0} \mid X \mid \sum_{i \in I} p_i.K_i \mid K \parallel M \mid \mathbf{rec}\ X.K & \textit{(Processes)}
\end{array}
$$

where e is an entity, X is a process variable, I is a finite nonempty set of indices, and $C \subseteq S$ is a possibly empty set of entities. The set of processes is denoted by **Proc** *and the set of prefixes by* **Pref**. *Processes are quotiented by a structural congruence relation ≡ satisfying: (1) $K \equiv M$ if they differ only by a renaming of local variables in recursive definitions (alpha-conversion); (2) $K \parallel M \equiv M \parallel K$; (3) $K \parallel (M \parallel N) \equiv (K \parallel M) \parallel N$; and (4) $K \parallel \mathbf{0} \equiv \mathbf{0} \parallel K \equiv K$.*

Let $D \subseteq S$ be a set of entities. The *prefix* C is always *enabled* in D and it *produces* the set of entities C. The prefix $c \quad \rightarrow \quad C$ is *enabled* if condition c holds in D. This happens when the function $check(c, D)$ evaluates to true where: $check(e, D) \triangleq e \in D$, $check(\overline{e}, D) \triangleq e \notin D$, and $check(c \bullet c', D) \triangleq check(c) \bullet check(c')$ with $\bullet \in \{\wedge, \vee\}$. For instance, the prefix $e_1 \wedge e_2 \wedge \overline{e_3} \wedge \overline{e_4} \rightarrow d$ is enabled in D if $\{e_1, e_2\} \subseteq D$ and $\{e_3, e_4\} \cap D = \emptyset$ (i.e., e_1, e_2 are present and e_3, e_4 absent). In that case, the entity d is produced.

The process $\mathbf{0}$ represents the end of a context sequence. The process X represents the *call* to a recursively defined process. The process $\sum_{i \in I} p_i.K_i$ nondeterministically chooses one of the enabled prefixes p_i and precludes the others from execution. To simplify the notation: when the set of indices I is a singleton, we omit the "$\sum_{i \in I}$" part and we write $p.K$; if $|I| = 2$, we write $p_1.K_1 + p_2.K_2$.

The parallel composition of two processes is represented by $K \parallel M$, where the outputs of K and M are merged together.

The process **rec** $X.K$ allows for recursive definitions. We assume that all occurrences of X in **rec** $X.K$ appear in the scope of a prefix (e.g., **rec** $X.p.X$ is guarded while **rec** $X.X$ is not). Moreover, all the occurrences of X in a process K must be bound by the binder "**rec** X." (i.e., there are no free variables in a well-formed process). These conditions are usually required in process algebras to guarantee, e.g., that the resulting transition system is finitely branching.

To avoid parentheses, the convention is that $\wedge$ binds tighter than $\vee$, and, in decreasing order of precedence, we have recursion, prefix, sums and parallel composition (e.g., **rec** $X.K \parallel p_1.K_1 + p_2.K_2$ must be understood as (**rec** $X.K$) $\parallel$ $(p_1.K_1 + p_2.K_2)$). Moreover, we write $!p$ instead of **rec** $X.p.X$, i.e., the unbounded repetition of the prefix p.

Operational Semantics. Now we define the input/output behavior of **ccReact** processes when interacting with a reaction system. For that, we consider three reduction relations:

1. $p \overset{D}{\rightsquigarrow} C$, determining that the prefix p produces the output C when the current state is D;
2. $K \xrightarrow{(D,C)} K'$, determining that the context specified by the process K reduces to K' on input D. On doing that, it produces C; and
3. $(D, K) \xRightarrow{(C)} (D', K')$, capturing the interaction of K with the RS where the process K outputs C on D. After that, the new *configuration* is (D', K'), where the state of the RS is D' and the context is provided by K'.

The next definition formalizes the relations in (1) and (2) above.

Definition 5 (Semantics of processes). $\rightsquigarrow$: **Pref** $\times S \times S$ *and* $\longrightarrow$: **Proc** $\times S \times S \times$ **Proc** *are the least relations satisfying, respectively, the rules* prefix *and* prefix-c*, and the rules* sum*,* sum$_\emptyset$*,* parallel *and* rec*, in Fig. 1.*

$$\frac{check(c,D) = true}{(c \rightarrow C) \overset{D}{\rightsquigarrow} C}\ \textsf{prefix-c} \qquad \frac{}{C \overset{D}{\rightsquigarrow} C}\ \textsf{prefix} \qquad \frac{p_j \overset{D}{\rightsquigarrow} C_j \quad \text{for some } j \in I}{\sum_I p_i.K_i \xrightarrow{(D,C_j)} K_j}\ \textsf{sum}$$

$$\frac{p_j \overset{D}{\not\rightsquigarrow} \quad \text{for all } j \in I}{\sum_I p_i.K_i \xrightarrow{(D,\emptyset)} \mathbf{0}}\ \textsf{sum}_\emptyset \qquad \frac{K \xrightarrow{(D,C_K)} K' \text{ and } M \xrightarrow{(D,C_M)} M'}{K \parallel M \xrightarrow{(D,C_K \cup C_M)} K' \parallel M'}\ \textsf{parallel}$$

$$\frac{K[\mathbf{rec}\ X.K/X] \xrightarrow{(D,C)} K'}{\mathbf{rec}\ X.K \xrightarrow{(D,C)} K'}\ \textsf{rec} \qquad \frac{K \xrightarrow{(D,C)} K' \text{ and } D' = res_{\mathcal{A}}(D \cup C)}{(D,K) \xRightarrow{(C)} (D',K')}\ \textsf{output}$$

Fig. 1. Input/output behavior of **ccReact** processes

The rule prefix produces the output C on input D when the prefix p is either C, or p is of the form $c \rightarrow C$ and c holds in D. The rule sum selects for execution one of the enabled prefixes and rules out the others. If none of the prefixes is enabled, then the empty set of entities is produced (rule sum$_\emptyset$). The rule parallel combines the outputs of K and M in the parallel composition $K \parallel M$. The rule rec unfolds the definition whenever the resulting process produces some output. The process $\mathbf{0}$ does not exhibit any transition (no rule for it), since it simply terminates a given process and halts the associated production of contexts.

The interaction of a process K with a reaction system $\mathcal{A}$ is a possibly infinite sequence of contexts produced by K together with the output (i.e. the result sets) generated by $\mathcal{A}$ (under the contexts produced by K).

Definition 6 (Interaction). *Let* $\mathcal{A} = (S, A)$ *be a RS.* $\Longrightarrow\ : S \times$ **Proc** $\times S \times S \times$ **Proc** *is the least relation satisfying the rule* output *in Fig. 1. A run* ρ *of* $K = K_0$ *on* $\mathcal{A}$ *is a sequence of the form* $(D_0, K_0) \xRightarrow{(C_0)} (D_1, K_1) \xRightarrow{(C_1)} \cdots$ *where* $D_0 = \emptyset$*. Moreover: the context sequence associated to* ρ*, notation* $\gamma(\rho)$*, is the sequence* $\{C_i\}_{i \geq 0}$*, the result sequence* $\delta(\rho)$ *is the sequence* $\{D_i\}_{i \geq 0}$ *and the state sequence associated to* ρ *is the sequence* $\{C_i \cup D_i\}_{i \geq 0}$*.*

Some **ccReact** processes may engage in an infinite sequence of interactions with a RS (take e.g., **rec** $X.C.X$). Given an infinite sequence s, we use $s \downarrow n$ to denote the first $n+1$ elements of s. Given a context sequence $\gamma = \{C_i\}_{i \in [0,n]}$, we use $P(\gamma)$ to denote the **ccReact** process $C_0.\cdots.C_n.\mathbf{0}$. Note that, given an arbitrary RS $\mathcal{A}$, there exists a unique run of length n of $P(\gamma)$ on $\mathcal{A}$.

The following result shows that n-step interactive processes (Definition 3) coincide with the associated state sequences of **ccReact** processes (Definition 6). This follows directly from the definition of the rule output (that coincides with the definition of D_i in Definition 3).

Theorem 1. *Let $\mathcal{A} = (S, A)$ be a RS. (1) Let $\gamma = \{C_i\}_{i \in [0,n]}$ be a context sequence, and ρ the unique run of $P(\gamma)$ on $\mathcal{A}$. Then, $\pi = (\gamma(\rho), \delta(\rho))$ is an n-step interactive process in $\mathcal{A}$; and (2) let K be a process and ρ a run of K on $\mathcal{A}$. If ρ is finite with length n, then $\pi = (\gamma(\rho), \delta(\rho))$ is an n-step interactive process in $\mathcal{A}$. If ρ is infinite, then for all $n \geq 0$, $\pi = (\gamma(\rho) \downarrow n\ ,\ \delta(\rho) \downarrow n)$ is an n-step interactive process in $\mathcal{A}$.*

4 Rewriting Logic Semantics for ccReact

This section gives a rewriting logic (RL) [30,31] semantics to **ccReact**. The proposed rewrite theory is executable in the Maude [22,24] system. This opens up the possibility of performing different analyses as shown later in Sect. 5. For presentation, we omit some details. The complete Maude specification can be found in the companion tool of this paper [11]. We start with a quick overview of RL and Maude (see [24,31] for further details).

Overview of Rewriting Logic and Maude. A *rewrite theory* [30] is a tuple $\mathcal{R} = (\Sigma, E, L, R)$ such that: (Σ, E) is an equational theory where Σ is a signature that declares sorts, subsorts, and function symbols; E is a set of (conditional) equations of the form $t = t' \mathbf{if} \psi$, where t and t' are terms of the same sort, and ψ is a conjunction of equations; L is a set of *labels*; and R is a set of labeled (conditional) rewrite rules of the form $l : q \longrightarrow r \mathbf{if} \psi$, where $l \in L$ is a label, q and r are terms of the same sort, and ψ is a conjunction of equations and *rewrite expressions* of the form $t \Rightarrow t'$. This latter expression is interpreted as a rewriting-based reachability goal in $\mathcal{R}$.

$T_{\Sigma,s}$ denotes the set of ground terms of sort s, and $T_\Sigma(X)_s$ denotes the set of terms of sort s over a set of sorted variables X. $T_\Sigma(X)$ and T_Σ denote all terms and ground terms, respectively. A substitution $\sigma : X \rightarrow T_\Sigma(X)$ maps each variable to a term of the same sort, and $t\sigma$ denotes the term obtained by simultaneously replacing each variable x in a term t with $\sigma(x)$.

A *one-step rewrite* $t \longrightarrow_{\mathcal{R}} t'$ holds if there is a rule $l : q \longrightarrow r \mathbf{if} \psi$, a subterm u of t, and a substitution σ such that $u = q\sigma$ (modulo equations), t' is the term obtained from t by replacing u with $r\sigma$, and $v\sigma = v'\sigma$ holds for each $v = v'$ in ψ. The reflexive-transitive closure of $\longrightarrow_{\mathcal{R}}$ is denoted as $\longrightarrow^*_{\mathcal{R}}$. Intuitively, terms in the equational theory (Σ, E) define system states, while the behavior of the system is given by local transitions between states described by rewrite rules.

Maude [22] is a language and tool supporting the specification and analysis of rewrite theories. A Maude module (`mod` M `is` ... `endm`) specifies a rewrite theory $\mathcal{R}$. Sorts and subsort relations are declared by the keywords `sort` and `subsort`; function symbols, or *operators*, are introduced with the `op` keyword: `op` f `:` s_1 ... s_n `->` s, where $s_1, \ldots, s_n$ are the sorts of the arguments of operator f, and s is its (value) sort. Operators can have user-definable syntax, with underbars '`_`' marking each of the argument positions (e.g., `_+_`). Some operators can have equational attributes, such as `assoc`, `comm`, and `id:` t, stating that the operator is, respectively, associative, commutative, and/or has identity element t. Equations are specified with the syntax `eq` t `=` t' or `ceq` t `=` t' `if` ψ; and rewrite rules as `rl [l] : u => v` or `crl [l] : u => v if ψ`. The mathematical variables in such statements are declared with the notations `var` and `vars` or, on the fly, with the notation `X:S` (variable `X` of sort `S`).

Maude provides a large set of analysis methods, including simulation by rewriting (`rew` t, that rewrites the term t) and explicit-state reachability analysis (`search` t `=>*` t' `such that` ψ, searching for a state reachable from t that matches the pattern t' and satisfies the condition ψ).

Reaction Systems as Maude's Terms. Entities in a RS are constants of `sort Entity`. For instance, `ops egf e p erbb1 : -> Entity` declares some of the entities in Example 1. The sort `SetEntity` represents "`,`"-separated sets of entities ("`,`" is an `associative`, `commutative` operator with `empty` as `identity`, and satisfies the equation for idempotency `eq X , X = X`). For instance, "`e , p`" is a term of sort `SetEntity`. Sets of reactions can be then specified as "`,`"-separated sets of terms of `sort Reaction`, built with the following constructor:

```
op [_;_;_] : SetEntity SetEntity SetEntity -> Reaction .
```

For instance, the reaction a_1 in Example 1 can be represented with the term `[egf ; (e , p) ; erbb1]` of sort `Reaction`.

ccReact processes are defined by the following operators:

```
sorts Condition Process PreProcess SumProcess .
subsort Qid < Process .          --- For processes variables
subsort PreProcess < SumProcess < Process .
subsort Entity < Condition .     --- An entity is a condition
op _^ : Entity -> Condition .    --- An entity that must be absent
op tt : -> Condition .           --- True condition
ops _and_ _or_ : Condition Condition -> Condition .
op {_}._   : SetEntity Process                      -> PreProcess .
op '[ _-->{_} '] . _   : Condition SetEntity Process -> PreProcess .
op 0       :                        -> Process .
op _+_     : PreProcess SumProcess -> SumProcess .
op _||_     : Process Process       -> Process [assoc comm id: 0] .
op rec_._  : Qid Process           -> Process .
```

The sort `Qid` in Maude denotes quoted identifiers as in `'X`. They are used here to denote process variables. Terms of sort `PreProcess` are *prefixed* processes of the form $p.K$ where $p \in \mathbf{Pref}$ and $K \in \mathbf{Proc}$. Summations $\sum_{i \in I} p_i.K_i$ are represented as terms of the sort `SumProcess`. For instance, the process **rec** $X.(\mathsf{ERBB1}.X + (\mathsf{e} \wedge \overline{\mathsf{p}} \rightarrow \mathsf{PLCG}).X)$ that either provides as input ERBB1 or PLCG (when e is present and p absent) is represented by the following term:

```
rec 'X . (erbb1 . 'X  +  [ e and p ^  --> { plcg } ] . 'X).
```

ccReact Semantics in Maude. The behavior of a **ccReact** process interacting with a RS is specified by: (1) a *deterministic* part, using equations, for checking *conditions* and computing the effects on the RS when a set of entities C is provided as input context; and (2) rewrite rules defining the non-deterministic behavior of **ccReact** processes.

The following specification defines the operator `eval`, that recursively computes (`evalRec`) the output D (a set of entities) from a set of reactions $SetR$ when the environment provides as stimulus the set of entities C:

```
vars C D R I P IN OUT : SetEntity . vars SetR : SetReaction .
op eval    : SetEntity SetReaction           -> SetEntity .
op evalRec : SetEntity SetReaction SetEntity -> SetEntity .
eq eval(C, SetR) = eval(C, SetR, empty) .
eq evalRec(C, empty, D) = D .
eq evalRec(C, ([R; I; P], SetR), D) =
   evalRec(C, SetR, (D, output(C, [R; I; P]))) .
op output : SetEntity Reaction -> SetEntity .
eq output(C, [R ; I ; P]) = --- Returns P if [R;I;P] is firable on C
 if (R subset C) and-then intersection(C,I) == empty then P else empty fi
```

As expected, the definition of `evalRec` checks every reaction against the current input C. A reaction (R, I, P) adds to the output D the products P, when the reactants R needed are a `subset` of the input C and $C \cap I = \emptyset$ (definition of `output`). In this specification, `if_then_else_fi` is the usual ternary operator.

Similarly, the operator `op check : Condition SetEntity -> Bool` is defined so that to check when a condition is true in a given state, as well as the operator `op firable : SetEntity SumProcess -> Bool` checks whether at least one of the prefixes in a summation is enabled.

States and Rules. We consider the following sorts and operators:

```
sorts PState IOState .      subsort PState IOState < State .
op { process:_ ; input:_ }  : Process SetEntity -> PState .
op { next:_; input:_; output:_ } : Process SetEntity SetEntity -> IOState
```

Terms of sort `PState` are used to capture the evolution of a process during the current interaction with the RS (relations $\rightsquigarrow$ and $\longrightarrow$). Terms of sort `IOState` specify the output of the system in the current interaction (relation $\Longrightarrow$). Note that both `PState` and `IOState` are subsorts of `State` which is defined in Maude's model-checker prelude [9] to represent the states of the transition system induced by a rewrite theory. Adding theses subsort relations allows us to seamlessly endow **ccReact** with verification capabilities (see Sect. 5).

A process $p.K$ (sort `PreProcess`) produces an output by using the union of the entities in the current input and those generated by the prefix p:

```
rl [prefix] : { process: ( { C } . K)  ; input: IN } =>
              { next: K ; input: IN ; output: eval((C, IN), reactions)} .
rl [prefixC] : { process: ( [ COND --> { C } ] . K)  ; input: IN } =>
   if check(COND, IN)
    then { next: K  ; input: IN ; output: eval( (C , IN), reactions) }
    else { next: 0  ; input: IN ; output: eval(      IN, reactions) } fi
```

In these rules, `reactions` is the set of reactions of the RS under consideration. The rule `prefix` produces (function `eval`) the output using the prefix C and the current input (set `(C , IN)`). In `prefixC`, if the condition $COND$ holds, the output is generated using the union of the entities in the current input and C. Otherwise, the output depends only on the current input ($\mathsf{sum}_\emptyset$ in Fig. 1).

In a summation $M = p.K + KS$, where $KS = \sum_{i \in I} p_i.K_i$, all the enabled (if any) prefixes must be considered. Hence, if p is enabled (i.e., `firable` on `IN`), the process $p.K$ can be scheduled for execution (rule `choiceL` below). Moreover, if there exists at least one enabled prefix in the continuation KS, KS is also considered for execution (rule `choiceR`).

```
crl [choiceL] : {process: Kp + KS ; input: IN} =>
    {process: Kp ; input: IN} if firable(IN, Kp) .
crl [choiceR] : {process: Kp + KS ; input: IN} =>
    {process: KS ; input: IN} if firable(IN, KS) .
```

The behavior of the parallel composition $K_1 \parallel K_2$ is specified by the rewrite rule `parallel` which exploits rewrite expressions in its condition to compute the outputs of both K_1 and K_2 (namely, `OUT1` and `OUT2`). Then, it combines `OUT1` and `OUT2` together by using the set union operator "`_,_`":

```
vars K1 K2 : Process . vars OUT1 OUT2 : SetEntity .
crl [parallel] : {process: K1 || K2 ; input: IN} =>
                 {next: K1' || K2' ; input: IN ; output: (OUT1 , OUT2)}
 if {process: K1; input: IN} => {next: K1'; input: IN ; output: OUT1} /\
    {process: K2 ; input: IN} => {next: K2'; input: IN ; output: OUT2} .
```

For recursive processes, the rule below unfolds the definition of the process with the aid of the function `subst` that computes the substitution $P[\mathbf{rec}\ X.P/X]$:

```
rl [rec] : { process: (rec X . K)              ; input: IN } =>
           { process: subst(X, (rec X . K), K) ; input: IN } .
```

Finally, the following rule allows us to observe the execution of the system across time-instants, where the entities produced in the current time-instant (i.e. `OUT`) are used as inputs in the next one:

```
rl [next] : {next: K; input: IN; output: OUT} => {process: K; input: OUT}
```

Adequacy. The representation of entities, reactions and processes as terms of the appropriate sorts is quite natural and allows for easily defining maps $[\![\cdot]\!]$ between RS components and their Maude counterparts, e.g., from sets of entities to terms of sort `SetEntity`, reactions to terms of sort `Reaction`, etc. In the following, given any syntactic entity ξ (entity, reaction, etc.), we shall use $[\![\xi]\!]$ to denote the corresponding term of the appropriate sort.

Theorem 2 (Adequacy). *For all processes K, K' and sets of entities C, D,* $(D, K) \xRightarrow{(C)} (D', K')$ *iff* $\{\mathtt{process}:\ [\![\mathtt{K}]\!]\ ;\ \mathtt{input}:\ [\![\mathtt{D}]\!]\} \longrightarrow^*_{\mathcal{R}}$

$$\{\mathtt{next}: [\![\mathtt{K'}]\!]\ ;\ \mathtt{input}: [\![\mathtt{D}]\!]\ ; \mathtt{output}:\ [\![\mathtt{C}]\!]\} \longrightarrow_{\mathcal{R}} \{\mathtt{process}:\ [\![\mathtt{K'}]\!]\ ;\ \mathtt{input}:\ [\![\mathtt{D'}]\!]\}.$$

5 Case Studies and Analyses

In this section we show how different analysis methods including simulation by rewriting, reachability analysis, and LTL/CTL model checking are available within our framework. We showcase the use of **ccReact** for the analysis of a RS that models the case study presented in Sect. 2. For this system, we verify results already found in the literature [27], and verify new hypotheses. It is worth noting that **ccReact** shows excellent performance in all the considered experiments (e.g., model-checking the properties under examination takes few milliseconds). However, further investigation is needed to benchmark system scalability on more complex RSs.

Simulation and State Reachability. Our Maude specification provides an executable model for **ccReact** that can be directly used to simulate interactive processes of RSs. This can be achieved by using the built-in Maude command `rew [n]` that performs a rewrite sequence involving `n` rewrite steps. For instance, consider the RS in Example 1. The following specification declares the entities and input context to represent this system.

```
ops egf  e  p  erbb1  erk12  p70s6k : -> Entity .
eq reactions = [(egf) ; (e ,p) ; (erbb1)] ,
   [(egf) ; ( empty ) ; (erk12)], [(erk12) ; ( empty ) ; (p70s6k)] .
op proc : -> Process . eq proc = {(egf,e)}.{empty}.{egf}.{empty}.0 .
op init : -> IOState  .
eq init = { next: proc ; input: empty  ; output: empty } .
```

Note that the **ccReact** process `proc` specifies the context sequence $\gamma = C_0, \ldots, C_3$ of the interactive process $\pi = (\gamma, \delta)$ in Example 1. The initial state `init` includes `proc`, the `input` representing D_0 (i.e., the first element of the result sequence δ) and the `output` field which is initially set to empty. Now, the following `rew` command generates a two-step rewrite sequence[1] that rewrites `init` and exactly simulates the computation of the next result set D_1 of δ. Indeed, the `output` in the resulting state is `erk12` as in Example 1.

```
Maude> rew [2] init . --- 2 steps of rewriting
result IOState: {next: {empty}. {egf}. ...; input: empty; output: erk12}
```

More interesting, we can search for all the `IOStates` reachable from `init`, thus obtaining the full result sequence $\delta = D_0, \ldots, D_3$ of π [2]:

```
Maude> search init =>* S:IOState .
Solution 1 (state 0)
S:IOState --> {next: {(egf, e)}. ... ; input: empty ; output: empty}
Solution 2 (state 2)
S:IOState --> {next: {empty}. {egf}. ... ; input: empty ; output: erk12}
```

[1] Some outputs are simplified (...) and comments (---) are added to explain the output. The complete specification and outputs are available in [11].

[2] Maude outputs an "extra" configuration D_4 (Solution 5), that corresponds to $D_4 = res_A(D_3 \cup C_3)$. "(`state n`)" is the number of the state in the search tree.

```
...
Solution 5 (state 8)
S:IOState --> {next: 0 ; input: erbb1, erk12 ; output: p70s6k}
No more solutions.
```

The Maude search capability can be used to answer more complex reachability queries. For instance, we can fully reproduce the results in [27] (see Table 1). We start by defining a predicate `isState` to check whether the steady state has been reached. In Maude, predicates (or atomic propositions) are terms of the sort `Prop` whose meaning is defined by the operator `op _|=_ : State Prop -> Bool`, specifying when a state satisfies the proposition. We then identify steady states by inspecting the `input` and `output` fields of any `IOState` as follows:

```
op isSteady : -> Prop . --- Is it a steady state?
eq { next: K ; input: IN ; output: OUT } |= isSteady   =   IN == OUT .
eq STATE:PState |= isSteady = false .
```

Note that the predicate `isState` (and similarly for other predicates introduced below) does not hold in states of sort `PState`. Recall that such states are only instrumental to decompose **ccReact** process and no output is really produced in them.

We also define the following predicates:

```
subsort SetEntity < Prop . --- Sets of entities as propositions
eq STATE:PState |= C = false .
eq { next: K ; input: IN ; output: OUT } |= C = C subset OUT .
op io-state : -> Prop .  --- Is it an IOState?
eq STATE:PState  |= io-state  = false .
eq STATE:IOState |= io-state  = true .
```

Due to the subsort relation `SetEntity` < `Prop` and its associated equations, a set of entities C is also an atomic proposition that holds in a `IOState` S iff C is a `subset` of the `OUTput` in S.

The results in Table 1 can be obtained by checking the presence of the attractors (e.g., AKT), when considering a **ccReact** process of the form ! (S, D). This process provides a context sequence including the stimulus S ={EGF,HRG} together with the drugs, e.g., D = {e, t}. For instance, for the BT474 cancer, the combination of erlotinib and trastuzumab is effective, whereas BT474 is resistant to the treatment with the drugs erlotinib and pertuzumab:

```
Maude> search [1] { process: ! (hrg, egf, e, t) ; input: empty } =>*
       STATE:IOState such that STATE:IOState |= (akt)  /\
                               STATE:IOState |= isSteady .
No solution.
Maude> search [1] { process: ! (hrg, egf, e, p) ...  } =>*  ...
Solution 1 (state 8)
STATE:IOState --> {... output: akt, p70s6k, mtor, plcg, erk12, pkca}
```

Model-Checking. The Unified Maude Model-Checking tool (`umaudemc`) [34] is a general interface to different model checkers for Maude models. Here we show

how `umaudemc` can be used to formally verify LTL as well as CTL temporal formulas against our models for HER2-positive cancers.

LTL and CTL temporal formulas are build from atomic `Proposition`, Boolean connectives and the usual temporal operators. In the case of LTL, we have $\square$ (always), $\Diamond$ (eventually), $\circ$ (next), **U** (until) and **R** (release). These operators are quantified in CTL over the paths of the computation tree (there **E**xists or for **A**ll paths). Temporal logic formulas can then be model-checked using the command `check` of `umaudemc` whose parameters include the file with the Maude specification, the initial state (a term of sort `State`) and the temporal formula.

The results of [27] that we reproduced using the `search` commands above can alternatively be obtained via model-checking. More concretely the presence/absence of the attractor AKT in steady states of the BT747 cell line can be checked by feeding `umaudemc` with the LTL formula $\Diamond(\square(\texttt{io-state} \rightarrow \texttt{akt}))$ and an initial state that specifies the stimulus and drug combination to be considered:

```
$> umaudemc check BT474 "{ process: ! (hrg, egf, e, t) ; input: empty }"
   "<> [] (io-state ->  akt)" --- Stimulus: HRG+EGF Drug combination: e+t

The property is not satisfied in the initial state ... counterexample ...
```

Using temporal formulas, we can go beyond the results in [27] and test other hypotheses. For instance, we can check that it is always the case that once the growth factor receptor ErbB-2 is present in a time-instant t, then it is also present in the time-instant $t+1$, thus highlighting a persistent activity of the ErbB-2 receptor across the pathway. We cannot model check directly the formula $\square(\texttt{erbb2} \rightarrow \circ\texttt{erbb2})$ since the *next* state of a term of sort `IOState` is a term of sort `PState`. This can be remediated by model checking, instead, the formula $\square(\texttt{erbb2} \rightarrow \circ(\neg\texttt{io-state}\ \mathbf{U}\ \texttt{erbb2}))$, where `erbb2` holds in the next `IOState` (after a sequence of `PStates`).

We can use CTL to check the behavior of a drug treatment on different runs. For instance, in order to observe the interactions when either erlotinib or pertuzumab are supplied, it suffices to consider the process `hrg,egf,e)(hrg, egf,e).(rec 'X . (e .'X) + (p .'X))`. For this process, Maude reports that the formula $\mathbf{E}\diamond(\texttt{isSteady} \wedge \neg\texttt{akt})$ is satisfied (there **E**xists at least one path where that treatment is successful) but $\mathbf{A}\diamond(\texttt{isSteady} \wedge \neg\texttt{akt})$ is not (not **A**ll paths avoid a steady state where AKT is present).

We can also check properties about the transient states, describing the evolution of the RS before reaching the steady state. Consider e.g., the process

```
! (hrg, egf)) || (rec 'X . ( ({e} . 'X) + ({ p } . 'X) + ({ t } . 'X) )))
```

Maude returns that, for BT474, the following two formulas are satisfied: $\square(\texttt{pdk1} \rightarrow (\diamond\square(\texttt{io-state} \rightarrow \texttt{akt})))$ and $\texttt{erbb1}\ \mathbf{R}\ \neg\texttt{pdk1}$. This (positively) answers the following questions: (i) regardless the drug used, once PDK1 is present, inevitably the steady state includes AKT; and (ii) PDK1 never appears before ErbB-1 is produced (which basically means that PDK1 is a product of the activation of the ErbB-1 receptor).

The results for the short-term experiments in [27] show that by *permanently* providing the drug erlotinib and the stimulus (EGF and HRG), the attractor AKT is never produced. By using **ccReact** conditional prefixes we can ask whether the drug needs to be provided in *all* the time-instants or can be provided only when an EGF receptor is activated. For that, consider the following process:

```
rec 'X . ([ erbb1   or  erbb2   --> { e } ]      . 'X) +
         ([ erbb1 ^ and erbb2 ^ --> { empty } ] . 'X)
```

Maude checks that the property $\mathbf{E}\diamond\mathbf{E}\square(\texttt{io-state} \rightarrow \neg\texttt{akt})$ is satisfied, i.e., the production of AKT can be also inhibited by providing erlotinib only when receptor ErbB-1 and ErbB-2 are active.

6 Related Work

There are some frameworks implemented in Maude that can be compared to ours. [10] provides a Maude formalization of RSs which is suitable for trace slicing, a technique used to simplify computation traces. The methodology in [10] needs context sequences to be manually given by the user, whereas our approach uses processes to generate (conditional as well as nondeterministic) contexts, automatically.

Pathway Logic (PL) [35] is a Maude framework that allows signaling pathways to be simulated and formally verified. However, PL is limited to the analysis of cell models, while our framework is more abstract and can be applied to multiple application domains (even beyond the biological context).

As for the formal verification of RSs, [29] introduces a CTL logic specifically tailored to context-restricted RSs, i.e., a computation model that generalizes standard RSs by controlling their context sequences. [32] proposes an LTL logic for a version of RSs with discrete concentrations, where collections of biological entities are modeled as multisets. Differently to our work, [29,32] use ad-hoc implemented model checkers, while we verify specifications by means of the standard model checkers. In [11] we have reproduced some of the experiments in [29,32] within our framework.

Nondeterministic contexts have already appeared in the literature [20,21,28], but here we are able to represent a collection of experiments within a single semantic object, which is easier to analyze and model-check. On the other hand, the RS semantics in [20,21] is compositional and hence well-suited to modular analysis, while our semantics is not, since we handle RSs as monolithic objects.

7 Conclusions and Future Work

Modeling and analyzing biological phenomena are complex tasks which require the support of adequate languages and tools. In this paper, we presented **ccReact**, a language of processes that formally captures the dynamic behavior of RSs. **ccReact** is endowed with a rich process algebra that allows for the definition of multiple external contexts that can drive the execution of a given RS w.r.t. different environment stimuli. Our language has been implemented in Maude, a rewriting-based language that offers both fast performance and an easy interconnection with several analysis tools. To this respect, we show how Maude built-in reachability capability and the Unified Maude Model-Checking tool can be directly employed to efficiently check properties related to resistance of different combinations of therapeutics in the HER2-positive breast cancer subtype.

As future work we want to consider other case studies, such as the rare disease Alkaptonuria [17], for which we are interested to analyze the evolution of the diseases under different treatments. We want also to investigate a possible integration of static analysis and slicing techniques [3–6, 14–16] within our framework, to study an extension to a distributed system [19], and the extension of the set of entities in the reactions with more structured values, exploiting some ideas from [1,2].

References

1. Alpuente, M., Ballis, D., Cuenca-Ortega, A., Escobar, S., Meseguer, J.: ACUOS^2: a high-performance system for modular ACU generalization with subtyping and inheritance. In: Calimeri, F., Leone, N., Manna, M. (eds.) JELIA 2019. LNCS (LNAI), vol. 11468, pp. 171–181. Springer, Cham (2019). https://doi.org/10.1007/978-3-030-19570-0_11
2. Alpuente, M., Ballis, D., Escobar, S., Sapiña, J.: Variant-based equational anti-unification. In: Villanueva, A. (ed.) LOPSTR 2022. LNCS, vol. 13474, pp. 44–60. Springer, Cham (2022). https://doi.org/10.1007/978-3-031-16767-6_3
3. Alpuente, M., Ballis, D., Frechina, F., Sapiña, J.: Combining runtime checking and slicing to improve Maude error diagnosis. In: Martí-Oliet, N., Ölveczky, P.C., Talcott, C. (eds.) Logic, Rewriting, and Concurrency. LNCS, vol. 9200, pp. 72–96. Springer, Cham (2015). https://doi.org/10.1007/978-3-319-23165-5_3
4. Alpuente, M., Ballis, D., Frechina, F., Sapiña, J.: Assertion-based analysis via slicing with ABETS. Theory Pract. Logic Program. **16**(5–6), 515–532 (2016). https://doi.org/10.1017/S1471068416000375
5. Alpuente, M., Ballis, D., Romero, D.: A rewriting logic approach to the formal specification and verification of web applications. Sci. Comput. Program. **81**, 79–107 (2014). https://doi.org/10.1016/j.scico.2013.07.014
6. Alpuente, M., Ballis, D., Sapiña, J.: Static correction of Maude programs with assertions. J. Syst. Softw. **153**, 64–85 (2019). https://doi.org/10.1016/j.jss.2019.03.061
7. Azimi, S.: Steady states of constrained reaction systems. Theor. Comput. Sci. **701**(C), 20–26 (2017). https://doi.org/10.1016/j.tcs.2017.03.047
8. Azimi, S., Iancu, B., Petre, I.: Reaction system models for the heat shock response. Fundam. Informaticae **131**(3–4), 299–312 (2014). https://doi.org/10.3233/FI-2014-1016
9. Bae, K., Meseguer, J.: The linear temporal logic of rewriting Maude model checker. In: Ölveczky, P.C. (ed.) WRLA 2010. LNCS, vol. 6381, pp. 208–225. Springer, Heidelberg (2010). https://doi.org/10.1007/978-3-642-16310-4_14
10. Ballis, D., Brodo, L., Falaschi, M.: Modeling and analyzing reaction systems in Maude. Electronics **13**(6), 1139 (2024). https://doi.org/10.3390/electronics13061139

11. Ballis, D., Brodo, L., Falaschi, M., Olarte, C.: Process calculi and rewriting techniques for analyzing reaction systems. Companion tool (2024). https://depot.lipn.univ-paris13.fr/olarte/reaction-systems-maude.git
12. Barbuti, R., Gori, R., Levi, F., Milazzo, P.: Investigating dynamic causalities in reaction systems. Theor. Comput. Sci. **623**, 114–145 (2016). https://doi.org/10.1016/j.tcs.2015.11.041
13. Barbuti, R., Gori, R., Milazzo, P.: Encoding boolean networks into reaction systems for investigating causal dependencies in gene regulation. Theor. Comput. Sci. **881**, 3–24 (2021). https://doi.org/10.1016/j.tcs.2020.07.031. Special Issue on Reaction Systems
14. Bodei, C., Brodo, L., Degano, P., Gao, H.: Detecting and preventing type flaws at static time. J. Comput. Secur. **18**(2), 229–264 (2010). https://doi.org/10.3233/JCS-2010-0361
15. Bodei, C., Brodo, L., Focardi, R.: Static evidences for attack reconstruction. In: Bodei, C., Ferrari, G.-L., Priami, C. (eds.) Programming Languages with Applications to Biology and Security. LNCS, vol. 9465, pp. 162–182. Springer, Cham (2015). https://doi.org/10.1007/978-3-319-25527-9_12
16. Bodei, C., Brodo, L., Gori, R., Levi, F., Bernini, A., Hermith, D.: A static analysis for brane calculi providing global occurrence counting information. Theor. Comput. Sci. **696**, 11–51 (2017). https://doi.org/10.1016/J.TCS.2017.07.008
17. Braconi, D., Millucci, L., Spiga, O., Santucci, A.: Cella and tissue models of alkaptonuria. Drug Discov. Today Dis. Model. **31**, 3–10 (2020). https://doi.org/10.1016/j.ddmod.2019.12.001
18. Brijder, R., Ehrenfeucht, A., Main, M., Rozenberg, G.: A tour of reaction systems. Int. J. Found. Comput. Sci. **22**(07), 1499–1517 (2011). https://doi.org/10.1142/S0129054111008842
19. Brodo, L.: On the expressiveness of the π-calculus and the mobile ambients. In: Johnson, M., Pavlovic, D. (eds.) AMAST 2010. LNCS, vol. 6486, pp. 44–59. Springer, Heidelberg (2011). https://doi.org/10.1007/978-3-642-17796-5_3
20. Brodo, L., Bruni, R., Falaschi, M.: Enhancing reaction systems: a process algebraic approach. In: Alvim, M.S., Chatzikokolakis, K., Olarte, C., Valencia, F. (eds.) The Art of Modelling Computational Systems: A Journey from Logic and Concurrency to Security and Privacy. LNCS, vol. 11760, pp. 68–85. Springer, Cham (2019). https://doi.org/10.1007/978-3-030-31175-9_5
21. Brodo, L., Bruni, R., Falaschi, M.: A logical and graphical framework for reaction systems. Theoret. Comput. Sci. **875**, 1–27 (2021). https://doi.org/10.1016/j.tcs.2021.03.024
22. Clavel, M., et al.: All About Maude - A High-Performance Logical Framework. LNCS, vol. 4350. Springer, Heidelberg (2007). https://doi.org/10.1007/978-3-540-71999-1
23. Corolli, L., Maj, C., Marinia, F., Besozzi, D., Mauri, G.: An excursion in reaction systems: from computer science to biology. Theor. Comput. Sci. **454**, 95–108 (2012). https://doi.org/10.1016/j.tcs.2012.04.003
24. Durán, F., et al.: Programming and symbolic computation in Maude. J. Log. Algebraic Methods Program. **110** (2020). https://doi.org/10.1016/J.JLAMP.2019.100497
25. Ehrenfeucht, A., Main, M.G., Rozenberg, G.: Combinatorics of life and death for reaction systems. Int. J. Found. Comput. Sci. **21**(3), 345–356 (2010). https://doi.org/10.1142/S0129054110007295

26. Ehrenfeucht, A., Main, M.G., Rozenberg, G.: Functions defined by reaction systems. Int. J. Found. Comput. Sci. **22**(1), 167–178 (2011). https://doi.org/10.1142/S0129054111007927
27. der Heyde, S.V., Bender, C., Henjes, F., Sonntag, J., Korf, U., Beißbarth, T.: Boolean ErbB network reconstructions and perturbation simulations reveal individual drug response in different breast cancer cell lines. BMC Syst. Biol. **8**(1), 75 (2014). https://doi.org/10.1186/1752-0509-8-75
28. Kleijn, J., Koutny, M., Mikulski, Ł, Rozenberg, G.: Reaction systems, transition systems, and equivalences. In: Böckenhauer, H.-J., Komm, D., Unger, W. (eds.) Adventures Between Lower Bounds and Higher Altitudes. LNCS, vol. 11011, pp. 63–84. Springer, Cham (2018). https://doi.org/10.1007/978-3-319-98355-4_5
29. Męski, A., Penczek, W., Rozenberg, G.: Model checking temporal properties of reaction systems. Inf. Sci. **313**, 22–42 (2015). https://doi.org/10.1016/j.ins.2015.03.048
30. Meseguer, J.: Conditioned rewriting logic as a united model of concurrency. Theor. Comput. Sci. **96**(1), 73–155 (1992). https://doi.org/10.1016/0304-3975(92)90182-F
31. Meseguer, J.: Twenty years of rewriting logic. J. Log. Algebraic Methods Program. **81**(7–8), 721–781 (2012). https://doi.org/10.1016/J.JLAP.2012.06.003
32. Meski, A., Koutny, M., Penczek, W.: Verification of linear-time temporal properties for reaction systems with discrete concentrations. Fundam. Informaticae **154**(1–4), 289–306 (2017). https://doi.org/10.3233/FI-2017-1567
33. Okubo, F., Yokomori, T.: The computational capability of chemical reaction automata. Nat. Comput. **15**(2), 215–224 (2016). https://doi.org/10.1007/s11047-015-9504-7
34. Rubio, R., Martí-Oliet, N., Pita, I., Verdejo, A.: Strategies, model checking and branching-time properties in maude. J. Log. Algebraic Methods Program. **123**, 100700 (2021). https://doi.org/10.1016/J.JLAMP.2021.100700
35. Talcott, C.: Pathway logic. In: Bernardo, M., Degano, P., Zavattaro, G. (eds.) SFM 2008. LNCS, vol. 5016, pp. 21–53. Springer, Heidelberg (2008). https://doi.org/10.1007/978-3-540-68894-5_2

BNClassifier: Classifying Boolean Models by Dynamic Properties

Nikola Beneš[1], Luboš Brim[1], Ondřej Huvar[1], Samuel Pastva[2], and David Šafránek[1](✉)

[1] Faculty of Informatics, Masaryk University, Brno, Czech Republic
{xbenes3,brim,safranek}@fi.muni.cz, xhuva@fi.muni.czr
[2] Institute of Science and Technology Austria (ISTA), Klosterneuburg, Austria
samuel.pastva@ist.ac.at

Abstract. Partially Specified Boolean Networks (PSBNs) represent a family of Boolean models resulting from possible interpretations of unknown update logics. Hybrid extension of CTL (HCTL) has the power to express complex dynamical phenomena, such as oscillations or stability. We present BNClassifier to classify Boolean Networks corresponding to a given PSBN according to criteria specified in HCTL. The implementation of the tool is fully symbolic (based on BDDs). The results are visualised using the machine-learning-based technology of decision trees.

1 Introduction

Boolean Networks (BNs) represent an essential qualitative modelling framework in computational systems biology [17]. A BN is given by a set of update Boolean functions. In reality, however, the modeller quite often does not know all the update functions exactly. In this paper, we provide a tool to explore how the lack of information about the update functions can influence the model dynamics.

The boolean model with (partially) unknown update functions can be seen as a set of fully specified BNs, where each model in the set represents one possible interpretation of unknown aspects. Such a set can be succinctly represented using the so-called *partially specified Boolean networks* (PSBNs) [7]. For a model-based understanding of biological reality, it is crucial to sort out or *classify* the models by diverse dynamic properties.

A proper framework for expressing dynamic properties is the hybrid extension of the branching-time logic CTL (HCTL) [9,10,12]. HCTL extends the standard CTL with hybrid operators that quantify and reference model states using state variables. This gives the logic the power to express complex phenomena, such as oscillations or stability, that would otherwise be impossible in (pure) CTL [5].

The work has been supported by the Czech Science Foundation grant No. GA22-10845S. This project has received funding from the European Union's Horizon 2020 research and innovation programme under the Marie Sklodowska-Curie Grant Agreement No. 101034413.

R. Gori et al. (Eds.): CMSB 2024, LNBI 14971, pp. 19–26, 2024.
https://doi.org/10.1007/978-3-031-71671-3_2

We present the tool BNClassifier that classifies a set of BNs given by a PSBN with respect to the properties expressed in HCTL. It is based on the *coloured HCTL model-checking method* introduced in [7, Supplement]. A distinguished feature of the classifier is that it handles the entire set of networks at once. To support large models, the implementation is fully symbolic, based on binary decision diagrams (BDD). For the visualisation part, we use the machine-learning-based technology of decision trees [4] to explore the model classes satisfying the same sets of HCTL formulas. It allows the user to comprehend the exhaustive information computed by the tool. We demonstrate the tool's main features on a model of MAPK signalling. BNClassifier (with benchmarks) is available at https://github.com/sybila/biodivine-bn-classifier under MIT licence.

Related Tools. We are not aware of any other tool that provides model classification with respect to HCTL formulas. In principle, a classifier can be built on top of an HCTL model checker that can deal with partially specified models. To the best of our knowledge, the only relevant tool is Antelope [2], which performs HCTL model checking on BNs and also works—to some extent—with PSBNs. However, there are two critical limitations: (i) the model-checking algorithm verifies the formula universally for all interpretations, i.e., it is not possible to compute only the satisfying interpretations, making it unfit for model classification; (ii) while the CTL part of the algorithm is symbolic, the computation regarding the hybrid operators is enumerative: the formulas employing the state variable quantifiers are solved by explicitly calling the labelling algorithm once for each possible state. This causes a considerable decrease in efficiency for larger models or when hybrid operators are nested.

In [3], we have presented the tool AEON for the symbolic identification of attractors in PSBNs. Subsequently, AEON has been enriched with decision trees [4] to ease the interpretation of produced results. BNClassifier is a completely novel tool that implements, for the first time, coloured HCTL model checking over PSBNs. However, in the architecture of BNClassifier we employ our general library for manipulating BDDs (`lib_bdd`) that is also used in AEON. For the GUI, we adopt the AEON-based concept of decision trees that is substantially re-implemented in BNClassifier to determine model classes satisfying the same sets of HCTL formulas. The HCTL properties decision trees visual explorer has a completely new architecture based on Tauri framework [1].

2 Model Classification Problem

Let AP be a set of atomic propositions. We say that $\mathcal{K}$ is a *Kripke structure* (KS) over AP iff $\mathcal{K} = (S, R, L)$, where $S \neq \emptyset$ is a finite set of states, $R \subseteq S \times S$ is a total transition relation, and $L : AP \rightarrow 2^S$ is a labelling assigning, to each atomic proposition, the set of states where the proposition holds. HCTL extends CTL with the notion of *state variables* and adds new operators: $\exists x.\varphi$, $\forall x.\varphi$, $\downarrow x.\varphi$, and $@_x.\varphi$. The first two are the standard quantifiers, the *down-arrow binder* $\downarrow x.\varphi$ stands for "name the current state of evaluation x and then proceed to check whether φ holds under this assumption", and the *at operator* $@_x.\varphi$ means "continue evaluating φ at the state x". In the following, we use $\mathcal{K} \models \varphi$ to mean that all states of $\mathcal{K}$ satisfy the HCTL formula φ. For more details and a formal definition of the semantics, see e.g. [12].

A *Boolean network* consists of n Boolean variables $v_1, \ldots, v_n$. Each variable is equipped with a Boolean update function that updates the variable based on the values of all the variables. We utilise the widely used *asynchronous semantics* assuming that every network variable can be updated independently of the others. Under this assumption, the semantics of a BN is a Kripke structure in which the states are the valuation of the variables (i.e., n-dimensional Boolean vectors), and each transition represents the update of a single variable. For details, see [13].

A *partially specified Boolean network* is a BN enriched with a set of function symbols that can be used in the Boolean expressions defining the update functions, thus representing the unknown parts of the update mechanism. An *interpretation* is a function I that assigns to each function symbol a Boolean function of the same arity. By fixing a particular interpretation I for a given PSBN $\mathcal{B}$, we substitute a concrete Boolean function for each function symbol in its partially specified Boolean expressions, and thus obtain a standard BN, which we denote by $\mathcal{B}_I$. We call these BNs the *instances* of a PSBN. Given a PSBN $\mathcal{B}$ and a finite set $\mathcal{I}$ of interpretations, we obtain a set of Kripke structures $\{\mathcal{K}_{\mathcal{B}_I} \mid I \in \mathcal{I}\}$. These Kripke structures share the set of states, but differ in their transition relations.

We now define our main problem. In addition to *classification* properties, we may also specify a set of *required* properties. We only consider interpretations for which all the required properties are satisfied. Our goal is to partition these interpretations based on the satisfaction of the classification properties.

Model Classification Problem. Given a PSBN $\mathcal{B}$, a set of function symbol interpretations $\mathcal{I}$, a set Ψ of HCTL formulas representing *required properties* and a (non-empty) set Φ of HCTL formulas representing *classification properties*, compute the *classification mapping class* : $2^{\Phi} \to 2^{\mathcal{I}}$ satisfying for any $X \subseteq \Phi$ the following property:

$$\forall I \in class(X).\, \mathcal{K}_{\mathcal{B}_I} \models \Big[\bigwedge_{x \in X} x \wedge \bigwedge_{y \in \Phi \setminus X} \neg y \wedge \bigwedge_{z \in \Psi} z\Big].$$

It follows from the definition that the classification mapping induces an equivalence relation on $\mathcal{I}$ that associates all interpretations that satisfy exactly the same subset of classification properties. The resulting quotient of $\mathcal{I}$ thus clusters BN models based on Φ. Since every class is unique for some particular set of formulas $X \subseteq \Phi$, we use X to denote the corresponding class.

The number of possible interpretations can be enormous and simultaneously high-dimensional. To make the classification practical, we utilise the notion of bifurcation decision trees (BDT) [6] that show the user how the interpretations of the function symbols lead to the particular classes. The BDT consists of two types of nodes. A *leaf node* represents interpretations whose corresponding PBSN instances satisfy the same combination of classification properties, i.e. a subset of a class $X \subseteq \Phi$. A *decision node* represents a choice on the interpretation of a particular function symbol. It has two outgoing edges, which lead to the subtrees where the corresponding interpretations meet the condition or not, respectively. An example of a BDT is given in Sect. 4, Fig. 2.

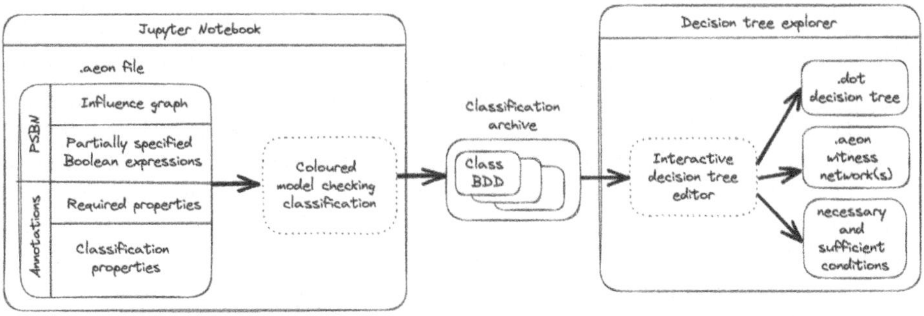

Fig. 1. Visual overview of the BNClassifier components and functionality.

3 Tool Components and Implementation

The operation of BNClassifier consists of two phases, which are depicted visually in Fig. 1. In the first phase, the user executes the model classification algorithm (typically using a Python script or a Jupyter notebook [14]) to obtain the *classification mapping* established in the previous section. In the second phase, the user constructs the BDT in an interactive manner using a desktop GUI, being able to choose which tree nodes to expand. The procedure behind this second phase is explained in more detail in [6].

The model classification phase relies internally on repeatedly calling the *coloured HCTL model-checking* subroutine, described in more detail in [7, Supplement]. This subroutine takes a PBSN $\mathcal{B}$ and a HCTL formula φ. The result is the set of interpretations I such that $\mathcal{K}_{\mathcal{B}_I} \models \varphi$. In our tool, the algorithm is implemented symbolically using binary decision diagrams (BDD).

To solve the model classification problem, the model-checking subroutine is first run with the input formula set to the conjunction of all the *required* properties ($\bigwedge_{z \in \Psi} z$). We call the resulting set of interpretations the *permissible* interpretations I_P. In further invocations of the model-checking subroutine, the interpretations of $\mathcal{B}$ are restricted to this permissible subset. Following that, the subroutine is executed once per each *classification* property $y \in \Phi$, obtaining a set of interpretations $I_y \subseteq I_P$ from each run. The classification mapping is then constructed incrementally from combinations of the valid and invalid properties using standard BDD set intersections and complements.

Technologically, the first phase is realised by our symbolic HCTL model-checking library written in Rust[1]. The library implements several state-of-the-art

[1] https://crates.io/crates/biodivine-hctl-model-checker.

optimizations. In particular, every HCTL formula is transformed into a canonical form, so that the computation of logically equivalent sub-formulas can be cached and reused. Furthermore, we use the *saturation* technique [8] to speed up the **EU** and **EF** fixed-point computations.

To make the classification workflow more accessible, all functionality is additionally available through Python bindings that can be integrated with Jupyter notebooks and other similar environments. This ensures interoperability within the larger ecosystem of systems biology tools, such as the ones maintained by the CoLoMoTo consortium [15].

The results of phase one are exported as a *classification archive*. This file contains the BDDs for all non-empty classes discovered by the classification algorithm, as well as the original PSBN@. The classification archive is then further investigated using the *decision tree explorer*.

Architecturally, the *decision tree explorer* consists of a Rust backend and an HTML/JS frontend that are packaged together as a single multi-platform desktop application using the framework Tauri [1]. This architecture maintains the flexibility and interoperability of a typical web application but also allows to perform any non-trivial computations locally within the application, without requiring a server or other distributed infrastructure.

The main element of the application is an interactive decision tree editor. Here, the user can partition the classification map based on various attributes of the PSBN interpretations into a decision tree. Such a decision tree helps to explain the relationships between classes and enables user-guided search for conditions leading to a particular class. The user can randomly sample the fully specified BNs belonging to the class. The explorer can also automatically generate a decision tree which is optimal with respect to the *information gain*, a standard metric in machine learning [16]. Finally, the user can export the resulting tree in the GraphViz `.dot` format.

4 Evaluation

For demonstration, we consider a well-known model [11] of one of the most important signalling mechanisms in human cells – the MAPK pathway. The model investigates the influence of several extracellular stimuli on cancer cell fate decisions. It consists of 18 variables. Four of these variables have update functions characterized by function symbols with zero arity. For simplicity, we refer to such variables as *inputs*. They represent the extracellular stimuli (`TGFBR_stimulus`, `EGFR_stimulus`, `FGFR3_stimulus`, and `DNA_damage`). All possible settings of these inputs result in 16 different BN instances. Another three variables represent possible modes of the cell behaviour: `Apoptosis` (programmed cell death), `Growth_Arrest` (temporary stop of growth), and `Proliferation` (cell reproduction).

We are interested in the relationship among the individual settings of the inputs and the long-term behaviour of the model (attractors). Moreover, we investigate the possible presence of biologically unrealistic behaviour to assess

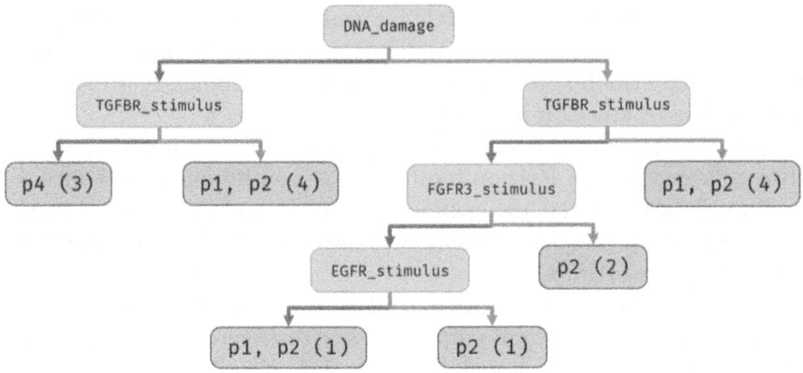

Fig. 2. Decision tree of the MAPK model produced by BNClassifier. It shows the effect of input variables on the classification properties. Each decision node contains the name of a variable used for the decision. Leaf node labels show the number of instances represented by the node (in parentheses) and the list of properties satisfied by these interpretations ($\mathtt{p_i}$ stands for φ_i).

the overall soundness of the model. To that end, we define the set of required properties $\Psi = \{\psi\}$ and the set of classification properties $\Phi = \{\varphi_1, \varphi_2, \varphi_3, \varphi_4\}$:

$\psi = \exists x.\, @_x\, ((\mathbf{AGEF}x) \wedge (\mathtt{Apoptosis} \vee \mathtt{Growth\_Arrest} \vee \mathtt{Proliferation}))$
There exists an attractor state where some cell mode is active.

$\varphi_1 = \forall x.\, @_x\, ((\mathbf{AGEF}x) \Rightarrow (\mathtt{Apoptosis} \wedge \mathbf{AX}x))$
Every attractor state is a fixed point with Apoptosis active (the system always converges to programmed cell death; *healthy* behaviour).

$\varphi_2 = \exists x.\, @_x\, (\mathbf{AG}\mathtt{Growth\_Arrest})$, $\varphi_3 = \exists x.\, @_x\, (\mathbf{AG}\mathtt{Proliferation})$
Growth_Arrest (φ_2) or Proliferation (φ_3) remain active in some attractor (programmed cell death may not be achieved; possibly *problematic* behaviour).

$\varphi_4 = \exists x.\, @_x\, ((\mathbf{AGEF}x) \wedge \mathtt{Apoptosis} \wedge \mathtt{Proliferation})$
Both Proliferation and Apoptosis are simultaneously active in an attractor state (biologically *unrealistic* behaviour).

We obtained classification results in less than two minutes, revealing several interesting model characteristics: (i) One instance (all stimuli turned off) does not satisfy the required property ψ. In this case, the model does not manifest any known biological cell mode. (ii) Nine instances satisfy φ_1, i.e. the cell is guaranteed to undergo the programmed death. All these instances also satisfy φ_2 (impeded cell growth). (iii) There is no instance where φ_3 holds (Proliferation cannot remain active indefinitely). However, (iv) three instances admit intermittent simultaneous activity of both Apoptosis and Proliferation (φ_4).

Although we found that φ_4 holds for some instances, it does not necessarily make the model unsound. Biologically unrealistic conditions can be admissible, but it is important to *understand* when and how they occur to interpret the model correctly. To do that, we construct a decision tree (Fig. 2) where we explore the conditions leading to the validity of individual properties. In particular, we see that the biologically unrealistic property φ_4 occurs if and only if both `DNA_damage` and `TGFBR_stimulus` are inactive. The decision tree also guides us towards conditions that ensure the programmed cell death. For example, we see that the presence of `TGFBR_stimulus` is a sufficient condition for φ_1.

Scalability. For presentation purposes, we intentionally demonstrate the tool on a relatively simple PSBN. Nevertheless, we also tested BNClassifier using similar properties (cyclic oscillations, bi-stable behaviour, etc.) and a variety of larger models. In particular, we were successful in obtaining results for models admitting up to 104 variables and $\sim 2^{75}$ interpretations in a manner of minutes.

References

1. Tauri apps. https://tauri.app/. Accessed 30 Sept 2023
2. Arellano, G., et al.: "Antelope": a hybrid-logic model checker for branching-time boolean GRN analysis. BMC Bioinform. **12**(1), 490 (2011)
3. Beneš, N., Brim, L., Kadlecaj, J., Pastva, S., Šafránek, D.: AEON: attractor bifurcation analysis of parametrised boolean networks. In: Lahiri, S.K., Wang, C. (eds.) CAV 2020. LNCS, vol. 12224, pp. 569–581. Springer, Cham (2020). https://doi.org/10.1007/978-3-030-53288-8_28
4. Beneš, N., Brim, L., Pastva, S., Šafránek, D.: Aeon 2021: bifurcation decision trees in boolean networks. In: Cinquemani, E., Paulevé, L. (eds.) CMSB 2021. LNCS, vol. 12881, pp. 230–237. Springer, Cham (2021). https://doi.org/10.1007/978-3-030-85633-5_14
5. Beneš, N., Brim, L., Demko, M., Pastva, S., Šafránek, D.: A model checking approach to discrete bifurcation analysis. In: Fitzgerald, J., Heitmeyer, C., Gnesi, S., Philippou, A. (eds.) FM 2016. LNCS, vol. 9995, pp. 85–101. Springer, Cham (2016). https://doi.org/10.1007/978-3-319-48989-6_6
6. Beneš, N., Brim, L., Pastva, S., Poláček, J., Šafránek, D.: Formal analysis of qualitative long-term behaviour in parametrised boolean networks. In: Ait-Ameur, Y., Qin, S. (eds.) ICFEM 2019. LNCS, vol. 11852, pp. 353–369. Springer, Cham (2019). https://doi.org/10.1007/978-3-030-32409-4_22
7. Beneš, N., Brim, L., Huvar, O., Pastva, S., Šafránek, D.: Boolean network sketches: a unifying framework for logical model inference. Bioinformatics **39**(4), btad158 (2023). https://doi.org/10.1093/bioinformatics/btad158
8. Ciardo, G., Lüttgen, G., Siminiceanu, R.: Saturation: an efficient iteration strategy for symbolic state—space generation. In: Margaria, T., Yi, W. (eds.) TACAS 2001. LNCS, vol. 2031, pp. 328–342. Springer, Heidelberg (2001). https://doi.org/10.1007/3-540-45319-9_23
9. Goranko, V.: Temporal logic with reference pointers. In: Gabbay, D.M., Ohlbach, H.J. (eds.) ICTL 1994. LNCS, vol. 827, pp. 133–148. Springer, Heidelberg (1994). https://doi.org/10.1007/BFb0013985
10. Goranko, V.: Temporal logics with reference pointers and computation tree logics. J. Appl. Non-Classical Logics **10**(3–4), 221–242 (2000)
11. Grieco, L., Calzone, L., Bernard-Pierrot, I., Radvanyi, F., Kahn-Perlès, B., Thieffry, D.: Integrative modelling of the influence of MAPK network on cancer cell fate decision. PLOS Comput. Biol. **9**(10), 1–15 (2013). https://doi.org/10.1371/journal.pcbi.1003286

12. Kernberger, D., Lange, M.: Model checking for hybrid branching-time logics. J. Logical Algebraic Methods Program. **110**, 100427 (2020)
13. Klarner, H., Streck, A., Šafránek, D., Kolčák, J., Siebert, H.: Parameter identification and model ranking of thomas networks. In: Gilbert, D., Heiner, M. (eds.) CMSB 2012. LNCS, pp. 207–226. Springer, Heidelberg (2012). https://doi.org/10.1007/978-3-642-33636-2_13
14. Kluyver, T., et al.: Jupyter notebooks–a publishing format for reproducible computational workflows. In: Positioning and Power in Academic Publishing: Players, Agents and Agendas, vol. 2016, pp. 87–90. IOS Press (2016)
15. Naldi, A., et al.: The colomoto interactive notebook: accessible and reproducible computational analyses for qualitative biological networks. Front. Physiol. **9** (2018). https://doi.org/10.3389/fphys.2018.00680
16. Quinlan, J.R.: Induction of decision trees. Mach. Learn. **1**, 81–106 (1986)
17. Schwab, J.D., Kühlwein, S.D., Ikonomi, N., Kühl, M., Kestler, H.A.: Concepts in Boolean network modeling: what do they all mean? Comput. Struct. Biotechnol. J. **18**, 571–582 (2020). https://doi.org/10.1016/j.csbj.2020.03.001

Enhancing Reaction Systems with Guards for Analysing Comorbidity Treatment Strategies

Juliana Bowles[1,2], Linda Brodo[3], Roberto Bruni[4], Moreno Falaschi[5], Roberta Gori[4], and Paolo Milazzo[4(✉)]

[1] School of Computer Science, University of St Andrews, St Andrews, UK
jkfb@st-andrews.ac.uk
[2] Software Competence Centre Hagenberg Softwarepark 32a, Hagenberg, Austria
[3] Dipartimento di Scienze Economiche e Aziendali, Università di Sassari, Via Muroni 25, Sassari, Italy
brodo@uniss.it
[4] Dipartimento di Informatica, Università di Pisa, Largo B. Pontecorvo 3, Pisa, Italy
{roberto.bruni,roberta.gori,paolo.milazzo}@unipi.it
[5] Department of Information Engineering and Mathematics, University of Siena, Via Roma 56, Siena, Italy
moreno.falaschi@unisi.it

Abstract. The coexistence of multiple medical conditions in an individual presents a complex challenge in healthcare. This research aims to enhance the analysis of comorbidity treatment plans by capturing nuanced interactions between different medical conditions and treatment interventions. Reaction systems provide a formal framework for modelling and analysing systems in which the dynamics is driven by promotion/inhibition mechanisms and external intervention of context processes. This study explores the integration of guards into context processes to increase the expressiveness of the formalism in order to analyse treatment plans for comorbidities. Through the analysis of combined clinical guidelines for atrial fibrillation and hypertension, we demonstrate the applicability and utility of this approach in optimizing patient care and improving health outcomes in the context of complex medical scenarios.

Keywords: Reaction systems · Clinical guidelines · SOS semantics

1 Introduction

Over time, it is common for people to develop long-term chronic conditions, such as diabetes, hypertension, cardiovascular diseases, chronic kidney disease,

Research supported by the Next Generation EU project PNRR ECS00000017 "THE - Tuscany Health Ecosystem" (Spoke 3 - CUP B63C22000680007, and Spoke 6 - CUP I53C22000780001), the Next Generation EU projects MEDICA (PRIN 2022, CUP_B53D23013170006 CUP_I53D23003720006), DELICE (PRIN PNRR 2022, P20223T2MF), RAP (PRIN PNRR 2022, P2022HXNSC), the INdAM GNCS project CUP_E53C22001930001, and the Austrian Science Fund (FWF) project Meitner M 3338-N.

R. Gori et al. (Eds.): CMSB 2024, LNBI 14971, pp. 27–44, 2024.
https://doi.org/10.1007/978-3-031-71671-3_3

cancer, chronic obstructive pulmonary disease, among many others. Patients diagnosed with such conditions are inevitably required to take medications as part of their treatment and/or to alleviate individual symptoms. Clinical guidelines were devised as evidence-based care plans to help manage the treatment of a wide range of health conditions [16,22]. They detail the essential steps followed when caring for patients with a specific clinical problem and play an important role in improving the healthcare provision for people with long-term conditions. These guidelines include recommendations for (group of) medications to be given at different stages of the treatment plan as well as alternatives, and are revised regularly. However, for patients with two or more chronic conditions (aka *multimorbidities*), these guidelines are insufficient and it is hard to have an oversight of the problems underlying following several care plans at the same time. Patients with multimorbidities follow several treatment plans simultaneously, which easily leads to *polypharmacy* (i.e., prescribing 5 or more medications) without clear guidance on how best to prioritise recommendations [17]. The risk of medication harm is exacerbated, that is, it is possible for patients to take medications that lead to adverse drug reactions, or for particular combinations of drugs to be less effective if administered at the same time.

Our study is motivated by the need to search for treatment paths that respect the preferences of patients with multimorbidities. This includes *looking ahead* for paths that will lead to fewer complications later on. Making the right choice as early as possible in the therapy (given the conditions of patients and the effects of their progression) can avoid worse outcomes in later stages.

In this paper, we use formal methods to model guidelines, different patients profiles and known adverse drug reactions. Our analysis intends to help doctors choose between alternative treatment options by pointing out the possible risks in choosing a particular therapy (also in the long term). As a side effect, our analysis may also identify missing conditions that should be considered in future revised guidelines (e.g., certain groups of drugs offered in one treatment should not be offered to patients with a particular comorbidity due to risky interactions between medications). To this aim, we chose to use Reaction Systems (RSs) [7,15], a computational framework that describes the dynamics of a system in terms of a set of reactions, each with some reactants, inhibitors and products. The *facilitation* and *inhibition* mechanisms inherent to RSs allow us to formalise medical guidelines in a modular way. The execution of a RS starts when an external environment, known as *context*, provides an initial set of entities, all the enabled reactions are fired and the union of their products are released. This step is iterative: next, the context provides another set of entities that are combined with the products from the previous step to fire enabled reactions again.

To model the scenario at hand, we must enhance the expressiveness of contexts so that they can supply different entities depending on those that are already present. We adopt the process algebra version of RS [9], which is easier to extend as already shown in [10,12], and we add a guarded prefix operator. Exploiting the same promotion/inhibition mechanisms of reactions, guarded contexts can check some properties of the actual state before providing new entities. Using this new functionality, we are able to encode guidelines as contexts that are driven by patient profiles, their conditions, and their current medications.

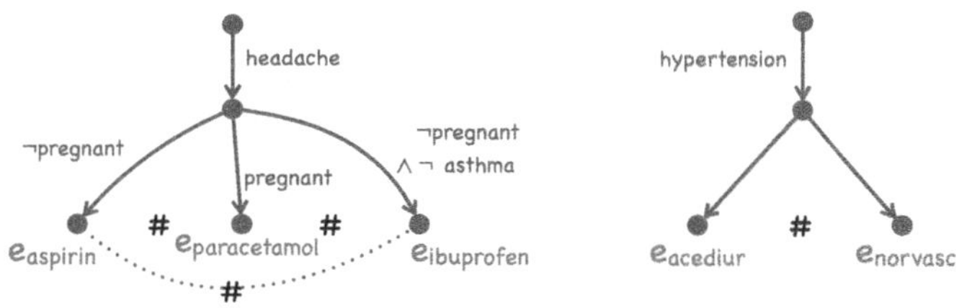

Fig. 1. Treatments for headache (on the left) and hypertension (on the right).

Our main result is the systematic encoding of guidelines, patient profiles and adverse drugs reactions in the enhanced version of RSs. Our encoding relies on an unambiguous representation of medical guidelines. In particular, we adopt the event structure modelling of therapies introduced in [3], paired with a tabular representation of adverse drug reactions which are the result of known side effects of drug combinations. The synthesised model is then used to derive a graph that describes, for each patient profile, all possible evolutions of patient therapies in accordance to clinical guidelines. Finally, we can derive a table that illustrates, for each patient profile, the choices that could lead into a critical medical situation.

Plan of the Paper. In Sect. 2 we introduce clinical guidelines with a toy example. Section 3 introduces the basics of RSs, that we extend with guards in Sect. 4. Section 5 outlines our modelling approach step by step. Section 6 presents our main clinical case study. Section 7 shows the efficacy of our approach on the case study. Section 8 presents related work and draws some conclusion.

2 Concise Guidelines Description

Medical guidelines are provided as long and complex textual documents. The analysis of guidelines with computational methods requires them to be expressed in a more schematic representation, suitable for computer processing. In [3], a representation is proposed based on the formalism of *event structures* paired with a table of adverse drug reactions for known side effects of combined usage of drugs. Turning a real medical guideline into such a representation and maintaining it updated over time requires human work and curation, that would be feasible if effective computational analysis methods were available. As we aim to propose one such analysis methods, we adopt the approach proposed in [3] and assume guidelines to be already available in such a concise form. For the guidelines used in this paper as case studies, the event structure representations were derived manually from the real and updated guideline text.

To illustrate the formal description of guidelines, we introduce two toy examples that will also be used to motivate the introduction of guarded contexts in RSs (cf. Sect. 4) as well as to exemplify our general methodology in Sect. 5.

The first therapy describes the prescription process of a painkiller for migraine, that can be either aspirin or paracetamol or ibuprofen. The second therapy represents the prescription process of two different drugs for hypertension: acediur contains the active ingredients captopril and hydrochlorothiazide (a diuretic drug) while norvasc is a medicine containing the active substance amlodipine besylate. The two therapies are depicted as event structures in Fig. 1. Each node represents an event and each directed arc an (immediate) causal dependency (the source precedes the target). The symbol # denotes (immediate) conflicting events. Events not related by causality or conflict are called concurrent. Arc labels represent guards that must be satisfied to take that path. Node labels show the drug prescriptions associated with that event.

For example, in the case of migraine, the top event represents the start of the treatment that leads to the prescription of one drug among aspirin or paracetamol or ibuprofen (one conflicting event for each choice). Guards such as pregnant, ¬pregnant and (¬pregnant ∧ ¬asthma) expose some factors that must be considered when choosing the right painkiller: patients with asthma are generally told to avoid ibuprofen while pregnant patient must take paracetamol only.

When combining the two therapies, some adversarial effects may arise. Indeed, taking aspirin and diuretics (that are contained in acediur) may lead to a (moderate) risk of renal toxicity. We represent potential drug conflicts as entries in a companion table called, Adverse Drug Reaction (ADR) table. Guarded events structures and ADR tables form the input of our method that aims to model, analyse and possibly improve medical guidelines within the RS framework. Patient profiles, patient conditions, medications and conflicts between adverse drugs will be modelled directly by reactions. Since reactions are deterministic, guarded event structures must necessarily be modelled by RS contexts. Since different therapies can be guided by guards on patient conditions, such as in the example of the painkiller above, we need to extend RS contexts in such a way that the choice to take can be driven by the current state.

Adverse Drug Reactions			
Medication	Medication	Reaction	Side effects & risks include
aspirin	acediur	moderate	renal toxicity

3 Background on Reaction Systems

First, we briefly account for the set theoretic definition of Reaction Systems (RSs) [7,15]. Then, we focus on their process algebraic version [9].

RS Basics. Let S be a finite set of *entities*. A *reaction* in S is a triple $a = (R, I, P)$, where $R, I, P \subseteq S$ are finite, non-empty sets and $R \cap I = \emptyset$. The sets R, I, P are the sets of *reactants*, *inhibitors*, and *products*, respectively. Without loss of generality, we admit the use of empty sets of reactants and inhibitors.

A Reaction System is a pair $\mathcal{A} = (S, A)$, where S is the set of entities, and A is a finite set of reactions in S. Given the current state $W \subseteq S$, a reaction $a =$

(R, I, P) is enabled in W if its *enabling predicate* $en_a(W) \triangleq R \subseteq W \wedge I \cap W = \emptyset$ is satisfied. The *result* of the reaction a on the current state W is defined as $res_a(W) \triangleq P$ if $en_a(W)$, and $res_a(W) \triangleq \emptyset$ otherwise. The result of all reactions A on the current state W, is $res_A(W) \triangleq \cup_{a \in A} res_a(W)$.

The no-permanency principle of RSs dictates that entities disappear if not sustained by some reaction. Thus, the current state $W = D \cup C$ is determined by the result D of all reactions on the previous state, together with additional entities C provided by the *context*. The dynamics of a RS is formalised in terms of *interactive processes*. They are pairs $\pi = (\gamma, \delta)$, where $\gamma \triangleq \{C_i\}_{i \in [0,n]}$ is the *context sequence* and $\delta = \{D_i\}_{i \in [0,n]}$ is the *result sequence*, s.t.: $D_0 = \emptyset$; and $D_{i+1} = res_A(D_i \cup C_i)$ for any $i \in [0, n-1]$. The sequence γ represents external interventions, while the sequence δ is entirely determined by γ and A. We call $\tau \triangleq \{W_i\}_{i \in [0,n]}$ the *state sequence*, with $W_i \triangleq C_i \cup D_i$ for any $i \in [0, n]$.

Example 1. Let $S \triangleq \{\mathsf{a}, \mathsf{b}, \mathsf{c}\}$, $\mathcal{A} \triangleq (S, A)$ and $\gamma \triangleq \{\mathsf{a}, \mathsf{b}\}, \{\mathsf{a}\}, \emptyset$, where $A \triangleq \{a_1\}$ contains a unique reaction $a_1 \triangleq (\{\mathsf{a}, \mathsf{b}\}, \{\mathsf{c}\}, \{\mathsf{b}, \mathsf{c}\})$. Then, the result sequence is $\delta \triangleq \emptyset, \{\mathsf{b}, \mathsf{c}\}, \emptyset$ and the state sequence is $\tau = \{\mathsf{a}, \mathsf{b}\}, \{\mathsf{a}, \mathsf{b}, \mathsf{c}\}, \emptyset$, because, e.g., $res_A(\{\mathsf{a}, \mathsf{b}\}) = \{\mathsf{b}, \mathsf{c}\}$ and $res_A(\{\mathsf{a}, \mathsf{b}, \mathsf{c}\}) = \emptyset$.

Process Algebraic RSs. Inspired by Plotkin's Structural Operational Semantics (SOS) approach [20] and process algebras such as CCS [19], we derive a Labelled Transition System (LTS) semantics for RSs by means of inductive inference rules. LTS states are terms of an algebra, each transition defines a computation step of the RS and its label records the entities involved in that step. Following [9], we also admit nondeterministic and recursively defined contexts.

Definition 1 (RS processes). RS processes *are defined by the grammar below:*

$$\mathsf{P} := [\mathsf{M}] \qquad \mathsf{M} := (R, I, P) \mid D \mid \mathsf{K} \mid \mathsf{M}|\mathsf{M} \qquad \mathsf{K} ::= \mathbf{0} \mid C.\mathsf{K} \mid \mathsf{K} + \mathsf{K} \mid X$$

where R, I, P, C, and D are sets of entities (with $P \neq \emptyset$ and $R \cap I = \emptyset$) and X is a context identifier drawn from a family of (recursive) definitions $\Delta \triangleq \{X_j = \mathsf{K}_j\}_{j \in J}$, called the environment.

A RS process P embeds a *mixture* process M obtained as the parallel composition of some reactions (R, I, P), some available entities D (if any), and some *context* K. We write $\prod_{i \in I} \mathsf{M}_i$ for the parallel composition of all M_i with $i \in I$.

A context process K is either: the nil context $\mathbf{0}$ that stops the computation; the prefixed context $C.\mathsf{K}$ that makes the entities C available to the reactions, and then will behave as K at the next step; the non-deterministic choice $\mathsf{K}_1 + \mathsf{K}_2$ that can behave as either K_1 or K_2; the context identifier X that behaves as K for $X = \mathsf{K} \in \Delta$. For example, if $A = \{\mathsf{a}\}.A + \emptyset.A \in \Delta$ and $B = \{\mathsf{b}\}.B + \emptyset.B \in \Delta$, then the context $A|B$ can offer any combination of a and b at each step.

We say that P and P' are structurally equivalent, written $\mathsf{P} \equiv \mathsf{P}'$, when they denote the same term up to the laws of Abelian monoids (unit, associativity and commutativity) for parallel composition $\cdot|\cdot$, with $\emptyset$ as the unit, and the laws of idempotent Abelian monoids for choice $\cdot + \cdot$, with $\mathbf{0}$ as the unit. We also assume $D_1|D_2 \equiv D_1 \cup D_2$ for any $D_1, D_2 \subseteq S$.

Definition 2 (RSs as RS processes). *Let $\mathcal{A} = (S, A)$ be a RS, and $\pi = (\gamma, \delta)$ an interactive process, with $\gamma = \{C_i\}_{i\in[0,n]}$ and $\delta = \{D_i\}_{i\in[0,n]}$. For any step $i \in [0, n]$, the corresponding RS process $[\![\mathcal{A}, \pi]\!]_i$ is defined as follows:*

$$[\![\mathcal{A}, \pi]\!]_i \triangleq [\ \textstyle\prod_{a\in A} a \mid D_i \mid C_i.C_{i+1}.\cdots.C_n.\mathbf{0}\ ]$$

The SOS semantics of RS processes is defined by the SOS rules in Fig. 2.

A transition label ℓ, written $\langle (D, C) \rhd R, I, P\rangle$, records: the sets D of entities currently in the system; the set C of entities provided by the context; the set R of entities whose presence justifies some reactions; the set I of entities whose absence justifies some reactions; and the set P of reaction products. The SOS rules guarantee that if $\mathsf{P} \xrightarrow{\langle (D,C)\rhd R,I,P\rangle} \mathsf{P}'$ it holds $en_{(R,I,P)}(D \cup C)$.

$$\frac{}{D \xrightarrow{\langle (D,\emptyset)\rhd\emptyset,\emptyset,\emptyset\rangle} \emptyset}\ (\mathit{Ent}) \qquad \frac{}{C.\mathsf{K} \xrightarrow{\langle (\emptyset,C)\rhd\emptyset,\emptyset,\emptyset\rangle} \mathsf{K}}\ (\mathit{Cxt})$$

$$\frac{\mathsf{K}_1 \xrightarrow{\ell} \mathsf{K}_1'}{\mathsf{K}_1 + \mathsf{K}_2 \xrightarrow{\ell} \mathsf{K}_1'}\ (\mathit{Suml}) \qquad \frac{\mathsf{K}_2 \xrightarrow{\ell} \mathsf{K}_2'}{\mathsf{K}_1 + \mathsf{K}_2 \xrightarrow{\ell} \mathsf{K}_2'}\ (\mathit{Sumr}) \qquad \frac{X = \mathsf{K} \in \Delta \quad \mathsf{K} \xrightarrow{\ell} \mathsf{K}'}{X \xrightarrow{\ell} \mathsf{K}'}\ (\mathit{Rec})$$

$$\frac{}{(R, I, P) \xrightarrow{\langle (\emptyset,\emptyset)\rhd R,I,P\rangle} (R, I, P) \mid P}\ (\mathit{Pro}) \qquad \frac{J \subseteq I \quad Q \subseteq R \quad J \cup Q \neq \emptyset}{(R, I, P) \xrightarrow{\langle (\emptyset,\emptyset)\rhd J,Q,\emptyset\rangle} (R, I, P)}\ (\mathit{Inh})$$

$$\frac{\mathsf{M}_1 \xrightarrow{\ell_1} \mathsf{M}_1' \quad \mathsf{M}_2 \xrightarrow{\ell_2} \mathsf{M}_2' \quad \ell_1 \frown \ell_2}{\mathsf{M}_1 \mid \mathsf{M}_2 \xrightarrow{\ell_1\cup\ell_2} \mathsf{M}_1' \mid \mathsf{M}_2'}\ (\mathit{Par}) \qquad \frac{\mathsf{M} \xrightarrow{\langle (D,C)\rhd R,I,P\rangle} \mathsf{M}' \quad R \subseteq D \cup C}{[\mathsf{M}] \xrightarrow{\langle (D,C)\rhd R,I,P\rangle} [\mathsf{M}']}\ (\mathit{Sys})$$

Fig. 2. SOS semantics of the RS processes.

The rule (*Ent*) records the set of current entities D. By rule (*Cxt*), a prefixed context process $C.\mathsf{K}$ makes available the entities in C and then reduces to K. Rules (*Suml*) and (*Sumr*) select a move of either the left or the right context, resp., discarding the other process. By rule (*Rec*), a context identifier X behaves according to its defining process K. The rule (*Pro*) assumes the reaction (R, I, P) is enabled: it records its reactants, inhibitors, and products in the label, and leaves the reaction available at the next step, together with its products P. The rule (*Inh*) records in the label the reasons why the reaction (R, I, P) should not be executed: possibly some inhibiting entities ($J \subseteq I$) are present or some reactants ($Q \subseteq R$) are missing, with $J \cup Q \neq \emptyset$, as at least one cause is needed. The rule (*Par*) puts two processes in parallel by pooling their labels and joining all labels components. We write $\ell_1 \cup \ell_2$ for the component-wise union of labels:

$$\textstyle\bigcup_{i=1,2} \langle (D_i, C_i) \rhd R_i, I_i, P_i\rangle \triangleq \langle (D_1 \cup D_2, C_1 \cup C_2) \rhd R_1 \cup R_2, I_1 \cup I_2, P_1 \cup P_2\rangle\ .$$

The sanity check $\ell_1 \frown \ell_2$ is required to guarantee that labels of reactants and inhibitors are consistent (see definition below):

$$\langle (D_1, C_1) \rhd R_1, I_1, P_1\rangle \frown \langle (D_2, C_2) \rhd R_2, I_2, P_2\rangle \triangleq (\textstyle\bigcup_{i=1,2} D_i \cup C_i \cup R_i) \cap (I_1 \cup I_2) = \emptyset.$$

Finally, the rule (*Sys*) checks that all the needed reactants are available in the system ($R \subseteq D \cup C$). Checking the absence of inhibitors ($(D \cup C) \cap I = \emptyset$) is not necessary, thanks to the sanity check in rule (*Par*). Notably, the SOS semantics matches the set-theoretic dynamics of RSs (as made precise by [9, Th. 19]).

Example 2. The encoding of the RS in Example 1, is as follows:

$$[\![\mathcal{A}, \pi]\!]_0 \triangleq [\ (\mathsf{ab}, \mathsf{c}, \mathsf{bc}) \mid \emptyset \mid \{\mathsf{a}, \mathsf{b}\}.\{\mathsf{a}\}.\emptyset.\mathbf{0}\ ] \equiv [\ a_1 \mid \{\mathsf{a}, \mathsf{b}\}.\{\mathsf{a}\}.\emptyset.\mathbf{0}\ ].$$

Then, from the SOS rules in Fig. 2, we derive, e.g.:

$$[\![\mathcal{A}, \pi]\!]_0 \xrightarrow{\langle(\emptyset,\mathsf{ab})\rhd\mathsf{ab},\mathsf{c},\mathsf{bc}\rangle} [a_1|\{\mathsf{b},\mathsf{c}\}|\{\mathsf{a}\}.\emptyset.\mathbf{0}] \xrightarrow{\langle(\mathsf{bc},\mathsf{a})\rhd\mathsf{c},\emptyset,\emptyset\rangle} [a_1|\emptyset.\mathbf{0}] \xrightarrow{\langle(\emptyset,\emptyset)\rhd\emptyset,\mathsf{ab},\emptyset\rangle} [a_1|\mathbf{0}]\ .$$

$$\frac{}{D \xrightarrow{\langle\langle D\rhd\emptyset,\emptyset,\emptyset\rangle\rhd\emptyset,\emptyset,\emptyset\rangle} \emptyset}\ (Ent) \qquad \frac{}{(R,I,C).\mathsf{K} \xrightarrow{\langle\langle\emptyset\rhd R,I,C\rangle\rhd\emptyset,\emptyset,\emptyset\rangle} \mathsf{K}}\ (Cxt)$$

$$\frac{}{(R,I,P) \xrightarrow{\langle\langle\emptyset\rhd\emptyset,\emptyset,\emptyset\rangle\rhd R,I,P\rangle} (R,I,P)|P}\ (Pro) \qquad \frac{J \subseteq I \quad Q \subseteq R \quad J \cup Q \neq \emptyset}{(R,I,P) \xrightarrow{\langle\langle\emptyset\rhd\emptyset,\emptyset,\emptyset\rangle\rhd J,Q,\emptyset\rangle} (R,I,P)}\ (Inh)$$

$$\frac{\mathsf{M} \xrightarrow{\langle\langle D\rhd R',I',C\rangle\rhd R,I,P\rangle} \mathsf{M}' \quad R' \subseteq D \quad R \subseteq D \cup C}{[\mathsf{M}] \xrightarrow{\langle\langle D\rhd R',I',C\rangle\rhd R,I,P\rangle} [\mathsf{M}']}\ (Sys)$$

Fig. 3. Updated rules to handle guarded contexts (rules (*Suml*), (*Sumr*), (*Rec*) and (*Par*) are left unchanged from Fig. 2 and thus omitted).

4 Reaction Systems with Guards

We enrich the expressiveness of context processes by introducing the possibility to provide some entities *whenever the current state satisfy certain conditions*. To this aim, we introduce guarded prefixes $(R, I, C).\mathsf{K}$.

Definition 3 (Guarded contexts). *We update Definition 1 by letting:*

$$\mathsf{K} ::= \mathbf{0} \mid (R, I, C).\mathsf{K} \mid \mathsf{K} + \mathsf{K} \mid X$$

where $R, I, C \subseteq S$ are possibly empty sets of entities such that $R \cap I = \emptyset$ and X is a context identifier drawn from a predefined environment Δ.

The similarity with the syntax of reactions is intended: the key difference is that in $(R, I, C).\mathsf{K}$ the presence of reactants R and the absence of inhibitors I is checked w.r.t. the set of current entities D. More importantly, the products C are made available immediately from the context, not at the next step. We admit R and I to be possibly empty, and abbreviate $(\emptyset, \emptyset, C).\mathsf{K}$ as $C.\mathsf{K}$. Note that a conditional prefixed process that is not enabled behaves as the $\mathbf{0}$ process.

The SOS rules are updated as shown in Fig. 3. To handle guarded contexts we need a more sophisticated kind of labels: we write ℓ for a tuple of the form

$$\langle\langle D \rhd R', I', C\rangle \rhd R, I, P\rangle$$

whose novel component $\langle D \rhd R', I', C\rangle$ records the available entities D together with those provided by the guarded contexts, assuming all entities in R' are present and those in I' are absent. The SOS rules guarantee that whenever $\mathsf{P} \xrightarrow{\langle\langle D \rhd R', I', C\rangle \rhd R, I, P\rangle} \mathsf{P}'$ it holds that $en_{(R',I',C)}(D)$ and $en_{(R,I,P)}(D \cup C)$. Consequently, we overload the notation $\ell_1 \frown \ell_2$ and $\ell_1 \cup \ell_2$ by letting:

$$\langle\langle D_1 \rhd R'_1, I'_1, C_1\rangle \rhd R_1, I_1, P_1\rangle \frown \langle\langle D_2 \rhd R'_2, I'_2, C_2\rangle \rhd R_2, I_2, P_2\rangle$$
$$\triangleq (\textstyle\bigcup_{i=1,2} D_i \cup R'_i) \cap (I'_1 \cup I'_2) = \emptyset \wedge (\bigcup_{i=1,2} D_i \cup C_i \cup R_i) \cap (I_1 \cup I_2) = \emptyset$$

$$\langle\langle D_1 \rhd R'_1, I'_1, C_1\rangle \rhd R_1, I_1, P_1\rangle \cup \langle\langle D_2 \rhd R'_2, I'_2, C_2\rangle \rhd R_2, I_2, P_2\rangle$$
$$\triangleq \langle\langle D_1 \cup D_2 \rhd R'_1 \cup R'_2, I'_1 \cup I'_2, C_1 \cup C_2\rangle \rhd R_1 \cup R_2, I_1 \cup I_2, P_1 \cup P_2\rangle$$

Example 3. The context $X = (\{\mathsf{a}\}, \emptyset, \{\mathsf{b}\}).X + (\{\mathsf{b}\}, \emptyset, \{\mathsf{a}\}).X + (\emptyset, \{\mathsf{a}, \mathsf{b}\}, \emptyset).X$ ensures that whenever a or b are present in the current state, then both a and b are available as reactants. The last option $(\emptyset, \{\mathsf{a}, \mathsf{b}\}, \emptyset).X$ handles the case where the first two guards fail: if omitted, when both a and b are absent the context X would behave as $\mathbf{0}$, blocking the computation.

5 RS Models for Clinical Guidelines

In this section, we outline our methodology for the modelling of clinical guidelines using RSs. Given the concise formal description of therapies in terms of guarded event structures and ADR tables, we distil a RS process model. To illustrate our construction, we take the toy guidelines from Sect. 2 as a running example.

The Void Process. We always include an auxiliary process $\mathsf{Void} = \emptyset.\mathsf{Void}$ that recursively provides the empty set of entities at any step. It will eventually lead to some attractor, so to model the long terms effect of the therapy.

Patient Profile. Therapies are often based on the health conditions of the patient. In the toy example, the patient profile is determined by the conditions that trigger the treatment (headache, hypertension) and by the conditions that appear in the arc labels (pregnant, asthma). We call them *features*. The patient profile is determined by any combination of features. In the experimentation, there will be exactly one context $\mathsf{K}_f = \emptyset.\mathsf{Void} + \{f\}.\mathsf{Void}$ for each feature f. When they are composed in parallel, their synchronous execution will account for any possible combination of features. In the example, the patient profile will be modelled by the context process: $\mathsf{K}_{\mathsf{headache}} \mid \mathsf{K}_{\mathsf{hypertension}} \mid \mathsf{K}_{\mathsf{pregnant}} \mid \mathsf{K}_{\mathsf{asthma}}$. Alternatively, specific profiles can be investigated by removing one alternative from each K_f. Once the profile is determined by the context, it is preserved during the rest of the computation by reactions of the form $(\{f\}, \emptyset, \{f\})$, one for each feature. In the example, we have, e.g., $(\{\mathsf{headache}\}, \emptyset, \{\mathsf{headache}\}) \mid \cdots \mid (\{\mathsf{asthma}\}, \emptyset, \{\mathsf{asthma}\})$.

Drugs. For each drug d that appears in the therapies, we consider three corresponding entities $\mathsf{get\_}d$, $\mathsf{stop\_}d$ and d: the first represents the prescription of d by the doctor, the second the removal of d from the current treatment and the third the intake of the drug by the patient. Typically, drugs are grouped into classes c and we introduce two more entities c and $\mathsf{stop\_}c$ for each drug class c when needed. Entities $\mathsf{get\_}d$ and $\mathsf{stop\_}d$ will be provided by the context that models the guideline. For each drug, d there will be the following reactions: $(\{\mathsf{get\_}d\}, \{\mathsf{stop\_}d, \mathsf{stop\_}c\}, \{d, c\})$ modelling the intake of the drug d as for doctor prescription, and $(\{d\}, \{\mathsf{stop\_}d, \mathsf{stop\_}c\}, \{d, c\})$ modelling the prosecution of the therapy. Note the presence of $\mathsf{stop\_}d$ and $\mathsf{stop\_}c$ as inhibitors. In the toy example we have five drugs (aspirin, paracetamol, ibuprofen, acediur and norvasc), but omit their classes for simplicity. Hence, we introduce reactions such as

$$(\{\mathsf{get\_aspirin}\}, \{\mathsf{stop\_aspirin}\}, \{\mathsf{aspirin}\}) \mid (\{\mathsf{aspirin}\}, \{\mathsf{stop\_aspirin}\}, \{\mathsf{aspirin}\}) \mid \cdots$$

ADR and Alerts. A generic entry of the ADR table consists of two drugs d_1 and d_2 and some side effects m. For simplicity, we assume that side effects are divided in just three categories: major, moderate and minor. To each entry, we thus introduce a reaction $(\{d_1, d_2\}, \emptyset, \{m\})$. In the toy example we have the reaction (aspirin acediur, $\emptyset$, moderate). Moreover, other kinds of alert can be raised in an analogous way, for example, since a drug can cure different symptoms, a particular alert will be raised in the case study of Sect. 7 when two different drugs of the same class are prescribed as a result of combined therapies.

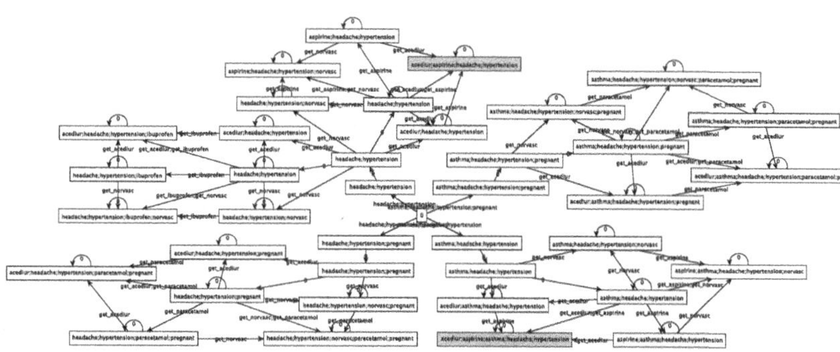

Fig. 4. LTS for (toy) headache/hypertension therapies.

Therapies. We assume that therapies are described as guarded event structures, which we translate to context processes. Roughly, we introduce an identifier E_e for each event e and define it as a sum of processes, one for each outgoing arc of e. If some guard is attached to the arc, then the corresponding alternative is also guarded. When some drug prescription for d is present in a node, then the corresponding choice produces $\mathsf{get\_}d$. Similarly, if the therapy requires stopping some drug d, the entity $\mathsf{stop\_}d$ is produced. Concurrent events are activated separately. For example, the first guideline of our toy example is modelled as

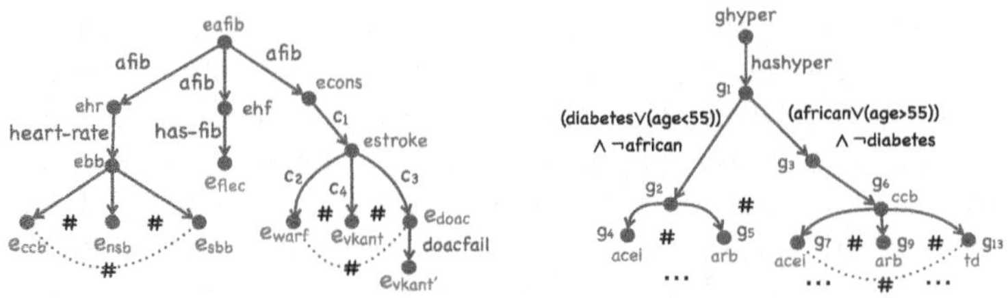

Fig. 5. Line of treatment for patients with atrial fibrillation (left) and first steps of drug treatment for hypertension (right).

$$\mathsf{E_{headache}} \triangleq (\emptyset, \{\mathsf{pregnant}\}, \emptyset).\mathsf{E_a} + (\{\mathsf{pregnant}\}, \emptyset, \emptyset).\mathsf{E_p} + (\emptyset, \{\mathsf{pregnant}, \mathsf{asthma}\}, \emptyset).\mathsf{E_i}$$

where we let $\mathsf{E_a} \triangleq \{\mathsf{get\_aspirin}\}.\mathsf{Void}$, $\mathsf{E_p} \triangleq \{\mathsf{get\_paracetamol}\}.\mathsf{Void}$, etc.

Timing Issues and Asynchronous Execution. The computational model of RSs is based on a fully synchronous semantics. This implies some degree of unmotivated synchronization in the simulation of concurrent therapies. E.g., in two concurrent therapies each consisting of two drugs to be administered in sequence, the first drug of the first therapy would be administered together with the first drug of the second therapy, and the same would hold for the two second drugs. This would happen in the simulation, regardless of the actual timing of the involved drugs. In order to abstract away from all timing issues, we design an asynchronous model of therapies. The idea is that passing from our drug to another is the result of a prescription by a doctor who evaluates the status of the patient and decides how to proceed on the basis of the guidelines. This can happen at any time. Enforcing asynchronicity in the model allows the analysis to provide general results that hold independently of the specific prescription timings. To introduce asynchrony in the simulation of a therapy, we can slightly modify the above definitions to account for arbitrary time lapses before the actual drug prescription, e.g., we can set $\mathsf{E_a} \triangleq \emptyset.\mathsf{E_a} + \{\mathsf{get\_aspirine}\}.\mathsf{Void}$, etc. In what follows, we will assume therapies to be modelled in an asynchronous way.

The Whole Model. The above fragments are assembled in parallel to form a single RS process. In this paper, we will exploit the following features of our RS toolkit:

1. automatic LTS generation in different formats (.dot, .graphml, .pdf, etc.) with customizable appearance: typically we label nodes with features, drugs and side effects and arcs with all entities supplied by the context; moreover, by using different colours for side effects (red for major, orange for moderate) we highlight the risks that patients may undergo (see Fig. 4);
2. automatic synthesis of patient profiles that are more at risk;
3. dynamic guideline revision: by refining guarded context to prevent severe effects for specific patients, we can readily check the efficacy of the changes.

Our framework has high flexibility. By tuning a few parameters, we can experiment with the same medical therapies under different circumstances (selected patient profiles, synchronous or asynchronous treatments, etc.).

6 A Clinical Case Study: AFib and Hypertension

Consider a diagnosis of atrial fibrillation (AFib), a heart condition which causes the heart rate to be irregular and often abnormally too fast. Treatment for AFib normally involves medications to *control the heart rate*, *restore normal heart rhythm* and *reduce the risk of stroke*. To control the heart rate patients can take: either standard beta blockers (SBB) (e.g., atenolol and bisoprolol), non-standard ones (NSBB) (e.g. carvedilol and propranolol) or a (rate limiting) calcium channel blocker (CCB) (e.g., verapamil and diltiazem). AFib patients may also have to take flecainide to restore the normal rhythm. Moreover, patients following one of the two previous therapies are at risk of blood clots forming in the heart chambers, which can cause a stroke. Hence, they can take either warfarin or a direct-acting anticoagulant (DOAC) (e.g., dabigatran and apixaban).

The guideline for AFib is shown in Fig. 5 (left) where a patient with AFib, if needs it, should be treated for medications to control heart rate, to restore the heart rate and to reduce the risk of stroke. The guard c_1 tests if the patient has at least a heart rate or a fibrillation problem and if he agrees to take a drug to reduce the stroke risks. In that case, the choice of the drug class is made on the basis of complex guards: if a patient is younger than 75 and does not have other comorbidities such diabetes or hypertension (guard c_2) the doctor should prescribe warfarin. Otherwise, the suggested therapy includes a DOAC drug unless the patient is intolerant or there are evidences that therapy will not work (guard c_3). In that case (guard $doacfail$) the cure is switched to vitamin K antagonist, the same for intolerance to DOAC drugs (guard c_4).

Now consider that the patient has also been diagnosed with hypertension[1]. whose first treatment steps are shown in Fig. 5 (right). Patients without type 2 diabetes but younger than 55 follow the same treatment as patients with diabetes type 2. In this case, the treatment choice consists of an ACEI or an ARB. Patients

Table 1. Interactions between drugs used for treating AFib and Hypertension.

Adverse Drug Reactions			
Medication (class)	Medication	Reaction	Side effects & risks include
apixaban (doac)	any ccb	moderate	bleeding issues
dabigatran (doac)	diltiazem (ccb)	moderate	bleeding issues
dabigatran (doac)	verapamil (ccb)	major	anaemia, bleeding issues
dabigatran (doac)	carvedilol (nsbb)	moderate	bleeding issues
warfarin	benazepril (acei)	minor	
warfarin	any td	minor	
warfarin	propranolol(nsbb)	minor	
flecainide	any ccb	major	chest pain, coma, ...
flecainide	any sbb	moderate	fainting, palpitations
flecainide	any nsbb	moderate	fainting, palpitations

[1] See details in the NICE guideline NG136 at www.nice.org.uk.

older than 55 or black African or African-Caribbean family origin (any age) are offered a different first treatment with *ccb*. Further steps, (not shown) due to intensification, involve adding one by one new medications not chosen before.

Drugs with clear adverse reactions are shown in Table 1.

7 Modelling and Analysis of the Clinical Case Study

We now apply the toolkit outlined in Sect. 5 to the case study in Sect. 6. For each of the two guidelines in Sect. 6, we detail the entities, reactions and guarded contexts representing patient features, drugs and therapies. The whole RS model, consisting of 123 reactions and 50 context process declarations and written in the syntax of the analysis tool BioReSolve [2], is available online [14].

AFib Guideline. *Patient profile.* By observing the guards in the event structure in Fig. 5, it emerges that patient features of interest are: afib (has to enter the AFib guideline), heart_rate (has hearth rate problem), has_fib (has fibrilation), consensus_acei (gives consensus to treatment for stroke risk reduction), over75 (is above 75 years old), diabetes (has diabetes), hyper (has hypertension), doac_int (is intolerant to DOAC). Context processes for feature setting and reactions for their maintenance are as explained in Sect. 5. Moreover, a guard in the right branch of the event structure from state e_doac is based on the result of the DOAC therapy prescribed in such a state. Hence, we include in the RS model two more entities, doac_ok and doac_fail, representing the success and the failure of such a therapy, respectively. These two entities are comparable to patient features (they contribute to describe the health status of the patient), but they are not set at the beginning of the analysis. As soon as the DOAC therapy is prescribed for the first time, either doac_ok or doac_fail is non-deterministically produced by the RS model in order to explore both possibilities in the analysis.

Drugs. Drugs administered in the AFib guideline together with their classes are modelled by the following entities: atenolol and bisoprolol (class sbb), carvedilol and propranolol (class nsbb), verapamil and diltiazem (class ccb), warfarin (no class specified), dabigatran and apixaban (class doac), vkant (vitamin K antagonist). Reactions for drug prescription and interruption are as explained in Sect. 5.

Therapies. Since the AFib guideline (see Fig. 5) activates three concurrent events ehr, ehf and econs, the initial context process includes three parallel processes eafib1, eafib2 and eafib3 leading to ehr, ehf and econs, respectively, if afib is present as a patient feature. Other context processes are defined as in Sect. 5.

Hypertension Guideline. *Patient profile.* According to the event structure in Fig. 5, patient features of interest for the hypertension guideline are: hyper, diabetes, below55 (is below 55 years old), and origin (has African origins).

Drugs. Drugs in the hypertension guideline are represented by the following entities: captopril and benazepril (class acei), irbesartan and olmesortan (class arb), verapamil and diltiazem (class ccb), chlorothiazide and indapamide (class td).

Therapies. We just apply the context processes encoding defined in Sect. 5.

Combined Guidelines Model. The models of the AFib and hypertension guidelines share some entities because of common patient features (hyper and diabetes) used in guards of both guidelines, and because of drugs that could be prescribed for both diseases (verapamil and diltiazem, in class ccb). Moreover, the two patient features over75 and below55 describe conflicting age classes. Reactions and context processes handling the shared entities are included only once in the combined model, while the two context processes related to age classes are replaced by the single process $\mathsf{K}_{\mathsf{age}} = \emptyset.\mathsf{Void} + \{\mathsf{over75}\}.\mathsf{Void} + \{\mathsf{below55}\}.\mathsf{Void}$.

ADR Table. All the adverse drug reactions listed in Table 1 are included in the combined model, as explained in Sect. 5.

Implementation of the Analyser. In order to analyse the case study, we exploited BioReSolve [2], a tool implementing the SOS semantics of RSs. BioReSolve has been developed in SWI-Prolog [21] by some of the authors of this paper, according to the specifications in [8,10]. In addition, we extended BioReSolve by implementing the support for RS with Guards proposed in this paper, by following the definition given in Sect. 4.

The analysis of the AFib and hypertension guidelines is conducted by a Python script developed on purpose, which uses BioReSolve to generate the LTS that describes all possible treatments (according to the guidelines) for all possible patients. Python-Prolog interaction is made possible by the `swiplserver` Python package. The LTS obtained from BioReSolve is then processed by exploiting the Python package `networkx` to identify attractors (i.e., loops) and determine which patient profiles could lead to the prescription of a combination of drugs at risk for adverse reactions. The patient profiles characterized in this way are then expresses as a minimal logic formula in disjunctive normal form, suitable for tabular representation. Minimization of the logic formula is performed by the `logicmin` Python package that implements the Quine-McCluskey method. The Python script we developed, along with the BioReSolve specification of the RS model to be analysed in this paper, are freely available at [14].

Analysis of the Guidelines. The analysis of the RS model of the two guidelines with BioReSolve and the Python script described above allows us to characterize the classes of patient profiles that could reach a state of major, moderate or minor risk of adverse reaction development. The LTS describing all treatment combinations for all patient profiles consists of 310k states and is constructed in 5 h by our (non optimized) Python script running on a single core of a Intel i7 laptop. Logical formulas characterizing the three classes are represented in tabular form in Fig. 6. In the tables, columns correspond to patient features, and rows to patterns of patient profiles. A green cell (TRUE) in a row means that the corresponding feature is present, a red cell (FALSE) that the corresponding feature is absent, and a white cell that the presence of such a feature is irrelevant.

In order to show how the results obtained from our analysis method can contribute to optimize patient treatments and can suggest potential guidelines improvements, let us consider a couple of patient profiles identified as possibly

Patient profiles possibily leading to a "major" adverse reaction										
	afib	has_fib	heart_rate	consensus_acei	over75	below55	diabetes	doac_int	hyper	origin
1	TRUE		TRUE	TRUE	FALSE			FALSE	TRUE	
2	TRUE		TRUE	TRUE	FALSE		TRUE	FALSE		
3	TRUE		TRUE	TRUE	TRUE	FALSE		FALSE		
4	TRUE	TRUE		TRUE				FALSE	TRUE	
5	TRUE	TRUE							TRUE	
6	TRUE	TRUE	TRUE							

Patient profiles possibily leading to a "moderate" adverse reaction										
	afib	has_fib	heart_rate	consensus_acei	over75	below55	diabetes	doac_int	hyper	origin
1	TRUE		TRUE	TRUE	FALSE			FALSE	TRUE	
2	TRUE		TRUE	TRUE	FALSE		TRUE	FALSE		
3	TRUE		TRUE	TRUE	TRUE	FALSE		FALSE		
4	TRUE	TRUE		TRUE				FALSE	TRUE	
5	TRUE	TRUE		TRUE				FALSE	TRUE	
6	TRUE	TRUE	TRUE							

Patient profiles possibily leading to a "minor" adverse reaction										
	afib	has_fib	heart_rate	consensus_acei	over75	below55	diabete	doac_int	hyper	origin
1	TRUE		TRUE	TRUE	FALSE		FALSE		FALSE	

Fig. 6. Summary of patient profiles that could lead to treatments associated with a major, moderate and minor risk for adverse drug reactions, resp. Note that, by design, over75 and below55 cannot be co-present (TRUE) in the same profile.

leading to major adverse reactions. In particular, we consider the patient profile constituted by features afib, heart_rate, consensus_acei and diabetes. This patient profile matches the pattern in row 2 of the table for major in Fig. 6, as well as the one in row 2 of the table for moderate. In Fig. 7 we show the LTS for such a profile in which red and orange states confirm that both situations of moderate and major risks for adverse reactions can be reached. By inspecting the LTS it is possible to deduce that one of the ways to avoid these risk states is to prevent prescribing the carvedilol, diltiazem and verapamil drugs, in favour of atenolol, bisoprolol or propranolol. Indeed, if we slightly change the RS model by adding extra guards preventing the administration of the identified drugs, we obtain the LTS in Fig. 8, in which dangerous states can no longer be reached.

This patient profile is a case in which only the AFib guideline is followed. Our analysis can be seen as a decision support approach for physicians, but also as a method to propose guideline refinements in a *precision medicine* style [1].

Let us now consider the patient profile constituted by the features afib, heart_rate, consensus_acei and hyper. This profile matches both the pattern in row 1 of the table for major in Fig. 6, and the one in row 1 of the table for moderate. This time the patient profile involves features from both guidelines, and this makes the corresponding LTS much more complex (more than 1.000 nodes and 2.300 arcs, see Fig. 9, left). The figure shows that several red states (associated to major risks for adverse reactions) can be reached. Nevertheless, a careful inspection of the LTS suggests that a cause of danger could be the administration of dabigatran. Again, a modification of the RS model preventing the prescription of such a drug in favour of apixaban (of the same class) allows

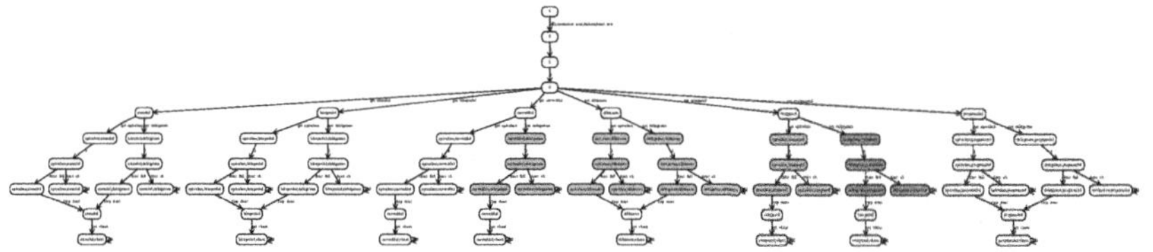

Fig. 7. LTS for patient profile: afib, heart_rate, consensus_acei, diabetes.

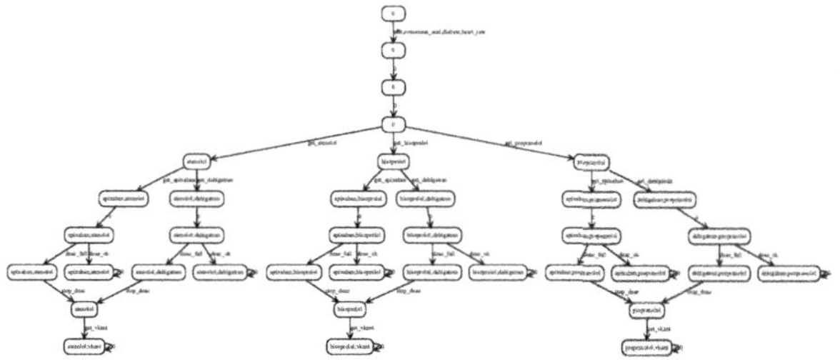

Fig. 8. LTS for the patient profile considered in Fig. 7, but in which the RS model has been modified to prevent prescribing carvedilol, diltiazem and verapamil.

states of major risk not to be reached, as shown in Fig. 9 (right), where red states are no longer present.

This second example of patient profile inspection shows that the proposed approach allows risks for adverse reactions to emerge before starting combinations of therapies from different guidelines for patients with comorbidities. This possibility opens to patient-targeted optimizations of combined treatments.

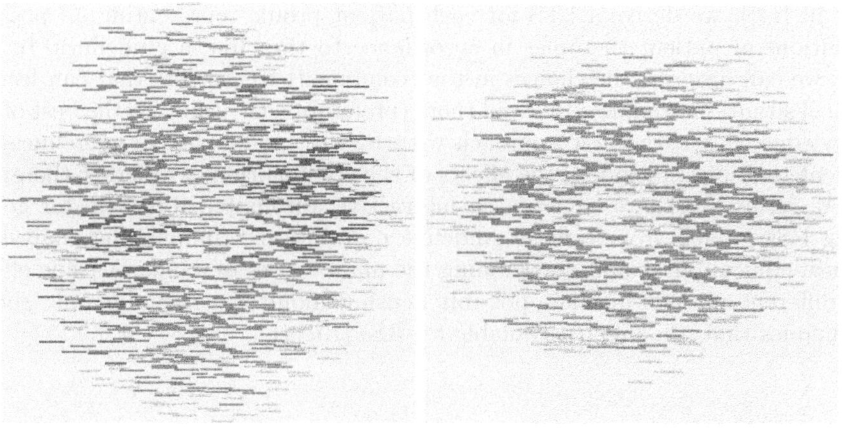

Fig. 9. Left: LTS for patient profile: afib, heart_rate, consensus_acei, hyper. Right: same profile, but with RS model modified to avoid dabigatran.

We remark that, although in this case study we resorted to visual inspection of the LTSs to identify causes of adverse reactions, the analysis could be automatized by applying, for instance, a slicing analysis method as we did in [13] for gene regulatory networks. Slicing analysis is currently able to find non-trivial causal relationships between reactions and produced entities in standard RSs. For the analysis of guidelines, the method should be extended to deal with guards and with the ability to consider as causes for the production of an entity also choices done in context processes. This extension is left for future work.

8 Related Work and Concluding Remarks

Earlier work (see, e.g., [3–6,18]) explored the combination of formal verification techniques, such as constraint solvers and theorem provers, to identify steps in different guidelines that cause problems if carried out together and to determine optimal treatment paths with respect to a particular parameter (e.g., drug efficacy, cost, etc.). However, none of them provided any visual aid to foresee and link current choices to future problems. They also lacked the ability to backtrack, to carry a more comprehensive analysis of treatment options (their advantages vs drawbacks) or to study different patient profiles. Indeed, earlier work focused entirely on the guidelines as given by the National Institute for Health and Care Excellence (NICE, see www.nice.org.uk) and the search through these in a more generic/idealised sense.

By contrast, here we model clinical guidelines directly with extended RSs, where contexts are guarded (recursive, nondeterministic) processes. Guarded contexts allow us to model (the administration of) different therapies, depending on the patients' profile. Despite our manual encoding of guidelines into RS processes, it is worth noting that their derivation can be fully automated. By running the RS specification in our toolkit (implemented by a prototype available in [14]), we derive a LTS for each patient profile representing all possible evolutions of patient therapies in accordance to the clinical guidelines. In this way, we can discover the choices in the (combined) treatments that can lead to critical situations and how to avoid them (preferably early on). To the best of our knowledge, this is the first approach with RSs to model clinical guidelines and patient therapies, including the effects of the combination of different therapies.

As future work, we envisage an interactive automated toolchain for generating RSs models from concise guideline descriptions, running the analysis of critical clinical paths and determining the main causes of dangerous side effects for different patient profiles, possibly reusing monitor-driven dynamic slicing techniques that are already available for RSs [10,11].

References

1. Ashley, E.A.: Towards precision medicine. Nat. Rev. Genet. **17**(9), 507–522 (2016). https://doi.org/10.1038/nrg.2016.86
2. BioReSolve web page, a prolog interpreter for Reaction Systems analysis. http://www.di.unipi.it/~bruni/LTSRS/. Accessed 3 May 2024
3. Bowles, J.K.F., Caminati, M.B.: A flexible approach for finding optimal paths with minimal conflicts. In: Duan, Z., Ong, L. (eds.) ICFEM 2017. LNCS, vol. 10610, pp. 209–225. Springer, Cham (2017). https://doi.org/10.1007/978-3-319-68690-5_13
4. Bowles, J.K.F., Caminati, M.B.: Balancing prescriptions with constraint solvers. In: Liò, P., Zuliani, P. (eds.) Automated Reasoning for Systems Biology and Medicine. CB, vol. 30, pp. 243–267. Springer, Cham (2019). https://doi.org/10.1007/978-3-030-17297-8_9
5. Bowles, J., Caminati, M.: Correct composition in the presence of behavioural conflicts and dephasing. Sci. Comput. Program. **185** (2020). https://doi.org/10.1016/j.scico.2019.102323
6. Bowles, J., Caminati, M., Cha, S., Mendoza, J.: A framework for automated conflict detection and resolution in medical guidelines. Sci. Comput. Program. **182**, 42–63 (2019). https://doi.org/10.1016/j.scico.2019.07.002
7. Brijder, R., Ehrenfeucht, A., Main, M., Rozenberg, G.: A tour of reaction systems. Int. J. Found. Comput. Sci. **22**(07), 1499–1517 (2011). https://doi.org/10.1142/S0129054111008842
8. Brodo, L., Bruni, R., Falaschi, M.: A process algebraic approach to reaction systems. Theoret. Comput. Sci. **881**, 62–82 (2021). https://doi.org/10.1016/j.tcs.2020.09.001
9. Brodo, L., Bruni, R., Falaschi, M.: A logical and graphical framework for reaction systems. Theoret. Comput. Sci. **875**, 1–27 (2021). https://doi.org/10.1016/j.tcs.2021.03.024
10. Brodo, L., Bruni, R., Falaschi, M.: Dynamic slicing of reaction systems based on assertions and monitors. In: Hanus, M., Inclezan, D. (eds.) PADL 2023. LNCS, vol. 13880, pp. 107–124. Springer, Cham (2023). https://doi.org/10.1007/978-3-031-24841-2_8
11. Brodo, L., Bruni, R., Falaschi, M.: A framework for monitored dynamic slicing of reaction systems. Nat. Comput. (2024). https://doi.org/10.1007/s11047-024-09976-3
12. Brodo, L., Bruni, R., Falaschi, M., Gori, R., Levi, F., Milazzo, P.: Quantitative extensions of reaction systems based on SOS semantics. Neural Comput. Appl. **35**(9), 6335–6359 (2023). https://doi.org/10.1007/s00521-022-07935-6
13. Brodo, L., Bruni, R., Falaschi, M., Gori, R., Milazzo, P.: Attractor and slicing analysis of a T Cell differentiation model based on reaction systems. In: Proceedings of DataMod 2023. LNCS. Springer, Berlin (in press)
14. Github repository with the RS model and Python script developed for this paper. https://github.com/Unipisa/AFib-Hyper-GuidelinesAnalysis. Accessed 3 May 2024
15. Ehrenfeucht, A., Rozenberg, G.: Reaction systems. Fundam. Inf. **75**(1-4), 263–280 (2007). http://content.iospress.com/articles/fundamenta-informaticae/fi75-1-4-15
16. Feder, G., Eccles, M., Grol, R., Griffiths, C., Grimshaw, J.: Using clinical guidelines. BMJ **318**(7185), 728–730 (1999). https://doi.org/10.1136/bmj.318.7185.728
17. Hughes, L., McMurdo, M.E.T., Guthrie, B.: Guidelines for people not for diseases: the challenges of applying UK clinical guidelines to people with multimorbidity. Age Ageing **42**, 62–69 (2013). https://doi.org/10.1093/ageing/afs100

18. Kovalov, A., Bowles, J.K.F.: Avoiding medication conflicts for patients with multimorbidities. In: Ábrahám, E., Huisman, M. (eds.) IFM 2016. LNCS, vol. 9681, pp. 376–390. Springer, Cham (2016). https://doi.org/10.1007/978-3-319-33693-0_24
19. Milner, R.: A Calculus of Communicating Systems. LNCS, vol. 92. Springer, Heidelberg (1980). https://doi.org/10.1007/3-540-10235-3
20. Plotkin, G.D.: A structural approach to operational semantics. J. Log. Algebraic Methods Program. **60–61**, 17–139 (2004)
21. SWI-Prolog home page. https://www.swi-prolog.org/. Accessed 3 May 2024
22. Woolf, S.H., Grol, R., Hutchinson, A., Eccles, M., Grimshaw, J.: Potential benefits, limitations, and harms of clinical guidelines. BMJ **318**(7182), 527–530 (1999). https://doi.org/10.1136/bmj.318.7182.527

Graphical Conditions Ensuring Equality Between Differential and Mean Stochastic Dynamics

Hugo Buscemi[1,2] and François Fages[2(✉)]

[1] ENS Paris-Saclay, Univ. Paris-Saclay, Gif-sur-Yvette, France
Hugo.Buscemi@ens-paris-saclay.fr
[2] Inria Saclay, Palaiseau, France
Francois.Fages@inria.fr

Abstract. Complex systems can be advantageously modeled by formal reaction systems (RS), a.k.a. chemical reaction networks in chemistry. Reaction-based models can indeed be interpreted in a hierarchy of semantics, depending on the question at hand, most notably by Ordinary Differential Equations (ODEs), Continuous Time Markov Chains (CTMCs), discrete Petri nets and asynchronous Boolean transition systems. The last three semantics can be easily related in the framework of abstract interpretation. The first two are classically related by Kurtz's limit theorem which states that if reactions are density-dependent families, then, as the volume goes to infinity, the mean reactant concentrations of the CTMC tends towards the solution of the ODE. In the more realistic context of bounded volumes, it is easy to show, by moment closure, that the restriction to reactions with at most one reactant ensures similarly that the mean of the CTMC trajectories is equal to the solution of the ODE at all time points. In this paper, we generalize that result in presence of polyreactant reactions, by introducing the Stoichiometric Influence and Modification Graph (SIMG) of an RS, and by showing that the equality between the two interpretations holds for the variables that belong to distinct SIMG ancestors of polyreactant reactions. We illustrate this approach with several examples. Evaluation on BioModels reveals that the condition for all variables is satisfied on models with no polymolecular reaction only. However, our theorem can be applied selectively to certain variables of the model to provide insights into their behaviour within more complex systems. Interestingly, we also show that the equality holds for a basic oscillatory RS implementing the sine and cosine functions of time.

Keywords: chemical reaction networks · Kurtz's theorem · continuous time markov chains · ordinary differential equations · oscillators · graphical conditions

R. Gori et al. (Eds.): CMSB 2024, LNBI 14971, pp. 45–61, 2024.
https://doi.org/10.1007/978-3-031-71671-3_4

1 Introduction

Inside cells, numerous chemical reactions can occur. It is valuable to understand the behaviour and state of each chemical species involved in these reactions. Such reactions can be represented by a Chemical Reaction Network (CRN), which describes the species involved, their resulting outcomes, and the specific rates at which they react. An Ordinary Differential Equation (ODE) can be associated with such a CRN [7]. Given a fixed initial state, each ODE has a unique deterministic solution.

On the other hand, a CRN can also be interpreted as a Continuous Time Markov Chain (CTMC) and simulated by the Stochastic Simulation Algorithm (SSA), a.k.a. Gillespie's Algorithm [8]. Simulations starting from the same initial state exhibit a variety of non-deterministic trajectories and a mean trajectory. This stochastic mean behaviour may or may not align with the deterministic solution of the ODE starting from the same initial state. Kurtz's limit theorem [9] asserts that, under certain circumstances, the mean behaviour converges to the deterministic one when the volume tends to the infinity. These circumstances involve rates that are density-dependent, possibly apart from a limited number of mode-switching transient states [11].

In the actual context of a cell however, the volume is bounded, and there is also a finite, and possibly small, number of molecules for some species, e.g. as an extreme, only one for genes. It is in this real-world model that we aim to study the relationship between the two continuous-time interpretations of CRNs, by ODE and by CTMC. Regarding the exact correspondence between the mean stochastic CTMC behaviours and the deterministic ODE behaviour, it is well known that if all reactions in the CRN have at most one reactant, then the equality holds. This can be easily by the calculus of moments derived from the Chemical Master Equation [1].

Our contribution is an extension of the unimolecular case, while still ensuring equality. Our result allows for polymolecular reactions with uncorrelated species, by guaranteeing the absence of correlations through graph-theoretic conditions on the structure of the network.

For the sake of generality of these results for dynamical systems beyond the molecular interaction scale, outside of chemistry, we prefer to speak of formal reaction systems (RS) over mathematical variables [3], rather than CRNs over molecular species. Apart from this update of terminology, there is no difference between RSs and CRNs, which are equipped with the same hierarchy of differential, stochastic, Petri net and Boolean semantics [6].

The rest of the paper is organized as follows. The next section illustrates the discrepancy there is between the differential and stochastic interpretations of a simple RS modeling the classical Lotka-Volterra prey-predator population dynamics system. The following section provides the necessary preliminaries on reaction systems with mass action kinetics and their differential and mean stochastic semantics, based respectively on ODE and CTMC interpretations. In Sect. 4, we introduce the Stoichiometric Influence and Modification Graph (SIMG) which somehow combines the classical notions of reaction hypergraph and influence graph of a reaction system [5]. That notion of SIMG is used in Sect. 5 to express and prove our main theorem of strong equality between the differential ODE and mean stochastic CTMC interpretations of RSs satisfying some structural conditions, independently of the kinetics and any limit considerations. In Sect. 6, we provide a low complexity algorithm to test our SIMG conditions. Then in the following sections, we provide several examples to illustrate these results. Interestingly, we also show that strong equality property for an oscillatory RS which implements the sine and cosine functions of time in both differential and stochastic interpretations. Then we report on the evaluation of our condition in the repository of RS models Biomodels [10]. Finally, we conclude on some perspectives for future work with three open questions.

2 Motivating Example

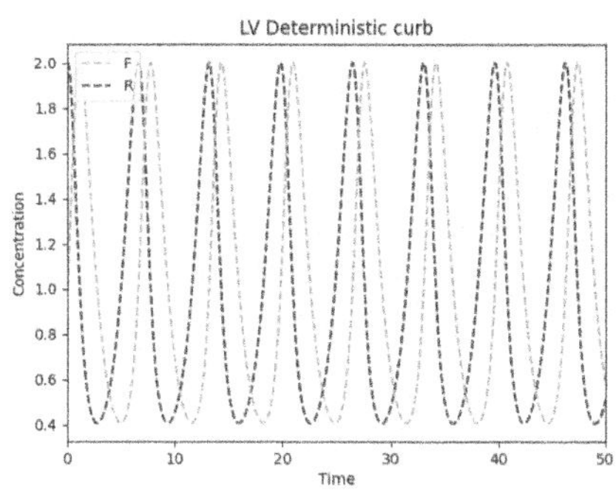

Fig. 1. Deterministic ODE simulation of Lotka-Volterra reaction system.

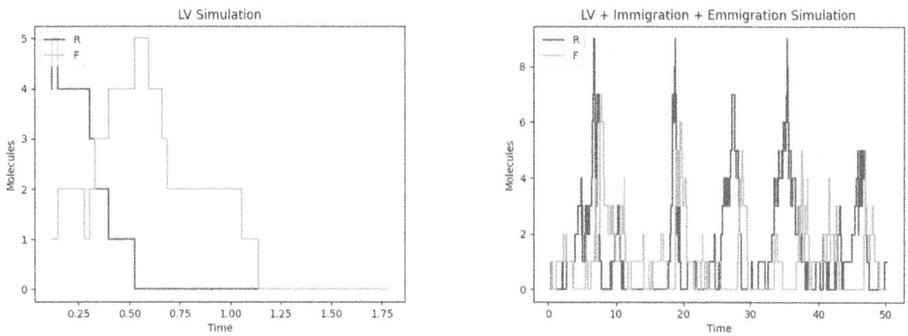

(a) Standard Lotka-Volterra system (b) plus immigration and emigration rules.

Fig. 2. Stochastic CTMC simulations of Lotka-Volterra reaction system.

The Lotka-Volterra reaction system, $F + R \to 2.F$, $R \to 2.R$, $F \to \emptyset$, is a classical model of interaction between two populations of preys (rabbits R) and predators (foxes F). When interpreted using ODEs, the system exhibits oscillatory behaviour, with the populations of rabbits and foxes fluctuating over time (see Fig. 1). However, when interpreted by a CTMC, the system leads almost surely to the extinction of the predator or both populations (see Fig. 2a). This is due to the presence in the stochastic semantics of an absorbing state 0 that is neither reached in the continuous semantics, nor by Kurtz's limit mean CTMC trajectory obtained by making the volume tends to infinity [9]. This is one reason why Kurtz's theorem should not be applied to this kind of counter-example.

On the other hand, one can remark that the addition of immigration and emigration reactions prevents the system from reaching an absorbing state where one population is completely extinct. By allowing for the possibility of individuals entering and leaving the system, there is no absorbing state anymore and the populations exhibit sustained stochastic oscillations (see Fig. 2b).

3 Preliminaries on Reaction Systems

We recall here the standard notions of Chemical Reaction Networks (CRN) over molecular species [7], by using a more general terminology of Reaction Systems (RS) over mathematical variables, in order to apply them to population dynamics or dynamical processes that have nothing to do with chemistry. Apart from this change of terminology to reflect the generality of that formalism of reaction systems, there is no difference between both formalisms of RSs and CRNs.

Definition 1. *A reaction ρ over a set of variables X is a triple (R, P, Λ), also noted*

$$\rho : \sum_i \ell_i R_i \Rightarrow^{\Lambda} \sum_j r_j P_j$$

where R, P are multisets of variables, $R_i, P_j \in X$ are the elements of R, P with non-zero coefficient multiplicity $\ell_i, r_j \in \mathbb{N}^*$, and $\Lambda : X \to \mathbb{R}^+$ is a rate function.

A Reaction System (RS) over X is a finite set of reactions over X.

The R_i's are called reactants and the P_i's products. A reaction with no reactant is called a *synthesis* reaction. A reaction with only one reactant and coefficient equal to 1 is called a *unireactant* reaction. Other reactions are *polyreactant* reactions. A polyreactant reaction with no product is called an *annihilation* reaction.

Notation 1. *For a reaction ρ and a variable $x \in X$, we write $\zeta_{\rho,x} = r_{\rho,x} - \ell_{\rho,x}$ the difference of coefficients between the production of product x and the consumption of reactant x in ρ.*

In this paper, we shall consider mass action law rate functions only.

Definition 2 (Mass-Action Law). *The mass action law rate of a reaction (R, P, Λ) is of the form $\Lambda = \lambda \prod_i \prod_{k=0}^{\ell_i - 1}(R_i - k)$ for some rate constant $\lambda \in \mathbb{R}$.*

Example 1. Mass action RS are written with just their rate constant, which can also be omitted if equal to 1, as in the reactions below:

$$A \Rightarrow^2 A + B$$
$$\_ \Rightarrow B$$
$$B + C \Rightarrow^6 B + D$$
$$D \Rightarrow \_$$

Definition 3 (Associated ODE). *The Ordinary Differential Equation (ODE) of an RS associates to each variable $x \in X$ the differential equation*

$$\frac{\mathrm{d}x}{\mathrm{d}t} = \sum_\rho \Lambda_\rho(X)\zeta_{\rho,x}$$

We denote by σ the unique solution of the system starting at the initial state $\sigma(0)$, and σ_x its projection on the variable x

Example 2. The ODE associated with the RS of Example 1 is

$$\frac{\mathrm{d}A}{\mathrm{d}t} = 0 \qquad \frac{\mathrm{d}B}{\mathrm{d}t} = 2A + 1$$
$$\frac{\mathrm{d}C}{\mathrm{d}t} = -6BC \quad \frac{\mathrm{d}D}{\mathrm{d}t} = 6BC - D$$

Definition 4 (Associated Continous Time Markov Chain). *The Continous Time Markov Chain (CTMC) associated to an RS is defined by the following transition probability*

$$\mathbb{P}\{X(t + \Delta t) = j | X(t) = i\} \sim_{\Delta t \to 0} q_{ij}\Delta t$$

where

$$q_{X(t),X(t)+\zeta} = \sum_{\rho: r_\rho - l_\rho = \zeta} \Lambda_\rho(X(t))$$

We shall make use of the following properties shown in [9] of the CTMC associated to an RS.

Proposition 1. *The mean of $X(t)$ at each point of time follows the equation*

$$\frac{\mathrm{d}\mathbb{E}\left[X(s)\right]}{\mathrm{d}t} = \sum_\rho \mathbb{E}\left[\Lambda_\rho(X(s))\right]\zeta_\rho$$

Definition 5 (Generator). *The generator of the CTMC is the function A defined for each function f by*

$$Af(x) = \sum_\rho \Lambda_\rho(x)(f(x + \zeta_\rho) - f(x))$$

Proposition 2. $f(X(t)) - f(X(0)) - \int_0^t Af(X(s))\mathrm{d}s$ *is a martingale and so* $\mathbb{E}\left[f(X(t))\right] = \mathbb{E}\left[f(X(0))\right] + \int_0^t \mathbb{E}\left[Af(X(s))\right]\mathrm{d}s.$ $\quad \dfrac{\mathrm{d}\mathbb{E}\left[f(X(s))\right]}{\mathrm{d}t} = \mathbb{E}\left[Af(X(t))\right]$

When $\zeta_{\rho,x} = 0$ the reaction ρ has no direct influence on the variable x, and one can focus on reactions where $\zeta_{\rho,x} \neq 0$ only.

4 Stoichiometric Influence and Modification Graph

In this section, we introduce a special kind of influence graph associated to a reaction system, which will be used to define our structural conditions ensuring the equality between the mean stochastic and differential interpretations of a given RS.

Definition 6 (SIMG). *The Stoichiometric Influence and Modification Graph (SIMG) of an RS is a bipartite multigraph with variables and reactions as vertices, with $\ell_{\rho,x}$ arcs from variable x to reaction ρ, and one additional arc from ρ to x if $\zeta_{\rho,x} \neq 0$.*

Example 3. THe SIMG of the RS of Example 1 is the following:

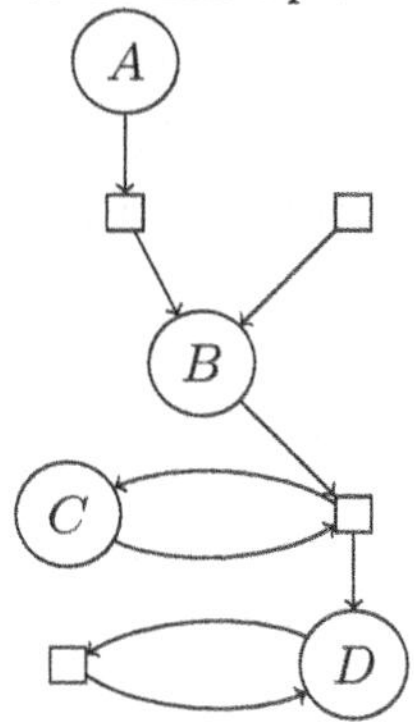

Once this directed graph is constructed, the directed acyclic graph (DAG) of its strongly connected components (SCCs) can be created. This DAG provides a hierarchy and depth for each node (reactions and variables) within the graph. In this graph, each node's ancestors are all the SCCs (including itself) higher in the hierarchy and leading to it.

Definition 7 (Ancestors). *A variable $y \in X$ is called an ancestor of a variable $x \in X$ if there exists a path in the SIMG from y to x. Especially, x is its own ancestor.*

Definition 8 (Depth). *The depth of a SCC is defined recursively as follows:*

- *All SCCs with no arc leading to them are of depth* 0.
- *Any other SCC is at depth $n+1$ where n is the maximum depth of the SCCs leading to it.*

By abuse of notation, we call the depth of a vertex y, the depth of the SCC of y.

Example 4. We show in the next section that the SIMG of the following RS satisfies graphical conditions that entail the equality between the deterministic ODE solution and the mean CTMC behaviour. Simulation traces are shown in Fig. 3a. The actual difference between the mean CTMC trace and the determinist ODE solution is shown in Fig. 3b, using a scale that shows that the difference is indeed neglectable.

$$A + B \Rightarrow A + B + C$$
$$A + B \Rightarrow A + B + D$$
$$C \Rightarrow D$$

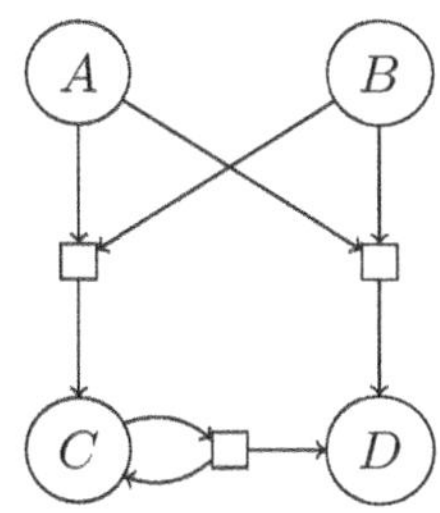

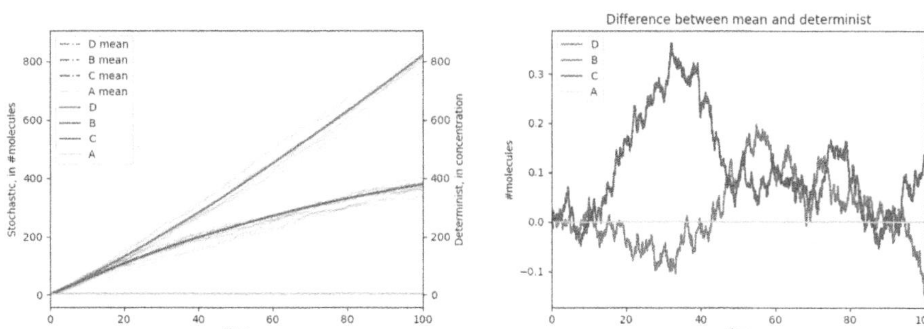

(a) Stochastic mean trace and deterministic solution

(b) Difference between mean stochastic and determinist solutions

Fig. 3. Simulation traces for Example 4 based on 10000 stochastic traces.

5 Main Theorem

Theorem 1. *For an RS with the mass-action law, if each reaction is*

1. *either at most unireactant,*
2. *or polyreactant and has all reactants with disjoint SIMG ancestors,*

then the mean stochastic behaviour, starting from a fixed initial state, is equal to the differential solution with the same initial state.

Proof. Let us examine the different types of reactions that can occur in the RS. As a reminder, they have the form:

$$\rho : \sum_i \ell_{\rho,R_i} R_i \Rightarrow^{\Lambda_\rho} \sum_j r_{\rho,P_j} P_j$$

A first observation is that each variable is its own ancestor, which means it cannot have multiple arcs leading to the same reaction. Hence, for any reaction ρ and variable x, we have $\ell_{\rho,x} \in \{0, 1\}$.

Therefore, for a specific variable x, reactions with $\zeta_{\rho,x} \neq 0$ can be categorized into the following five sets according to the reactants and their ancestors:

1. $\pi_0(x) = \{\rho : \_ \Rightarrow^{\Lambda_\rho} r_{\rho,x}x + \sum_j r_{\rho,P_j} P_j \mid r_{\rho,x} > 0\}$
2. $\pi_1(x) = \{\rho : x \Rightarrow^{\Lambda_\rho} r_{\rho,x}x + \sum_j r_{\rho,P_j} P_j \mid r_{\rho,x} \neq 1\}$
3. $\pi'_1(x) = \{\rho : R \Rightarrow^{\Lambda_\rho} r_{\rho,x}x + \sum_j r_{\rho,P_j} P_j \mid r_{\rho,x} > 0, R \neq x\}$
4. $\pi_2(x) = \{\rho : x + \sum_i R_i \Rightarrow^{\Lambda_\rho} r_{\rho,x}x + \sum_j r_{\rho,P_j} P_j \mid r_{\rho,x} \neq 1, R_i \neq x\}$ and x and all R_i have disjoint ancestors.
5. $\pi'_2(x) = \{\rho : \sum_i R_i \Rightarrow^{\Lambda_\rho} r_{\rho,x}x + \sum_j r_{\rho,P_j} P_j \mid r_{\rho,x} > 0, R_i \neq x\}$ and R_i have disjoint ancestors

It is worth noting that if $\rho \in \pi_2(x)$, then reactant R_0 is an ancestor of both x and R_0 itself. However, this violates the disjoint ancestors property condition. Therefore, we conclude that $\pi_2(x) = \emptyset$.

Now, we can express the set of differential equations for the mean value of each variable:

$$\begin{aligned}\frac{\mathrm{d}\mathbb{E}[x]}{\mathrm{d}t} &= \sum_{\rho\in\pi_0(x)} \lambda_\rho \zeta_{\rho,x} \\ &+ \sum_{\rho\in\pi_1(x)} \lambda_\rho \mathbb{E}[x]\,\zeta_{\rho,x} + \sum_{\rho\in\pi'_1(x)} \lambda_\rho \mathbb{E}[R]\,\zeta_{\rho,x} \\ &+ \sum_{\rho\in\pi'_2(x)} \lambda_\rho \mathbb{E}\left[\prod_i R_i\right] \zeta_{\rho,x}\end{aligned}$$

Now, by applying Lemma 2 proven below, the differential equation becomes:

$$\begin{aligned}\frac{\mathrm{d}\mathbb{E}[x]}{\mathrm{d}t} &= \sum_{\rho\in\pi_0(x)} \lambda_\rho \zeta_{\rho,x} \\ &+ \sum_{\rho\in\pi_1(x)} \lambda_\rho \mathbb{E}[x]\,\zeta_{\rho,x} + \sum_{\rho\in\pi'_1(x)} \lambda_\rho \mathbb{E}[R]\,\zeta_{\rho,x} \\ &+ \sum_{\rho\in\pi'_2(x)} \lambda_\rho \prod_i \mathbb{E}[R_i]\,\zeta_{\rho,x}\end{aligned}$$

This is exactly the same set of differential equations satisfied by the differential solution σ. Furthermore, we have $\mathbb{E}[X](0) = \sigma(0)$. Therefore, we conclude that $\mathbb{E}[X] = \sigma$.

Lemma 1. *Under the hypotheses of the theorem, a reaction is either alone in its SCC, or all the reactions of its SCC are unireatant reactions.*

Proof. By simple inspection of the different cases.

Lemma 2. *Under the hypotheses of the theorem, for all variables x_i's with no common ancestors, the following equality holds:*

$$\mathbb{E}\left[\prod_i x_i\right] = \prod_i \mathbb{E}[x_i]$$

Proof. Since the variables x_i's have no common ancestors, they are not modified by a same reaction. Therefore, for any reaction ρ, we have $\zeta_{\rho,x_i}\zeta_{\rho,x_j} = 0$ for any x_i and x_j with no common ancestor and no x_i and x_j are in the same SCC. We prove the result by induction on the maximum depth of the variables occurring in the product.

Base Case: At depth 0, for any i, the sets $\pi_0(x_i)$ and $\pi_2'(x_i)$ are empty otherwise the SIMG would have one of this patterns $\rho \rightarrow x_i$ or $R_1, R_2 \Rightarrow \rho \rightarrow x_i$ with ρ in its own SCC, using Lemma 1, and so x_i would have a depth > 0.

By using the equation from Proposition 2, we have:

$$\frac{\mathrm{d}\mathbb{E}\left[\prod_{i\in I} x_i\right]}{\mathrm{d}t} = \sum_{j\in I}\sum_{\rho\in\pi_1(x_j)} \lambda_\rho \mathbb{E}\left[\prod_{i\in I} x_i\right]\zeta_{\rho,x_j} + \sum_{\rho\in\pi_1'(x_j)} \lambda_\rho \mathbb{E}\left[R\prod_{i\neq j} x_i\right]\zeta_{\rho,x_j}$$

For the product of means, we have:

$$\frac{\mathrm{d}\prod_{i\in I}\mathbb{E}[x_i]}{\mathrm{d}t} = \sum_{j\in I}\sum_{\rho\in\pi_1(x_j)} \lambda_\rho \prod_{i\in I}\mathbb{E}[x_i]\,\zeta_{\rho,x_j} + \sum_{\rho\in\pi_1'(x_j)} \lambda_\rho \mathbb{E}[R]\prod_{i\neq j}\mathbb{E}[x_i]\,\zeta_{\rho,x_j}$$

All variables appearing in those equations are of depth at most 0. We can rewrite those equation for any variable in each SCC of the x_i's. This gives us a whole system of equations with one equation for each possible product of variables. Since only those products appear in the system, it has a unique solution at a fixed initial state. The system of equations for the mean of the product and the product of means have the same structure. Since they have the same initial conditions, they are equivalent. Therefore, $\mathbb{E}[\prod_i x_i] = \prod_i \mathbb{E}[x_i]$ holds for any product of variables of depth 0.

Inductive Step: Let us assume △ that the result holds for any product of variables with a maximum depth of n, and prove it for products of variables with a maximum depth of $n+1$. We use induction on the $n+2$-tuple $(d_{n+1}, \ldots, d_0)$ where d_i represents the number of variables of the product at depth i, with lexicographic order.

- ⋆ *Base case:* If there are no variable at depth $n+1$, then the induction hypothesis △ gives the result for products of variables with a maximum depth of n. Thus, the result holds.
- ⋆ *Induction step:* Assume that the result holds for any product with at most k variables of depth exactly $n+1$ (and all others at most n). Considering a product with exactly $k+1$ variables of depth exactly $n+1$. We denote its $n+2$-tuple by $\tau = (k+1, d_n, \ldots, d_0)$. Consider the differential equation system for the product $\prod_{i \in I} x_i$:

$$\begin{aligned}\frac{\mathrm{d}\mathbb{E}\left[\prod_{i\in I} x_i\right]}{\mathrm{d}t} &= \sum_{j\in I}\sum_{\rho\in\pi_0(x_j)} \lambda_\rho \mathbb{E}\left[\prod_{i\neq j} x_i\right] \zeta_{\rho,x_j} \\ &+ \sum_{\rho\in\pi_1(x_j)} \lambda_\rho \mathbb{E}\left[\prod_{i\in I} x_i\right] \zeta_{\rho,x_j} \\ &+ \sum_{\rho\in\pi_1'(x_j)} \lambda_\rho \mathbb{E}\left[R\prod_{i\neq j} x_i\right] \zeta_{\rho,x_j} \\ &+ \sum_{\rho\in\pi_2'(x_j)} \lambda_\rho \mathbb{E}\left[\prod_{\ell} R_\ell \prod_{i\neq j} x_i\right] \zeta_{\rho,x_j}\end{aligned}$$

  For $\rho \in \pi_0(x_j)$, the tuple $\tau_{j,\rho} = \tau[d(x_j) \leftarrow \tau(d(x_j)) - 1]$ is such that $\tau_{j,\rho} < \tau$. Then, the induction hypothesis applies to the product $\mathbb{E}\left[\prod_{i\neq j} x_i\right] = \prod_{i\neq j} \mathbb{E}[x_i]$.
  Since for $\rho \in \pi_2'(x_j)$, using Lemma 1, the R_ℓ are higher in the hierarchy than x_j, and they have no common ancestors with x_i for $i \neq j$. So we have $\tau_{j,\rho} < \tau$ and thus $\mathbb{E}\left[\prod_\ell R_\ell \prod_{i\neq j} x_i\right] = \prod_\ell \mathbb{E}[R_\ell] \prod_{i\neq j} \mathbb{E}[x_i]$. The same reasoning applies to $\rho \in \pi_1'(x_j)$ such that R is not from the SCC of x_j. In this case, R is higher and independent from the other variables, and $\mathbb{E}\left[R\prod_{i\neq j} x_i\right] = \mathbb{E}[R]\prod_{i\neq j}\mathbb{E}[x_i]$.
  It remains to consider $\rho \in \pi_1'(x_j)$ such that R is part of the SCC of x_j, and also $\rho \in \pi_1(x_j)$ which is the same but with $R = x_j$. In this case, we can write the system of differential equations for $R\prod_{i\neq j} x_i$ for all R of the SCC. Since we have the same number of equations as unknown functions and the initial values are fixed, we conclude that $\mathbb{E}\left[R\prod_{i\neq j} x_i\right](0) = \left(\mathbb{E}[R]\prod_{i\neq j}\mathbb{E}[x_i]\right)(0)$, which implies $\mathbb{E}\left[R\prod_{i\neq j} x_i\right] = \mathbb{E}[R]\prod_{i\neq j}\mathbb{E}[x_i]$ for the considered product.

6 SIMG Ancestor Condition Checking Algorithm

```
input : SIMG of an RS of size n (number of variables and reactions)
output: Is the graphical condition of Thm. 1 verified?
sccs() calculus ;                                              // in O(n)
Sort sccs() with depth(scc 1) ≤ depth(scc 2) ≤ ...;
foreach scc ∈ sccs do
    Check scc is unireactant ;                                 // in O(|scc|)
    ancestors ← scc;
    foreach variable x ∈ scc do
        foreach reaction ρ with x as SIMG successor do
            match reactants of ρ with :
                • ∅ : do nothing;
                • {R} : ancestors.add(R.ancestors);
                • {R_1 ... R_k} :
                    if ∀i < j, R_i.ancestors ∩R_j.ancestors = ∅ then
                        ancestors.add(∪_{i=1}^k R_i.ancestors)
                    else
                        return False
                    end
        end
    end
    foreach variable x ∈ scc do
        x.ancestors ← ancestors;
    end
end
return True
```

Algorithm 1: Checking of the graphical condition

The construction of the SIMG of an RS can be easily computed from the reaction graph and is of linear size. Now, Algorithm 1 provides a means to verify whether the graphical condition is fulfilled given the SIMG of an RS. By simple inspection of the loops, one can show:

Proposition 3. *(termination and complexity) Algorithm 1 terminates in* $O(nd^2)$ *steps where* n *is the number of variables and reactions and* d *is the maximal number of reactants for a reaction.*

Proposition 4. *(correctness) Algorithm 1 checks the condition of Theorem 1.*

Proof. Let us prove its correctness by first showing that for each variable x, $x.ancestors$ are indeed the real ancestors of x. Then we'll see why this implies the correctness of the algorithm.

First, it is clear that all members of a SCC have the same ancestors.

Now, let us prove by induction on the depth of SCCs that $x.ancestors$ is actually the set of ancestors of x.

- At depth 0:
  - $\supseteq$ The only ancestors of any x at depth 0 are the variables from its SCC. They are added in Alg. line 5
  - $\subseteq$ The only reactions leading to any x (in the SIMG) are the ones from its SCC. Then only Alg. line 10 adds ancestors but it is nothing else than the SCC itself.

  So ancestors at depth 0 are well computed.
- At depth n:
  - $\supseteq$ If A os an ancestor of x
    - Either A is in the same SCC than x and is added in Alg. line 5
    - Either it is not and there exists a path in the SIMG such that $A \rightarrow \ldots \rightarrow R \rightarrow \square \rightarrow x'$, with some R such that $depth(R) < n$ and x' is in the same SCC than x. Then by induction hypothesis, $A.ancestors$ is well formed. So $R \in A.ancestors$ and $A.ancestors$ is adding to $x'.ancestors = x.ancestors$ when considering x' in Alg. line 6.

    Then all ancestors of x are in $x.ancestors$.
  - $\subseteq$ If A is in $x.ancestors$.
    * Either it is added with Alg. line 13 and was in some $R_i.ancestors$ such that $R_i \rightarrow \square \rightarrow x'$ with x' in the same SCC than x. So $depth(R_i) < n$ and then by induction hypothesis A is an ancestor of R_i and so of x' and of x.
    * Either it is added with Alg. line 10 and so $R \rightarrow \square \rightarrow x'$ with x' in the SCC of x and A in $R.ancestors$. The only way to be adde is then with $R.ancestors \neq \emptyset$ which can't happend with R in the same SCC. So $depth(R) < n$ and the same reasoning as just before with induction hypothesis and transitivity of ancestors leads to A as an ancestor of x.

    So all variables in $x.ancestors$ are true ancestors of x.

  Therefore $x.ancestors$ is well computed for each x at depth n.

By the induction principle, it is the case for any depth. Finally, checking the condition of Theorem 1 is done by Alg. line 4 and line 12.

7 Examples

Example 5. One can check that the following RS satisfies the condition of Theorem 1 on all variables. Figure 4 shows the equality between the ODE trace and the mean CTMC trace, together with their high variance.

$$
\begin{aligned}
A + B &\Rightarrow A + B + C \\
C &\Rightarrow D \\
D &\Rightarrow \_ \\
D &\Rightarrow D + E \\
E &\Rightarrow E + F + G \\
E &\Rightarrow \_ \\
F &\Rightarrow \_ \\
F &\Rightarrow F + C \\
G &\Rightarrow H \\
E + I &\Rightarrow E + I + H \\
H &\Rightarrow \_ \\
D + J &\Rightarrow D + J + K \\
K &\Rightarrow \_ \\
K + L &\Rightarrow K + L + M \\
\_ &\Rightarrow M \\
M &\Rightarrow \_
\end{aligned}
$$

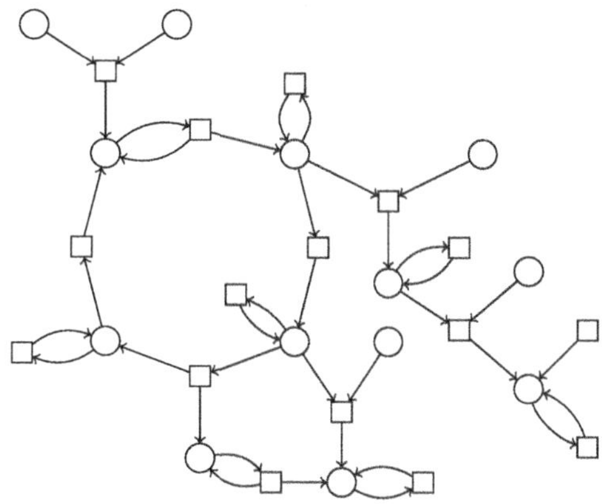

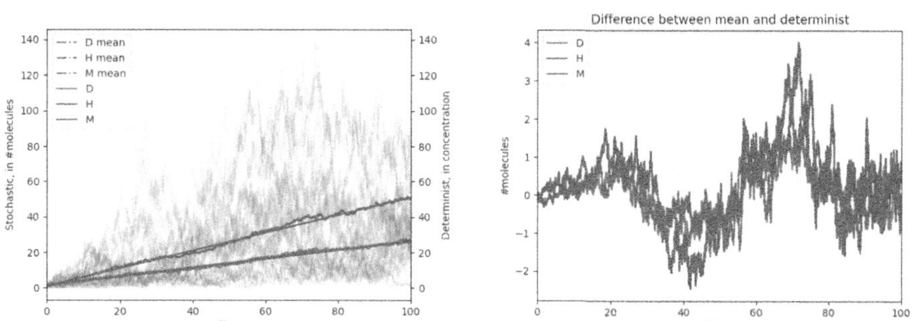

(a) Stochastic mean and determinist solutions.

(b) Difference between mean stochastic and determinist solutions.

Fig. 4. ODE trace and 100 CTMC traces with mean in Example 5.

Example 6. The following RS implements arithmetic multiplication [4]. Indeed, with A and B constant, in the ODE semantics we get at stable state, $\mathbb{E}[C] = AB$. Theorem 1 applies on all variables and thus entails the correctness of the result in the mean stochastic interpretation as well:

$$A + B \Rightarrow A + B + C$$
$$C \Rightarrow \_$$

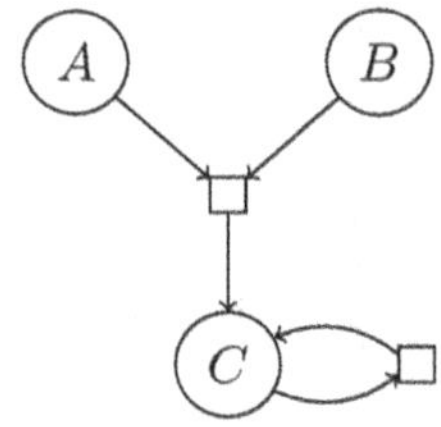

Example 7. Let us consider the following RS with SIMG violating the conditions:

$$A + B \Rightarrow A + B + C$$
$$C \Rightarrow \_$$
$$C \Rightarrow C + D$$
$$D \Rightarrow \_$$
$$C + D \Rightarrow C + D + E$$
$$E \Rightarrow \_$$

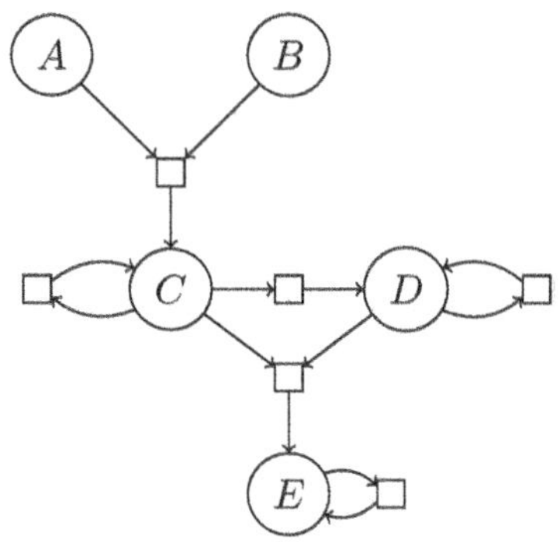

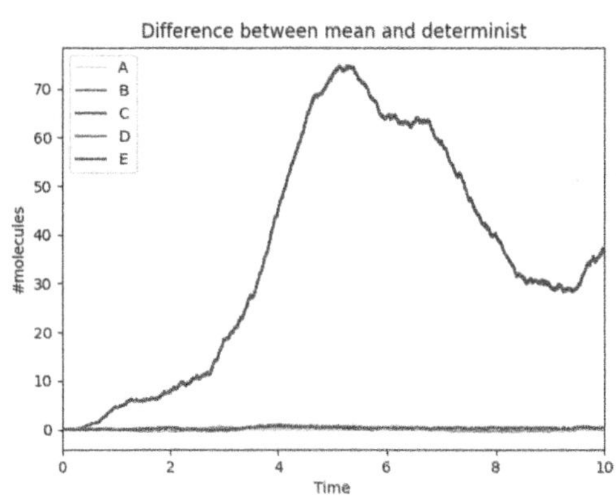

Fig. 5. Difference between the mean over 100 iterations and the deterministic solution for Example 7 with $A(0) = 5$ and $B(0) = 11$

The difference of behaviours between the stochastic and deterministic interpretations is plotted in Fig. 5 for the different variables. Interestingly, one can observe no difference on all variables except E. This can be proven on the SIMG as follows. Because the sub-graph until D verifies the graphical condition, we have $\mathbb{E}[A] = \sigma_A = a$, $\mathbb{E}[B] = \sigma_B = b$ and $\mathbb{E}[C] = \sigma_C$ and $\mathbb{E}[D] = \sigma_D$.

Then $\frac{d\sigma_C}{dt} = ab + \sigma_C$ so, with $\sigma_C(0) = 0$ we have $\sigma_C = ab(1 - e^{-t})$. Thus

$$\begin{aligned}\frac{d\xi}{dt}(t) = \frac{d(\mathbb{E}\left[C^2\right] - \sigma_C^2)}{dt}(t) &= \mathbb{E}\left[ab((C+1)^2 - C^2) + C((c-1)^2 - C^2)\right](t) \\ &\quad - (2(ab - \sigma_C)\sigma_C)(t) \\ &= (ab + \sigma_C - 2\xi)(t)\end{aligned}$$

and $\xi(t) = ab(1 - e^{-t})$.

The same reasoning leads for $\psi = \mathbb{E}[CD] - \sigma_C\sigma_D$ to $\psi' + 2\psi = \xi$ and so $\psi(t) = \frac{ab}{2}(1 - 2e^{-t} + e^{-2t})$.

Then $\phi = \mathbb{E}[E] - \sigma_E$ verifies $\phi' + \phi = \psi$ and so $\phi = \frac{ab}{2}(1 - 2te^{-t} - e^{-2t}) \neq 0$.

8 Equality Property for Cosine Oscillatory RS

The sine and cosine functions of time constitute a fundamental example of oscillator. A CRN implementing $sin(t)$ and $cos(t)$ with 4 molecular species was given in [4] for illustrating the Turing-completeness of continuous CRNs, and the compiler of computable mathematical functions in CRNs implemented in BIOCHAM[1] [2]. That automatically generated RS, called cosRS, uses the dual-rail encoding of negative values as the difference between the concentrations of two reactants with a fast annihilation reaction between both with rate f:

$$\begin{aligned} A &\Rightarrow A + B \\ B &\Rightarrow B + C \\ C &\Rightarrow C + D \\ D &\Rightarrow D + A \\ A + C &\Rightarrow^f \_ \\ B + D &\Rightarrow^f \_ \end{aligned}$$

With initial conditions $A(0) = 1$ and $B(0) = C(0) = D(0) = 0$, we have $A - C = \cos(t)$ and $D - B = \sin(t)$ in the ODE semantics of cosRS. We show that the same holds for $\mathbb{E}[A - C]$ and $\mathbb{E}[D - B]$ in the stochastic CTMC semantics, and does generalize to any initial conditions.

Theorem 2. *For any initial conditions, the mean stochastic traces of $A - C$ and $D - B$ are equal to their ODE traces in the ODE semantics of cosRS.*

Proof. We just have to show that the mean stochastic CTMC traces of $D - B$ and $A - C$ obey the differential equations of the ODE semantics of cosRS.

Variables A and C represent the positive and negative parts of the same function, $A - C$, while B and D represent similarly another function, $D - B$. By linearity, we have $\frac{d\mathbb{E}[A-C]}{dt} = \mathbb{E}[D - B]$ and $\frac{d\mathbb{E}[D-B]}{dt} = \mathbb{E}[C - A]$. Therefore, we get $\frac{d^2\mathbb{E}[A-C]}{dt^2} = -\mathbb{E}[A - C]$ and $\frac{d^2\mathbb{E}[D-B]}{dt^2} = -\mathbb{E}[D - B]$.

[1] http://lifeware.inria.fr/biocham.

9 Evaluation in Biomodels and Partial Approximations

Alg. 1 has been applied to the first 2000 models of the database BioModels [10]. The only models satisfying the condition did not contain polyomolecular reactions. As mentioned in the introduction, the property for at most unimolecular reaction models is well-known and our graphical condition for all variables does not apply to more models in Biomodels repository. Nevertheless, even when the conditions of Theorem 1 are not satisfied on the whole SIMG of the model, they can be used to identify a subset of reactant variables for which the ODE solution is indeed equal to the mean CTMC behaviour.

The key observation is that the condition is additive with respect to the variables ordering based on their depth in the SIMG. If variables do not violate the condition, then it is true for all previous variables considered in the algorithm. Therefore, the algorithm only needs to focus on the partial portion of the graph where all ancestors satisfy the condition. The result then holds for those variables and means nothing for the other variables.

A partial application of Theorem 1 can thus be used to identify RS input variables which may be non constant, but with provably equal differential and stochastic mean dynamics. Independently of the more complex behaviours of the other variables, the theorem guaranties that the ODE semantics provides the exact mean concentration of these input variables over time.

This is illustrated here in Example 7 where the partial condition applies to variables A, B, C and D. This explains why in Fig. 5, variables A, B, C and D follow the determinist ODE solution in mean, unlike variable E. In BioModels, the statistics are however not very significant since many reactions have no mass-action law kinetics, or have non well-formed kinetics [3].

10 Conclusion and Perspectives

Kurtz's theorem is the state-of-the-art result to relate the mean stochastic CTMC and deterministic ODE interpretations of a reaction system. As a limit theorem however, it does not apply to bounded systems and leads to wrong results if applied for instance to population dynamics models in finite space.

We have shown that the strong equality holds between the mean CTMC and deterministic ODE interpretations independently of the kinetics and without any limit approximation, provided some graphical conditions are satisfied on the structure the RS. The interest of our result is that it is not a limit theorem and can thus be applied to bounded dynamical systems with low values for the variables, such as models of gene regulatory networks if we think that one gene is in single copy in a cell.

The evaluation of our graphical conditions in the reaction models in Biomodels shows, on the one hand, that they are satisfied for all species, only in models containing no polymolecular reactions, but, on the other hand, that they can be used to identify in each model a subset of molecular species for which the property holds. Interestingly, we have also shown that the basic oscillatory RS implementing the sine and cosine functions enjoys the strong equality property for all species.

These results also leave some new questions open. First, back to our first example, can we determine the mean stochastic behaviour of the Lotka-Volterra system extended with immigration and emigration reactions for which we have shown the existence of sustained stochastic oscillations ? Second, can we generalize the reasoning done for the sine-cosine RS, to prove the property for general RSs similarly involving annihilation reactions as sole degradation reactions ? Third, our assumptions require no correlations between reactants, while some variables may be balanced and yield the same result, they do not meet our condition. These questions should deserve further investigations.

Acknowledgements. We are grateful to the reviewers of CMSB for their comments, which where useful to improve the presentation of our results, and to Mathieu Hemery for discussions, and for improving the parsing of SBML models in BIOCHAM. This research was partially funded by ANR-20-CE48-0002 project Difference.

References

1. Anderson, D.F., Kurtz, T.G.: Stochastic Analysis of Biochemical Systems, vol. 674. Springer, Heidelberg (2015). https://doi.org/10.1007/978-3-319-16895-1
2. Calzone, L., Fages, F., Soliman, S.: BIOCHAM: an environment for modeling biological systems and formalizing experimental knowledge. Bioinformatics **22**(14), 1805–1807 (2006). https://doi.org/10.1093/bioinformatics/btl172
3. Fages, F., Gay, S., Soliman, S.: Inferring reaction systems from ordinary differential equations. Theoret. Comput. Sci. **599**, 64–78 (2015). https://doi.org/10.1016/j.tcs.2014.07.032
4. Fages, F., Le Guludec, G., Bournez, O., Pouly, A.: Strong turing completeness of continuous chemical reaction networks and compilation of mixed analog-digital programs. In: Feret, J., Koeppl, H. (eds.) CMSB 2017. LNCS, vol. 10545, pp. 108–127. Springer, Cham (2017). https://doi.org/10.1007/978-3-319-67471-1_7
5. Fages, F., Martinez, T., Rosenblueth, D., Soliman, S.: Influence networks compared with reaction networks: semantics, expressivity and attractors. IEEE/ACM Trans. Comput. Biol. Bioinf. (2018). https://doi.org/10.1109/TCBB.2018.2805686
6. Fages, F., Soliman, S.: Abstract interpretation and types for systems biology. Theoret. Comput. Sci. **403**(1), 52–70 (2008). https://doi.org/10.1016/j.tcs.2008.04.024
7. Feinberg, M.: Mathematical aspects of mass action kinetics. In: Lapidus, L., Amundson, N.R. (eds.) Chemical Reactor Theory: A Review, chap. 1, pp. 1–78. Prentice-Hall (1977)
8. Gillespie, D.T.: Exact stochastic simulation of coupled chemical reactions. J. Phys. Chem. **81**(25), 2340–2361 (1977)
9. Kurtz, T.G.: Solutions of ordinary differential equations as limits of pure jump Markov processes. J. Appl. Probab. **7**(1), 49–58 (1970)
10. Malik-Sheriff, R.S., et al.: BioModels - 15 years of sharing computational models in life science. Nucleic Acids Res. **48**(D1), D407–D415 (2020). https://doi.org/10.1093/nar/gkz1055
11. Piho, P., Hillston, J.: Fluid approximation-based analysis for mode-switching population dynamics. ACM Trans. Model. Comput. Simul. **31**(2) (2021). https://doi.org/10.1145/3441680

Bio-STARK: A Tool for the Time-Point Robustness Analysis of Biological Systems

Valentina Castiglioni[1(✉)], Michele Loreti[2], and Simone Tini[3]

[1] Eindhoven University of Technology, Eindhoven, The Netherlands
v.castiglioni@tue.nl
[2] University of Camerino, Camerino, Italy
[3] University of Insubria, Como, Italy

Abstract. We present Bio-STARK, an extension of STARK for the simulation and analysis of biological systems. Specifically, to simulate the stochastic, dynamical, behaviour of these systems, Bio-STARK exploits the core simulation model of STARK, the evolution sequence model, and it extends it by refining the discrete step modelling into a time point modelling. We show how Bio-STARK allows us to verify robustness properties in systems biology, by capturing the effects of (unpredictable) perturbations on species in biochemical networks, as well as on the oscillatory behaviour of gene regulatory networks.

1 Bio-STARK: Motivation and Originality

Stochastic simulation is fundamental in systems biology [31], since it allows for capturing the effects of external interference and internal fluctuations on the behaviour of a system. These *perturbations* might be due to an unpredictable environment [25]; to a dynamic [13], noisy [1,18], behaviour of the system itself; or even to the lack of knowledge on the exact behaviour of, e.g., a disease [23]. Moreover, the complex behaviour of systems arises from the interaction of their numerous components: malfunctioning or corruption of these interactions may originate in severe diseases, such as cancer or diabetes. It is then crucial to study the effects of perturbations on a system, and whether the system maintains its nominal behaviour in spite of their presence. Hence, we need to provide the means to analyse the *robustness* of biological systems against perturbations.

To this end, we present an extension of the STARK tool [4,9,11], called *Bio*-STARK, for the stochastic simulation of biological systems and the verification of their robustness properties against internal and external perturbations.

The STARK Tool. STARK [11] is a Java tool for the specification and verification of robustness properties of Cyber-Physical Systems (CPS) operating under uncertainty. It builds on the *evolution sequence* model [6,8] for simulating systems behaviour, and *Robustness Temporal Logic* (*RobTL*) [7,10] for the specification of robustness properties.

The evolution sequence model follows a discrete-time, data-driven approach: the behaviour of the system is given by the modifications that the interaction of the agents (the cyber component) with the environment (the physical component) induces on a set

R. Gori et al. (Eds.): CMSB 2024, LNBI 14971, pp. 62–70, 2024.
https://doi.org/10.1007/978-3-031-71671-3_5

of application-relevant data, called *data state*. Due to the unpredictability of the environment and potential approximations in the specification of agents, those modifications are modelled as continuous *distributions* on the attainable data states. We call *evolution sequence* the sequence of the distributions over data states that are obtained at each step.

RobTL is a temporal logic for the specification of robustness properties of CPS-like systems, seen as temporal properties of *distances* between nominal and perturbed behaviours of a system. An atomic proposition $\Delta(\mathtt{exp},\mathtt{p}) \bowtie \eta$ is used to evaluate, at a given time step, the *distance*, specified by a *distance expression* exp, between an evolution sequence $\mathcal{S}$ and its perturbed version, obtained by applying some *perturbation* p to $\mathcal{S}$, and to compare it with the threshold η. Atomic propositions are combined with classic Boolean and temporal operators, to extend and compare these evaluations over a chosen time horizon. Perturbations are the effects of unpredictable, unspecified events on the behaviour of the system. We can imagine them as some random noise on data, and represent them as *perturbation functions* mapping each data state to a distribution over data states. This noise is propagated, in time, along the evolution sequence.

From Steps to Time Points: Bio-STARK. One can think of a direct application of STARK to the analysis of biological systems. For instance, for biochemical networks, the data state is the tuple of the number of molecules (or concentration) of each species in the mixture. Then, we use the environment for the stochastic simulation of reactions: by an implementation of (part of) Gillespie's algorithm [14], we can establish the likelihood of each reaction to be the next one to occur. Hence, a computation step by the environment gives the distribution over the data states at the next step, each obtained by applying a different reaction to the current data state. A similar *reaction-by-reaction* approach was proposed in [20], where the SPEAR tool [5] is extended to study robustness in biochemical networks [19] step-by-step. This is in advantage with respect to existing techniques that analyse it only at the steady state (cf. Sect. 3), as it allows us to account for the effects of random events on the networks, and to capture step-wise differences between them that are abstracted away in the analysis at steady state.

However, with a step-by-step approach we cannot take into account the time discrepancies in the dynamics of the networks: the stochastic behaviour is only related to the selection of the next reaction to occur, but the information on time jumps between reactions is disregarded. An analogous argument applies to the modelling of intracellular behaviour, regulatory networks, biomolecular interactions, etc. Moreover, when modelling higher level systems, we can make use of agents for controlling their dynamics. For instance, agents can be used to model a therapy scheduling. Hence, to obtain a more relevant and physically meaningful comparison of biological systems, we need to enhance the discrete-time approach of STARK, and transform the step-by-step modelling into a *time point* modelling. This is what we do in *Bio*-STARK, which is available at https://github.com/stark-tool/Bio-STARK/tree/Tony.

The special features of STARK, based on the uniqueness of the evolution sequence model and RobTL specifications, make Bio-STARK one of a kind. Not only the inherent stochastic behaviour can be easily included in the specification of the system, but it also allows for simulating the effects of arbitrary external perturbations that can be applied at any time to it. Moreover, given the complete freedom in the formalisation of the system and its robustness properties, the tool can be applied in several contexts, such as biochemical networks, gene regulatory networks, disease treatment, etc. The users can use the formalisation that suits best their application context and objectives, and to analyse systems behaviour with respect to several robustness properties at a time.

2 Bio-STARK: Implementation

Bio-STARK is mainly built on the Java library of STARK and offers:

1. A module for the time point simulation of (perturbed) evolution sequences.
2. A module for the evaluation of distances between evolution sequences.
3. A RobTL model checker, with either Boolean or three-valued evaluation.

Due to space constraints, we discuss only the modifications we made to the simulation module to obtain the time point modelling. See https://github.com/stark-tool/Bio-STARK/wiki for a detailed description of the three modules.

In (Bio-)STARK a system is described by the interface SystemState, that provides the methods used by the simulator and the model checker, and that is used to specify system configurations. In Bio-STARK the interface is implemented by the class TimedSystem that differs from the ControlledSystem class of STARK in that:

- We extend the DataState class, modelling a data state, with four special variables, tp, te, δ t, and gr, allowing us to manage time. Specifically: tp represents the *last time point that was reached*; te represents the *total time elapsed* since the beginning of the computation; δ t represents the *duration of the current computation step*; gr represents the chosen *granularity*, i.e., the time interval between observations.
- We add a DataStateExpression, used to quantify δ t for the current computation step. For instance, in the case studies presented in Sect. 3 this procedure is implemented using Gillespie's algorithm.

When simulating each computation step, the δ t is added to te, allowing us to keep track of the total time elapsed. Variable gr is used to state how frequently we will observe the system. Intuitively, the granularity is the time interval between time points at which we can observe the configuration of the system, and it disregards the number of computation steps that occur between two consecutive time points. An evolution sequence is then the sequence of distributions that are reached at the time points.

The TimedSystem class is smoothly integrated in the tool, in the sense that all other classes that are implemented in STARK (like those necessary to apply perturbations, or to evaluate distances and or formulae) remain unchanged in Bio-STARK, since they will automatically interact with the EvolutionSequence generated by the TimedSystem, i.e., only with the distributions reached at the time points.

3 Case Studies

We present two applications of Bio-STARK. Firstly, we consider robustness against perturbations that are mainly due to an unpredictable environment: we analyse the variations in the oscillatory behaviour of a gene regulatory network, the *repressilator* [13]. Secondly, we focus on the robustness of a biochemical network, the *isocitrate dehydrogenase regulatory system* (*IDHKP-IDH*) of the bacterium Escherichia coli [12].

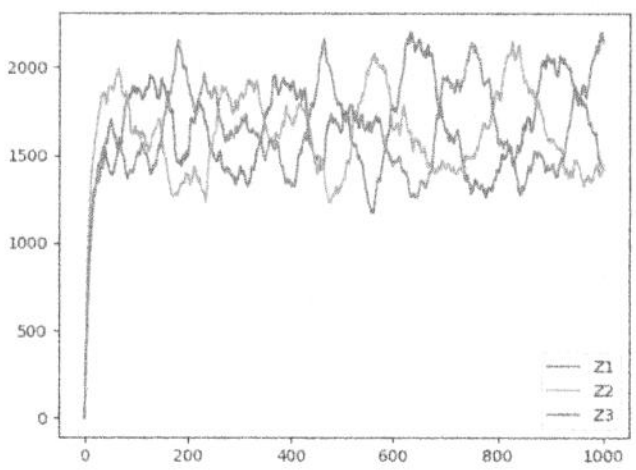

(a) Nominal evolution: avg. proteins level

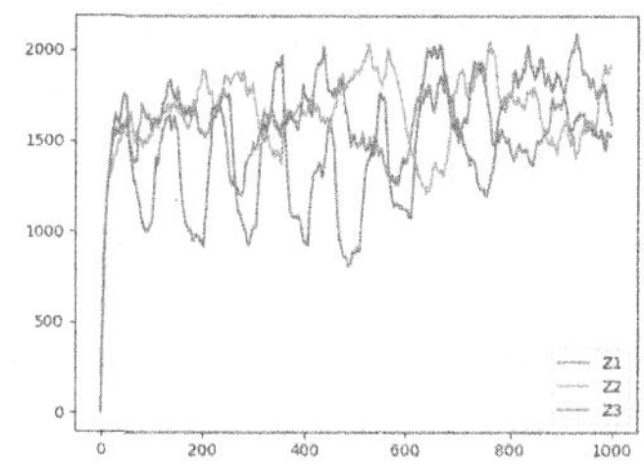

(b) Perturbed evolution: avg. proteins level

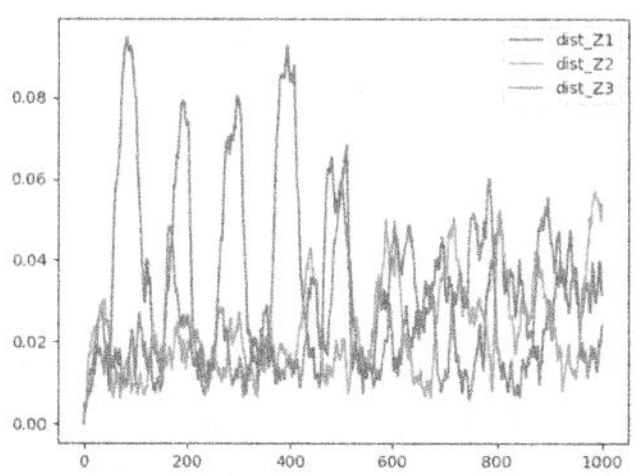

(c) Distances between Fig. 1a and Fig. 1b

(d) Evaluation of formula in Eq. 1

Fig. 1. Analysis of the repressilator network.

Repressilator. The repressilator network [13] consists in m genes forming a directed cycle of *negative interactions*. We model it following [15,16]: (i) we adopt the *two state model* of gene expression, where the gene promoter can be either active or inactive, (ii) we consider mRNA molecules, which can be transcribed only during the active period, and proteins, which are produced by mRNA molecules at a constant rate, (iii) we use chemical reactions for specifying the model, and (iv) we let $m = 3$.

For $i = 1, 2, 3$ we let AG_i denote the active promoter, G_i the inactive promoter, X_i the amount of mRNA molecules and Z_i the amount of proteins. The system is then modelled by the following reactions, with G_i and AG_i satisfying the conservation relation $\mathsf{G}_i + \mathsf{AG}_i = 1$, whereas X_i and Z_i are not conserved quantities:

$$\mathsf{G}_i \underset{k_{\mathtt{off},i}}{\overset{k_{\mathrm{on},i}}{\rightleftharpoons}} \mathsf{AG}_i \qquad \mathsf{AG}_i \xrightarrow{s_{0,i}} \mathsf{AG}_i + \mathsf{X}_i \qquad \mathsf{X}_i \xrightarrow{s_{1,i}} \mathsf{X}_i + \mathsf{Z}_i \qquad \mathsf{X}_i \xrightarrow{d_{0,i}} \emptyset \qquad \mathsf{Z}_i \xrightarrow{d_{1,i}} \emptyset$$

Experimental data [30] suggest a regime with $s_{0,i} \gg d_{0,i}$ and $k_{\mathtt{off},i} \gg k_{\mathrm{on},i}$, where mRNA is transcribed *in bursts*. Gene interaction is modelled by defining the burst frequency as a function of proteins levels [15,16]: $k_{\mathrm{on},i} = \frac{k_{0,i} + k_{1,i} \exp\left(\beta_i + \sum_{j=1}^{3} \theta_{j,i} \mathsf{Z}_j\right)}{1 + k_{1,i} \exp\left(\beta_i + \sum_{i=1}^{3} \theta_{j,i} \mathsf{Z}_j\right)}$, where $k_{0,i}$ and $k_{1,i}$ are the minimal and maximal burst frequency, respectively, and satisfy $k_{0,i} \ll k_{1,i}$, β_i encodes the basal gene activity, and the $\theta_{j,i}$ realise the gene interaction matrix. We consider the same values used in [15]: $k_{\mathtt{off},i} = 5$, $s_{0,i} = 250$, $s_{1,i} = 7$, $d_{0,i} = 1$, $d_{1,i} = 1$, $\theta_{j,i} = -10$ for $(j, i) = (1, 2), (2, 3), (3, 1)$.

We use Bio-STARK[1] to study system's robustness in the protein levels with respect to *protein translation deregulation*, namely the alteration of the normal process of protein synthesis, resulting in aberrant protein production or impaired regulation of protein expression [17]. This may be due to several factors, such as environmental and cellular stress conditions, like abnormal conditions of light for plants [3,24]. Here we abstract from the mechanisms originating such a phenomenon: we simply assume that it affects the translation rate, and express it through perturbations. We consider the perturbation $\mathtt{p}_{s11}(x, a, b, n)$, which affects the translation rate $s_{1,1}$ of gene 1. $\mathtt{p}_{s11}$ is cyclic: for n times, it has no effect for the first a time points, i.e., the system behaves regularly; in the subsequent b time points, $s_{1,1}$ is decremented by x, which impacts directly on the evolution of Z_1 and, through interactions, on Z_2 and Z_3. Using the Bio-STARK simulation module, we simulate a nominal (Fig. 1a) and a perturbed (Fig. 1b) evolution sequence, using perturbation $\mathtt{p}_{s11}(3, 50, 50, 5)$, using 100 sampled runs for each, used to sample the distributions. For readability, in the figures we represent each distribution with the average value assumed in it by Z_1, Z_2, Z_3. While in Bio-STARK the parameters x, a, b can be randomised at each cycle, we fixed them to make the results reproducible.

By comparing Fig. 1a and Fig. 1b we infer that, when active, the perturbation affects protein levels. We formalise it by using the Bio-STARK module for the evaluation of distances. We start by defining the rank function $\rho_{\mathsf{Z}_i}(\mathbf{d}) = \mathbf{d}(\mathsf{Z}_i)/\max(\mathsf{Z}_i)$, assigning to a data state $\mathbf{d}$ the normalised value of Z_i. Then, the atomic distance $<^{\rho_{\mathsf{Z}_i}}$ maps two distributions to the Wasserstein distance [32] between their ranks (as the ranks, this is a real in [0,1]). Pointwise evaluations of $<^{\rho_{\mathsf{Z}_1}}$, $<^{\rho_{\mathsf{Z}_2}}$ and $<^{\rho_{\mathsf{Z}_3}}$ are reported in Fig. 1c.

Finally, we employ the Bio-STARK model checker to verify a robustness property formalising that the perturbed system re-aligns with the nominal one, modulo a tolerance, after the perturbation terminates its effects. We define the distance expression $\mathtt{dMax} = \max(<^{\rho_{\mathsf{Z}_1}}, \max(<^{\rho_{\mathsf{Z}_2}}, <^{\rho_{\mathsf{Z}_3}}))$ giving the maximal value among $<^{\rho_{\mathsf{Z}_i}}$, $i = 1, 2, 3$. Then, the distance expression $\mathtt{intDMax}(u, v) = \mathtt{G}^{[u,v]}\,\mathtt{dMax}$ returns the maximum of $\mathtt{dMax}$ over the interval $[u, v]$. Finally, we give the RobTL atomic formula $\varphi_{(x,a,b,n,u,v)}$ expressing that the distance $\mathtt{intDMax}(u, v)$ between the nominal behaviour and its perturbation via $\mathtt{p}_{s11}(x, a, b, n)$ is below a threshold η:

$$\varphi_{(x,a,b,n,u,v)} = \Delta(\mathtt{intDMax}(u, v), \mathtt{p}_{s11}(x, a, b, n)) \leq \eta \quad (1)$$

The three-valued evaluations of $\varphi_{(3,50,50,5,800,900)}$ for several values of η are in Fig. 1d.

Isocitrate Dehydrogenase Regulatory System. The IDHKP-IDH system controls the biosynthesis of cell constituents of E. coli when it grows on substances that are low on carbon. It regulates the phosporylation level of the TCA cycle enzyme IDH: protein IDH, denoted by I, regulates the flow of carbon through the system, but it is inactive when in phosphorylated form, denoted by $\mathsf{I_P}$. Let E be the bifunctional enzyme that (de)phosphorylates I. The system is characterised by the following reactions:

$$\mathsf{E} + \mathsf{I_P} \underset{0.5}{\overset{0.02}{\rightleftharpoons}} \mathsf{EI_P} \qquad \mathsf{EI_P} + \mathsf{I} \underset{0.5}{\overset{0.02}{\rightleftharpoons}} \mathsf{EI_PI} \qquad \mathsf{EI_P} \xrightarrow{0.5} \mathsf{E} + \mathsf{I} \qquad \mathsf{EI_PI} \xrightarrow{0.1} \mathsf{EI_P} + \mathsf{I_P}$$

[1] Code available at https://github.com/stark-tool/Bio-STARK/tree/Tony/examples/repressilator.

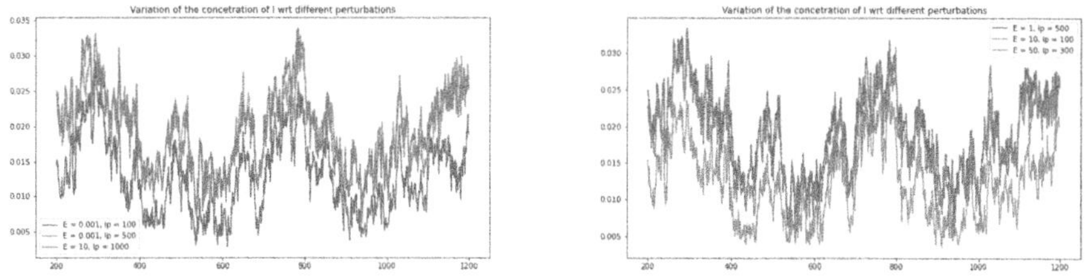

Fig. 2. Pointwise evaluation of $<^{\rho_\mathsf{I}}$ wrt various perturbations.

Robustness in biochemical networks is usually associated to a measure on the variation of species concentration at the steady state [2,26–29]. In [21,22], a network is α-*robust* if, by varying the initial concentration of some species, one observes, at steady state, a variation up-to α of the concentration of a species of interest. Here[2], we show how our time point analysis accounts for the differences in time between networks, abstracted away in the steady state approach, and the effects of perturbations on their behaviour. We study the robustness of the system in the species I with respect to variations in the species E and $\mathsf{I_P}$. We use perturbations to simulate the variations in the initial concentration of species: perturbation $\mathsf{p_I}(x, y) = \mathtt{EandIp}(x, y)@0$, where function $\mathtt{EandIp}(x, y)$ is applied at the same time 0, and it simulates the addition of x molecules of E and y molecules of $\mathsf{I_P}$ to the mixture. To measure the variation in the species I, we use the rank function $\rho_\mathsf{I}(\mathbf{d}) = \mathbf{d}(\mathsf{I})/\max(\mathsf{I})$ to collect the normalised number of molecules of I, and the distance expression $\exp_\mathsf{I} = \mathtt{G}^{[400,1000]} <^{\rho^\mathsf{I}}$ to evaluate the maximal difference in the concentration over the interval $[400, 1000]$. Robustness with respect to a specific variation is given by testing at time 0 the following RobTL formula

$$\varphi_{\mathsf{I},x,y} = \Delta(\exp_\mathsf{I}, \mathsf{p_I}(x, y)) \leq \zeta$$

where $\zeta \in [0, 1]$ is a chosen tolerance. In our experiments, we considered the following initial concentrations: $\mathsf{E} = 0.001$, $\mathsf{I} = 100$, $\mathsf{I_P} = 10000$, $\mathsf{EI_P} = 10$. In Fig. 2, we report the pointwise evaluation, in the interval $[200, 1200]$, of the atomic distance $<^{\rho_\mathsf{I}}$ between the nominal system and the ones obtained by applying, respectively, $\mathsf{p_I}(0.001, 100)$, $\mathsf{p_I}(0.001, 500)$ and $\mathsf{p_I}(10, 1000)$ in the left-hand figure, and $\mathsf{p_I}(1, 500)$, $\mathsf{p_I}(10, 100)$ and $\mathsf{p_I}(50, 300)$ in the right-hand figure. The first three perturbations are the same used to study α-robustness in [21], where it is argued that the system, at steady state, is 0-robust. Here, we capture the tiny discrepancies between the nominal system and its perturbations, and conclude that, for $\zeta \geq 0.04$, the system is robust.

4 Concluding Remarks

Bio-STARK is the first step towards the development of the TempOral aNalYsis module (TONY) of STARK. To obtain it, we will need to further refine the identification of

[2] Code available at https://github.com/stark-tool/Bio-STARK/tree/Tony/examples/Isocitrate.

time points: the current approach, based on granularity, works perfectly with biological systems and, in general, systems in which the time scale of events is large enough to disregard discrepancies between time points of the order of a millisecond. However, those become relevant in CPS, safety-critical systems, digital twins, etc. Moreover, we will need to extend (Bio-)STARK with ODE modelling. We also plan to provide a front-end module for Bio-STARK, similar to that of STARK, that will include a domain specific language, for the specification of systems and their properties, and a command line interpreter, to allow the user to easily interact with all the modules in the back-end.

Acknowledgements. This study received funding from the European Union - NextGenerationEU - National Recovery and Resilience Plan (NRRP) - MISSION 4 COMPONENT 2, INVESTMENT N. 1.1, CALL PRIN 2022 D.D. 104 02-02-2022 - MEDICA Project, CUP N. J53D23007180006.

This publication is part of the project *NODES* which has received funding from the MUR - M4C2 1.5 of PNRR with grant agreement no. ECS00000036.

References

1. Ballarini, P., Bentriou, M., Cournède, P.: A formal approach for tuning stochastic oscillators. In: Pang, J., Niehren, J. (eds.) CMSB 2023. LNCS, vol. 14137, pp. 1–17. Springer, Cham (2023). https://doi.org/10.1007/978-3-031-42697-1_1
2. Barkai, N., Leibler, S.: Robustness in simple biochemical networks. Nature **387**, 913–917 (1997). https://doi.org/10.1038/43199
3. Casal, J.J., Yanossky, M.J.: Regulation of gene expression by light. Int. J. Dev. Biol. **49**, 501–511 (2005). https://doi.org/10.1387/ijdb.051973jc
4. Castiglioni, V., Lanotte, R., Loreti, M., Tini, S.: Evaluating the effectiveness of digital twins through statistical model checking with feedback and perturbations. In: Haxthausen, A.E., Serwe, W. (eds.) FMICS 2024. LNCS, vol. 14952, pp. 21–39. Springer, Cham (2024). https://doi.org/10.1007/978-3-031-68150-9_2
5. Castiglioni, V., Loreti, M., Tini, S.: Measuring adaptability and reliability of large scale systems. In: Margaria, T., Steffen, B. (eds.) ISoLA 2020, Part II. LNCS, vol. 12477, pp. 380–396. Springer, Cham (2020). https://doi.org/10.1007/978-3-030-61470-6_23
6. Castiglioni, V., Loreti, M., Tini, S.: How adaptive and reliable is your program? In: Peters, K., Willemse, T.A.C. (eds.) FORTE 2021. LNCS, vol. 12719, pp. 60–79. Springer, Cham (2021). https://doi.org/10.1007/978-3-030-78089-0_4
7. Castiglioni, V., Loreti, M., Tini, S.: RobTL: a temporal logic for the robustness of cyber-physical systems. CoRR abs/2212.11158 (2022). https://doi.org/10.48550/arXiv.2212.11158
8. Castiglioni, V., Loreti, M., Tini, S.: A framework to measure the robustness of programs in the unpredictable environment. Log. Methods Comput. Sci. **19**(3) (2023). https://doi.org/10.46298/LMCS-19(3:2)2023
9. Castiglioni, V., Loreti, M., Tini, S.: STARK: a software tool for the analysis of robustness in the unknown environment. In: Jongmans, S.S., Lopes, A. (eds.) COORDINATION 2023. LNCS, vol. 13908, pp. 115–132. Springer, Cham (2023). https://doi.org/10.1007/978-3-031-35361-1_6
10. Castiglioni, V., Loreti, M., Tini, S.: RobTL: robustness temporal logic for CPS. In: Proceedings of CONCUR 2024. LIPIcs, Schloss Dagstuhl - Leibniz-Zentrum für Informatik (2024, to appear)

11. Castiglioni, V., Loreti, M., Tini, S.: STARK: a tool for the analysis of CPSs robustness. Sci. Comput. Program. **236**, 103134 (2024). https://doi.org/10.1016/j.scico.2024.103134
12. Dexter, J.P., Gunawardena, J.: Dimerization and bifunctionality confer robustness to the Isocitrate Dehydrogenase regulatory system in Escherichia coli. J. Biol. Chem. **288**(8), 5770–5778 (2013). https://doi.org/10.1074/jbc.M112.339226
13. Elowitz, M.B., Leibler, S.: A synthetic oscillatory network of transcriptional regulators. Nature **403**, 335–338 (2000). https://doi.org/10.1038/35002125
14. Gillespie, D.T.: Exact stochastic simulation of coupled chemical reactions. J. Phys. Chem. **81**(25), 2340–2361 (1977)
15. Herbach, U.: Harissa: stochastic simulation and inference of gene regulatory networks based on transcriptional bursting. In: Pang, J., Niehren, J. (eds.) CMSB 2023. LNCS, vol. 14137, pp. 97–105. Springer, Cham (2023). https://doi.org/10.1007/978-3-031-42697-1_7
16. Herbach, U., Bonnaffoux, A., Espinasse, T., Gandrillon, O.: Inferring gene regulatory networks from single-cell data: a mechanistic approach. BMC Syst. Biol. **11**, 335–338 (2017). https://doi.org/10.1186/s12918-017-0487-0
17. Jia, X., He, X., Huang, C., Li, J., Dong, Z., Liu, K.: Protein translation: biological processes and therapeutic strategies for human diseases. Signal Transduct. Target. Ther. **9**(44), 2786–2791 (2024). https://doi.org/10.1038/s41392-024-01749-9
18. Kim, J., Enciso, G.: Absolutely robust controllers for chemical reaction networks. J. R. Soc. Interface **17**, 20200031 (2020). https://doi.org/10.1098/rsif.2020.0031
19. Kitano, H.: Towards a theory of biological robustness. Mol. Syst. Biol. **3**(1), 137 (2007). https://doi.org/10.1038/msb4100179
20. Lanotte, R., Manicardi, D., Tini, S.: Step-by-step robustness for biochemical networks. In: Proceedings of ICTCS 2023. CEUR Workshop Proceedings, vol. 3587, pp. 299–313. CEUR-WS.org (2023). https://ceur-ws.org/Vol-3587/2906.pdf
21. Nasti, L.: Verification of robustness property in chemical reaction networks. Ph.D. thesis, University of Pisa, Italy (2020). https://etd.adm.unipi.it/theses/available/etd-02172020-165444/
22. Nasti, L., Gori, R., Milazzo, P.: Formalizing a notion of concentration robustness for biochemical networks. In: Mazzara, M., Ober, I., Salaün, G. (eds.) STAF 2018. LNCS, vol. 11176, pp. 81–97. Springer, Cham (2018). https://doi.org/10.1007/978-3-030-04771-9_8
23. Pernice, S., et al.: Computational modeling of the immune response in multiple sclerosis using epimod framework. BMC Bioinform. **21-S**(17), 550 (2020). https://doi.org/10.1186/S12859-020-03823-9
24. Petrillo, E., Herz, M., Barta, A., Kalyna, M., Kornblihtt, A.: Let there be light: regulation of gene expression in plants. RNA Biol. **11**(10), 1215–1220 (2014). https://doi.org/10.4161/15476286.2014.972852
25. Raun, W.R., et al.: Unpredictable nature of environment on nitrogen supply and demand. Agron. J. **111**(6), 2786–2791 (2019). https://doi.org/10.2134/agronj2019.04.0291
26. Rizk, A., Batt, G., Fages, F., Soliman, S.: A general computational method for robustness analysis with applications to synthetic gene networks. Bioinform. **25**(12) (2009). https://doi.org/10.1093/bioinformatics/btp200
27. Rizk, A., Batt, G., Fages, F., Soliman, S.: Continuous valuations of temporal logic specifications with applications to parameter optimization and robustness measures. Theor. Comput. Sci. **412**(26), 2827–2839 (2011). https://doi.org/10.1016/j.tcs.2010.05.008
28. Shinar, G., Feinberg, M.: Structural sources of robustness in biochemical reaction networks. Science **327**(5971), 1389–1391 (2010). https://doi.org/10.1126/science.1183372
29. Shinar, G., Feinberg, M.: Design principles for robust biochemical reaction networks: what works, what cannot work, and what might almost work. Mathe. Biosci. **231**(1), 39–48 (2011). https://doi.org/10.1016/j.mbs

30. Tunnacliffe, E., Chubb, J.R.: What is a transcriptional burst? Trends Genet. **36**(4), 288–297 (2020). https://doi.org/10.1016/j.tig.2020.01.003
31. Uri, A.: An Introduction to Systems Biology: Design Principles of Biological Circuits. Chapman and Hall/CRC (2006)
32. Vaserstein, L.N.: Markovian processes on countable space product describing large systems of automata. Probl. Peredachi Inf. **5**(3), 64–72 (1969)

BoNesis: a Python-Based Declarative Environment for the Verification, Reprogramming, and Synthesis of Most Permissive Boolean Networks

Stéphanie Chevalier[1,2], Déborah Boyenval[3], Gustavo Magaña-López[3], Théo Roncalli[3], Athénaïs Vaginay[3], and Loïc Paulevé[3(✉)]

[1] Translational Medicine, Servier, France
[2] LISN, Univ. Paris-Saclay, CNRS, Gif-sur-Yvette, France
[3] Univ. Bordeaux, CNRS, Bordeaux INP, LaBRI, UMR 5800, 33400 Talence, France
loic.pauleve@labri.fr

Abstract. BoNesis is a Python library which offers a declarative framework for the synthesis of Boolean networks from advanced dynamical properties, such as reachability, bifurcation, minimal trap spaces, stable states, and mutations. It combines recent theoretical advances on Boolean networks with the Most Permissive update mode and efficient resolution of logic programs expressed in Answer-Set Programming. Its main application domain is the inference of Boolean models from bulk and single-cell gene expression data of cell-fate, differentiation and reprogramming processes.

BoNesis is distributed under the GPLv3-compatible free software license CeCILL and is available at https://bnediction.github.io/bonesis.

1 Introduction

Boolean networks (BNs) are executable models with an extensive record of applications in biology and medicine, with the modeling of cellular processes such as cell cycle [9,10,21], lineage differentiation [14,20,22,32], fate decision [18,23,35], and reprogramming [1,8]. Yet, the design of BNs that are able to reproduce observed behaviors while satisfying desired structural properties (typically on its influence graph), and reflecting background and expert knowledge, is an outstanding challenge at different levels.

First, the model properties need to be specified, both in terms of structure and dynamics. In case of data-driven modeling, e.g. from multiscale sequencing data, a preliminary interpreting step is necessary in order to translate quantitative data into dynamic Boolean properties. This modeling task requires biological and statistical expertise, and an upstream well-defined goal about the BN reconstruction. Moreover, dealing with quantitative observations of the system typically involves uncertainty, e.g. measurement noise or lack of statistical significance for gene activity classification or cell clustering.

R. Gori et al. (Eds.): CMSB 2024, LNBI 14971, pp. 71–79, 2024.
https://doi.org/10.1007/978-3-031-71671-3_6

Then, provided a specification of what characterizes a good model arises the challenge of constructing such a model. Most of BNs described in systems biology literature have been designed by hand, either entirely, or by combining previously handmade models. This involves a huge workload, expertise, rules of thumb, trials and errors. Obviously, the automatic construction, also known as logical model inference or *synthesis*, is a long-lived goal [33]. However, the combinatorics of the search space and the complexity of BN verification form a strong bottleneck for realistic applications. Provided a specification and an algorithm for BN synthesis, one would potentially face an issue with a huge number of solutions. Indeed, the experimental observations of cellular dynamics are often derived from few initial conditions and few perturbations with regard to the number of genes, resulting in a largely under-specified synthesis problem. There is consequently a pressing need for efficient computational methods to explore the solution space of BNs.

Recently, there have been several noticeable advances to provide application-focused tools that are tailored for specific type of data and biological and dynamical interpretation of those data [2,11,16,17,19,25,27]. Efforts for providing more generic engines for BN synthesis led to notable frameworks, such as RE:IN [36], involving domain specific language (DSL) and Satisfiability-Modulo-Theory (SMT) technologies, or BRE:IN [13] and AEON [3,4], involving temporal logics and model-checking technologies.

We present the software BoNesis, which aims at providing a general logical modeling environment and solving engine for the synthesis of BNs from rich dynamical properties. It takes the form of a Python library and offers a declarative interface to specify the synthesis problem, together with various methods for browsing the solution space. The synthesis itself relies on logic Answer-Set Programming (ASP) solving. The flexibility of the framework enables to exploit different types of data (bulk/single-cell RNA-seq, perturbation data) with different interpretations, depending on the biological context of the observations. Moreover, BoNesis can also be employed for the verification and reprogramming of a given BN, to identify mutations enforcing a specified dynamical property. Compared to the prior mentioned softwares, one of the main specificity of BoNesis is that it accounts for the *Most Permissive* (MP) dynamics of BNs, which bring a strong connection with quantitative models and a lower complexity of analysis [29]. BoNesis is available at https://bnediction.github.io/bonesis, with tutorials and documentation. It is distributed as part of the CoLoMoto Docker image [26], as well as standard `pip` and `conda` packages. It can be tried online at https://bnediction.github.io/bonesis/online-demo.html.

2 Features

BoNesis is dedicated to the synthesis of BNs from the combination of two kinds of inputs (Fig. 1): the *domain* of candidate BNs, describing the model search space, and the *specification of dynamical properties* that must be verified. BoNesis offers various methods to browse the BNs which belong to the

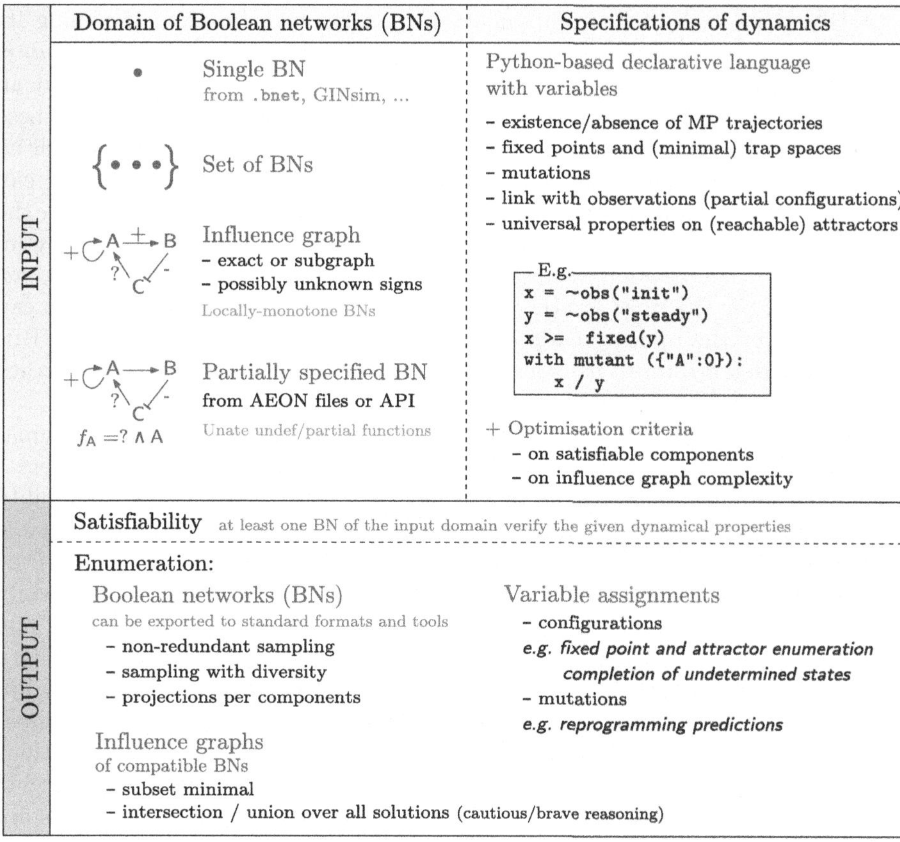

Fig. 1. Overview of BoNesis features

input domain and possess the declared dynamical properties: enumeration of solutions, sampling of diverse solutions, projections over subsets of components, enumeration of involved influence graphs, etc., with the possibility to specify optimization criteria, such as model size.

The input domain of BNs is typically defined from an influence graph which determines the components of the network and the allowed directed pairwise signed dependencies between them. BoNesis also supports partially defined BNs, thanks to the AEON framework [3]. In these cases, BoNesis will account for any BN matching with the influence graph and partially specified Boolean functions which are locally monotone, i.e. no component has both a positive and negative influence on another one. The domain can also be an explicit set of BNs, or a single BN, without any restriction. This latter case is useful for *verification*, i.e., for checking whether the input BN verifies the specified dynamical properties, and for *reprogramming*, i.e., the computation of permanent mutations ensuring desired dynamical properties [28].

A BN maps each of its components with a Boolean function modeling its target Boolean state with respect to the state of its regulators. A *configuration* of the BN associates to each component a Boolean state. Then, provided an *update mode*, one can compute possible transitions and trajectories between these configurations. *Attractors* are the smallest sets of configurations from which no transition can leave. They represent the long-term dynamics of the BN, and can either be a single configuration (fixed points), or a set of configurations. In the case of MP update mode, transitions and trajectories can be computed by an iterative process over subcubes. There, the attractors correspond to the *minimal trap spaces* of the BN, that are the smallest subcubes for which no transition can leave, whatever the update mode [24,29]. It has been show in [29] that MP is able to capture trajectories from any quantitative refinement of the BN, contrary to traditional update modes that can miss possible behaviors.

The expected dynamical properties are specified using predicates, summarized in Table 1. Currently, BoNesis supports properties related to observations, fixed points, minimal trap spaces, existence and absence of reachability between configurations, and universal properties on fixed points and reachable fixed points. The properties can be conditioned with mutations, that replace the function of specified components with a constant value. This enables to specify jointly wild-type and mutant properties. E.g., one can specify that, without any mutation, all the fixed points match with an observation A or B, and that in a mutant where g is knocked out, the fixed points match only with B. The properties can involve partially defined configurations, e.g., by requiring they match with observations over a subset of components, as well as placeholders mutations (`Some` objects). In these cases, BoNesis will search for satisfying assignments for complete configuration and mutations so that the dynamical properties are fulfilled. This is notably employed to identify reprogramming strategies: one can declare that under a `Some` mutation, all fixed points satisfy a given property. Then, BoNesis can enumerate all mutations that can replace `Some`.

In practice, BoNesis builds on the Python language to offer a declarative language for such properties. Figure 2 illustrates BoNesis usage for the modeling of a simple dynamical property: the reachability of a fixed point matching with a specific marker in the wild-type condition, and the absence of such behavior in a given mutant.

3 Implementation and Use Cases

BoNesis relies on Answer-Set Programming (ASP) logic framework and the solver clingo [12]. When requesting for solutions, BoNesis compiles the predicates in ASP using the encodings defined in [6,34] and relies on clingo's Python API to perform the different resolution mechanisms (sampling with diversity, projections, etc.). The ASP encodings are designed such that each solution of the logic program corresponds to a distinct BN verifying the input properties.

Currently, for each component, the Boolean functions to synthesize are internally represented in disjunctive normal form (DNF), i.e. by a list of clauses, each

Table 1. Overview of the main predicates to declare dynamical properties in the scope of a Python object `BoNesis(`*dom*`,`***data***`)`, where ***dom*** specifies the domain of BNs, including the set of components, and ***data*** defines the named observations. For a complete reference, see https://bnediction.github.io/bonesis/language.html.

<table>
<tr><th colspan="2">Objects</th></tr>
<tr><td colspan="2">Observation: maps a subset of components to 0 or 1</td></tr>
<tr><td><code>O = obs({a:0, b:1,..})</code></td><td>Obs from partial mapping of components to 0/1</td></tr>
<tr><td><code>O = obs("name")</code></td><td>Named obs defined in the data dictionnary</td></tr>
<tr><td colspan="2">Configuration: maps each component to 0 or 1</td></tr>
<tr><td><code>C = cfg()</code></td><td>Allocates a fresh configuration</td></tr>
<tr><td><code>C = +O</code></td><td>Fresh configuration matching with obs O</td></tr>
<tr><td><code>C = ~O</code></td><td>Default configuration matching with obs O</td></tr>
<tr><td><code>S = Some(max_size=k)</code></td><td>Represents a mutation of at most <code>k</code> components.
Use <code>S.assignments()</code> to get satisfying valuations.</td></tr>
<tr><th colspan="2">Properties</th></tr>
<tr><td colspan="2">Constraints on configurations</td></tr>
<tr><td><code>C != C'</code></td><td><code>C</code> and <code>C'</code> differ on at least one component</td></tr>
<tr><td><code>C[a] == C'[b] | 0 | 1</code></td><td>State of component a in <code>C</code> is equal (resp. different)</td></tr>
<tr><td><code>C[a] != C'[b] | 0 | 1</code></td><td>to state of component b in <code>C'</code> (resp. 0 and 1)</td></tr>
<tr><td><code>C != O</code></td><td><code>C</code> does not match with O</td></tr>
<tr><td colspan="2">Attractor properties</td></tr>
<tr><td><code>fixed(C)</code></td><td><code>C</code> is a fixed point</td></tr>
<tr><td><code>fixed(O)</code></td><td>there exists a trap space where O is fixed</td></tr>
<tr><td><code>in_attractor(C)</code></td><td><code>C</code> belongs to an attractor</td></tr>
<tr><td colspan="2">Reachability</td></tr>
<tr><td><code>C >= C'</code> or <code>reach(C,C')</code></td><td>Exists an MP trajectory from <code>C</code> to <code>C'</code></td></tr>
<tr><td><code>C >= O</code></td><td>... to a config matching with O (equiv to <code>C >= +O</code>)</td></tr>
<tr><td><code>C >= fixed(O)</code></td><td>... to a config in a trap space where O is fixed</td></tr>
<tr><td colspan="2">Note: can be composed, e.g., <code>C1 >= C2 >= fixed(C3)</code></td></tr>
<tr><td><code>C / □</code> or <code>nonreach(C,□)</code></td><td>Absence of MP traj from <code>C</code> to ... (same as <code>reach</code>)</td></tr>
<tr><td colspan="2">Universal properties</td></tr>
<tr><td><code>all_fixpoints({O,O',..})</code></td><td>All the fps match with at least one given obs</td></tr>
<tr><td><code>C >> "fixpoints" ^ {O,O',..}</code></td><td>All the fps reachable from <code>C</code> ...</td></tr>
<tr><th colspan="2">Contexts</th></tr>
<tr><td><code>with mutant({a:0,..} | S):</code>
<code>   ...</code></td><td>Properties within the with block are subject to
mutation specified by <code>{a:0,..}</code> (resp. S)</td></tr>
</table>

```
import bonesis
```

```
# use complete graph as domain
dom = bonesis.InfluenceGraph.complete(["A","B","C"])
# named observations
data = {
    "init": {"A": 0, "B": 0, "C": 0},
    "marker": {"A": 1, "C": 0}
}
```

```
bo = bonesis.BoNesis(dom, data)
x = ~bo.obs("init")
y = ~bo.obs("marker")
x >= bo.fixed(y)  # y is a fixed point and can be reached from x
with bo.mutant({"C": 0}):
    x / y
```

```
bo.is_satisfiable()
```

```
True
```

```
solutions = list(bo.boolean_networks(limit=3))
solutions[0] # show one solution
```

```
A <- A|(!B&C)
B <- !B|(A&!C)
C <- !B
```

```
# list attractors
import pandas as pd
pd.DataFrame(solutions[0].attractors())
```

	A	B	C
0	1	1	0

```
solutions[0].save("solution.bnet")
```

Fig. 2. Example use of BoNesis in a Jupyter session

clause being a conjunction of at most m literals, where m is the number of regulators of the component. Because the number of clauses grows exponentially with m, BoNesis allows setting an arbitrary upper bound on the number of clauses for Boolean functions. This enables synthesing functions that depend on numerous components, but have a limited number of clauses. As shown in [5,6], BoNesis can tackle the synthesis of BNs with thousands of components.

BoNesis has been employed for different concrete biological applications. In [7], authors revisited a published BN of cell fate in cancer (32 components) with ensembles of diverse alternative models synthetized by BoNesis from the same observations in different mutation conditions. In [30], authors included BoNesis in a pipeline to infer an ensemble of BNs over 232 genes from gene perturbation experiments for the study of epilypsy. To model early hematopoiesis aging, [15] employed BoNesis from scRNA-seq data and an expert influence graph (15 components) and results in the discovery of new explanatory regulatory mechanisms. [5] also integrated scRNA-seq with BoNesis but with a large influence graph from a public database (>1,000 nodes) to identify key genes involved in hematopoiesis and analyze variability of candidate BNs.

4 Conclusion

BoNesis aims at offering a flexible and efficient framework for BN synthesis from rich dynamical properties, for instance generated from experimental data. Thanks to its flexibility, BoNesis can also be employed for the verification and reprogramming of BNs. The defined predicates form a domain-specific language that enables to express a large variety of dynamical properties over BNs with a higher abstraction level than classical temporal logic. Moreover, the intertwining with Python greatly facilitates the programmatic generation of the dynamical properties (loops, variables), and the combination with other tools, both as preprocessing of input data and post-processing of obtained models. In future work, we plan to extend support for the synthesis of non-monotone BNs, and for further constraints over attractors and trajectories, notably by generalizing the universal reasoning over minimal trap spaces introduced in [31].

References

1. Abou-Jaoudé, W., et al.: Model checking to assess t-helper cell plasticity. Front. Bioeng. Biotechnol. **2** (2015). https://doi.org/10.3389/fbioe.2014.00086
2. Aghamiri, S.S., Delaplace, F.: Taboon boolean network synthesis based on tabu search. IEEE/ACM Trans. Comput. Biol. Bioinform. **19**(4), 2499–2511 (2021). https://doi.org/10.1109/TCBB.2021.3063817
3. Beneš, N., Brim, L., Kadlecaj, J., Pastva, S., Šafránek, D.: AEON: attractor bifurcation analysis of parametrised boolean networks. In: Lahiri, S.K., Wang, C. (eds.) CAV 2020. LNCS, vol. 12224, pp. 569–581. Springer, Cham (2020). https://doi.org/10.1007/978-3-030-53288-8_28
4. Beneš, N., Brim, L., Huvar, O., Pastva, S., Šafránek, D.: Boolean network sketches: a unifying framework for logical model inference. Bioinformatics **39**(4) (2023). https://doi.org/10.1093/bioinformatics/btad158
5. Chevalier, S.: Inférence logique de réseaux booléens à partir de connaissances et d'observations de processus de différenciation cellulaire. Ph.D. thesis, Université Paris-Saclay (2022). https://theses.hal.science/tel-03917566/file/106218_CHEVALIER_2022_archivage.pdf
6. Chevalier, S., Froidevaux, C., Paulevé, L., Zinovyev, A.: Synthesis of boolean networks from biological dynamical constraints using answer-set programming. In: 2019 IEEE 31st International Conference on Tools with Artificial Intelligence (ICTAI), pp. 34–41. IEEE (2019). https://doi.org/10.1109/ICTAI.2019.00014
7. Chevalier, S., Noël, V., Calzone, L., Zinovyev, A., Paulevé, L.: Synthesis and simulation of ensembles of boolean networks for cell fate decision. In: Abate, A., Petrov, T., Wolf, V. (eds.) CMSB 2020. LNCS, vol. 12314, pp. 193–209. Springer, Cham (2020). https://doi.org/10.1007/978-3-030-60327-4_11
8. Collombet, S., et al.: Logical modeling of lymphoid and myeloid cell specification and transdifferentiation. Proc. Natl. Acad. Sci. **114**(23), 5792–5799 (2017). https://doi.org/10.1073/pnas.1610622114
9. Davidich, M.I., Bornholdt, S.: Boolean network model predicts cell cycle sequence of fission yeast. PLoS ONE **3**(2), e1672 (2008). https://doi.org/10.1371/journal.pone.0001672
10. Fauré, A., Naldi, A., Chaouiya, C., Thieffry, D.: Dynamical analysis of a generic boolean model for the control of the mammalian cell cycle. Bioinformatics **22**(14), e124–e131 (2006). https://doi.org/10.1093/bioinformatics/btl210
11. Gao, S., Sun, C., Xiang, C., Qin, K., Lee, T.H.: Learning asynchronous Boolean networks from single-cell data using multiobjective cooperative genetic programming. IEEE Trans. Cybern. **52**(5), 2916–2930 (2022). https://doi.org/10.1109/TCYB.2020.3022430
12. Gebser, M., Kaminski, R., Kaufmann, B., Schaub, T.: Clingo = ASP + control: preliminary report. CoRR (2014). https://doi.org/10.48550/arXiv.1405.3694
13. Goldfeder, J., Kugler, H.: BRE:IN - a backend for reasoning about interaction networks with temporal logic. In: Bortolussi, L., Sanguinetti, G. (eds.) CMSB 2019. LNCS, vol. 11773, pp. 289–295. Springer, Cham (2019). https://doi.org/10.1007/978-3-030-31304-3_15
14. Hamey, F.K., Nestorowa, S., Kinston, S.J., Kent, D.G., Wilson, N.K., Göttgens, B.: Reconstructing blood stem cell regulatory network models from single-cell molecular profiles. Proc. Natl. Acad. Sci. **114**(23), 5822–5829 (2017). https://doi.org/10.1073/pnas.1610609114

15. Hérault, L., Poplineau, M., Duprez, E., Remy, É.: A novel Boolean network inference strategy to model early hematopoiesis aging. Comput. Struct. Biotechnol. J. **21**, 21–33 (2023). https://doi.org/10.1016/j.csbj.2022.10.040
16. Heydari, T., et al.: IQCELL: a platform for predicting the effect of gene perturbations on developmental trajectories using single-cell RNA-seq data. PLoS Comput. Biol. **18**(2), e1009907 (2022). https://doi.org/10.1371/journal.pcbi.1009907
17. Hung-Cuong, T., Yung-Keun, K.: CGA-BNI: a novel constrained genetic algorithm-based boolean network inference method from steady-state gene expression data. Bioinformatics **37**, i383–i391 (2021). https://doi.org/10.1093/bioinformatics/btab295
18. Ikonomi, N., Kühlwein, S.D., Schwab, J.D., Kestler, H.A.: Awakening the HSC: dynamic modeling of HSC maintenance unravels regulation of the TP53 pathway and quiescence. Front. Physiol. **11** (2020). https://doi.org/10.3389/fphys.2020.00848
19. Liu, X., Wang, Y., Shi, N., Ji, Z., He, S.: GAPORE: boolean network inference using a genetic algorithm with novel polynomial representation and encoding scheme. Knowl.-Based Syst. **228**, 107277 (2021). https://doi.org/10.1016/j.knosys.2021.107277
20. Martínez-Sosa, P., Mendoza, L.: The regulatory network that controls the differentiation of t lymphocytes. Biosystems **113**(2), 96–103 (2013). https://doi.org/10.1016/j.biosystems.2013.05.007
21. Meyer, P., et al.: A model of the onset of the senescence associated secretory phenotype after DNA damage induced senescence. PLoS Comput. Biol. **13**(12), e1005741 (2017). https://doi.org/10.1371/journal.pcbi.1005741
22. Moignard, V., et al.: Decoding the regulatory network of early blood development from single-cell gene expression measurements. Nat. Biotechnol. **33**(3), 269–276 (2015). https://doi.org/10.1038/nbt.3154
23. Montagud, A., et al.: Patient-specific boolean models of signalling networks guide personalised treatments. eLife **11** (2022). https://doi.org/10.7554/elife.72626
24. Moon, K., Lee, K., Paulevé, L.: Computational complexity of minimal trap spaces in boolean networks. arXiv e-prints (2022). https://doi.org/10.48550/ARXIV.2212.12756
25. Muñoz, S., Carrillo, M., Azpeitia, E., Rosenblueth, D.A.: Griffin: a tool for symbolic inference of synchronous boolean molecular networks. Front. Genet. **9** (2018). https://doi.org/10.3389/fgene.2018.00039
26. Naldi, A., et al.: The CoLoMoTo interactive notebook: accessible and reproducible computational analyses for qualitative biological networks. Front. Physiol. **9**, 680 (2018). https://doi.org/10.3389/fphys.2018.00680
27. Palli, R., Palshikar, M.G., Thakar, J.: Executable pathway analysis using ensemble discrete-state modeling for large-scale data. PLoS Comput. Biol. **15**(9), e1007317 (2019). https://doi.org/10.1371/journal.pcbi.1007317
28. Paulevé, L.: Marker and source-marker reprogramming of most permissive boolean networks and ensembles with BoNesis. Peer Commun. J. **3**, e30 (2023). https://doi.org/10.24072/pcjournal.255
29. Paulevé, L., Kolčák, J., Chatain, T., Haar, S.: Reconciling qualitative, abstract, and scalable modeling of biological networks. Nat. Commun. **11**(1), 4256 (2020). https://doi.org/10.1038/s41467-020-18112-5
30. Réda, C., Delahaye-Duriez, A.: Prioritization of candidate genes through boolean networks. In: Petre, I., Păun, A. (eds.) CMSB 2022. LNCS, vol. 13447, pp. 89–121. Springer, Cham (2022). https://doi.org/10.1007/978-3-031-15034-0_5

31. Riva, S., Lagniez, J.M., López, G.M., Paulevé, L.: Tackling universal properties of minimal trap spaces of boolean networks. In: CMSB 2023: Proceedings of the 21st International Conference on Computational Methods in Systems Biology (2023). https://doi.org/10.48550/arXiv.2305.02442
32. Schwab, J.D., Ikonomi, N., Werle, S.D., Weidner, F.M., Geiger, H., Kestler, H.A.: Reconstructing boolean network ensembles from single-cell data for unraveling dynamics in the aging of human hematopoietic stem cells. Comput. Struct. Biotechnol. J. **19**, 5321–5332 (2021). https://doi.org/10.1016/j.csbj.2021.09.012
33. Thieffry, D., Thomas, R.: Dynamical behaviour of biological regulatory networks-ii. immunity control in bacteriophage lambda. Bull. Math. Biol. **57**, 277–297 (1995). https://doi.org/10.1007/BF02460619
34. Trinh, V.G., Benhamou, B., Paulevé, L.: mpbn: a simple tool for efficient edition and analysis of elementary properties of Boolean networks. arXiv (2024)
35. Werle, S.D., et al.: Unraveling the molecular tumor-promoting regulation of cofilin-1 in pancreatic cancer. Cancers **13**(44), 725 (2021). https://doi.org/10.3390/cancers13040725
36. Yordanov, B., Dunn, S.J., Kugler, H., Smith, A., Martello, G., Emmott, S.: A method to identify and analyze biological programs through automated reasoning. NPJ Syst. Biol. Appl. **2**(1), 1–16 (2016). https://doi.org/10.1038/npjsba.2016.10

Computing Thermodynamically Consistent Elementary Flux Modes with Answer Set Programming

Emma Crisci[1,2], Maxime Mahout[3], and Sabine Peres[1,2](✉)

[1] UMR CNRS 5558, Laboratoire de Biométrie et de Biologie Évolutive, Université Claude Bernard Lyon 1, 69100 Villeurbanne, France
sabine.peres@univ-lyon1.fr

[2] ERABLE, INRIA Lyon Centre, 69100 Villeurbanne, France

[3] LIFEWARE, INRIA Saclay, Palaiseau, France

Abstract. Elementary Flux Modes (EFM) allow the description of the minimal sets of reactions in a metabolic network under steady-state conditions, representing unique and feasible pathways. They fully characterize the solution space but a combinatorial explosion prevents their calculation when the network is large. Furthermore, it is not necessary to calculate all of them, as many of them are not biologically relevant.

Therefore, the software *aspefm*, which combines the use of Answer Set Programming and Linear Programming, proposes to integrate different types of constraints in the EFM computation such as equilibrium constants, Boolean regulatory rules, growth yields and growth medium. The addition of constraints makes it possible to cut off research pathways that lead to non-relevant EFMs. The computation of the EFMs of interest significantly reduces the computational time and saves space. In this article, we have added thermodynamic constraints in terms of the Gibbs energy of reactions, which constrain metabolite concentrations within a chosen interval. This constraint is added as a theory propagator and it reduces the enumeration during the computation. We applied our tool to the central carbon metabolism of *E. coli* and showed that the Gibbs energy constraints suppress a large number of non-relevant EFMs.

Keywords: Metabolic Networks · Elementary flux modes · Logic programming · Thermodynamics

1 Introduction

A metabolic network is a set of successive chemical reactions, catalysed by enzymes, that result in the production of metabolites required by the cell. The concept of elementary flux mode (EFM) plays a crucial role in their analysis from a pathway-focused point of view. An EFM is defined as a smallest sub-network (with respect to reactions set inclusion) that enables the metabolic system to operate at steady state with all irreversible reactions proceeding in the appropriate direction [16]. A major challenge in the computation of EFMs is the large

R. Gori et al. (Eds.): CMSB 2024, LNBI 14971, pp. 80–88, 2024.
https://doi.org/10.1007/978-3-031-71671-3_7

number of EFMs associated with large biochemical networks. One approach to calculating EFMs is to add biological knowledge to restrict the search to biologically relevant EFMs. The idea is not to enumerate the whole set and then select the EFMs of interest, but to restrict the search to biologically feasible solutions. In the double description method [12] which is used by *efmtool*, only the support of monotone constraints such as negative transcriptomic Boolean regulations [9] thermodynamic equilibrium constants [14,15] and the Gibbs energy [7,8] could be integrated into the algorithm. Such constraints significantly reduce the complexity of enumerating EFMs, but may not be sufficient to aid computation in a genome-scale metabolic model. In previous works, we proposed constraint logic programming methods based on SAT Boolean satisfiability to enumerate subsets of EFM of biological interest [11,13], but their efficiency suffered from the complexity of minimizing the size of the solutions searched. Recently, we developed *aspefm* [10], a highly efficient software based on Answer Set Programming (ASP), to search for relevant EFMs respecting different kind of biological constraints, such as transcriptomic regulations, equilibrium constants, growth yields and growth medium.

In this article, we develop the implementation of a new *aspefm* extension, called *DeltaGChecker*. *DeltaGChecker* specifically checks whether the Gibbs free energies, or thermodynamics, of reactions are possible. With this new constraint type, we aim to reduce the total number of EFMs and the complexity of computations. We test our *DeltaGChecker* on one small-scale metabolic networks based on *Escherichia coli*. We observe that the number of EFMs retrieved is particularly dependent on the minimum and maximum bounds chosen for the metabolite concentration variables and that the extension does not lose significant amounts of computational time.

2 Computation of EFMs Compatible with Thermodynamic Constraints

A metabolic network is represented by a stoichiometric matrix $S \in \mathbb{R}^{|\mathcal{M}| \times |\mathcal{R}|}$, where $|\mathcal{M}|$ and $|\mathcal{R}|$ are the numbers of metabolites and reactions, respectively. Each element $S_{m,r}$ denotes the stoichiometric coefficient of metabolite m in reaction r. The steady-state condition is expressed as $S \cdot v = 0$, where $v \in \mathbb{R}^{|\mathcal{R}|}$ is a flux vector. A flux mode v satisfies two conditions $S \cdot v = 0$ and $v_i \geq 0$ for all $i \in irrev$, where $irrev$ is the set of indices corresponding to irreversible reactions. An Elementary Flux Mode (EFM) e is a flux mode with minimal support, where the support is defined as $supp(e) = \{r \in \mathbb{N} : e_r \neq 0\}$. Depending on the size of the metabolic network, there may be a combinatorial explosion in the search for EFMs. Even if it could be possible to enumerate billions of EFMs, for some networks, analysing in post-processing all the EFMs would be time-consuming. *aspefm* allow us to directly find the EFMs of interest, by exploring only solutions that satisfy the constraints. Several types of constraints can be added, including the so-called linear constraints, referring to the field of linear programming.

By analogy, our thermodynamics extension *DeltaGChecker* checks the Gibbs free energies of reactions by using linear programming. Linear programming allows the algorithm to check whether a given support of an EFM in construction (the set of active reactions in the EFM) satisfies these thermodynamic constraints. To determine if an EFM is thermodynamically feasible, the extension finds a set of metabolite concentrations that satisfies the thermodynamic constraints.

2.1 Thermodynamics of Chemical Reactions and Formulation in LP

Thermodynamics is the study of energy transformations that occur within mechanical systems. In thermodynamics, chemical reactions are analysed from an energetic perspective, in order to predict whether a given reaction is favorable or unfavorable.

In this article, we will focus on the second law of thermodynamics, expressed through the Gibbs free energy equation, which provides a criterion for determining the spontaneity and direction of chemical reactions [1].

The Gibbs free energy of formation ($\Delta_F G'$) of metabolites is calculated using the Gibbs free energies of formation in their standard states ($\Delta_F G'^{\circ}$), the molar gas constant, R and the temperature, T. The standard state is thus at the metabolite level. The free energy of formation denoted $\Delta_F G'$ is the Gibbs energy associated with the formation of one mole of a substance from its constituents in their standard states at a reference pressure and temperature. In other words, it is the energy required to form one mole of a substance from its most stable components. The sign of $\Delta_R G_r$ gives the direction of the reaction r. If the calculated $\Delta_R G_r$ is strictly negative, the reaction is thermodynamically favorable under current conditions. Conversely, if $\Delta_R G_r$ is positive, the reaction is unfavorable or requires additional energy to make it possible. Thus, no EFM can therefore have reactions in that direction. Finally, if the energy $\Delta_R G_r$ is null, this means that the reactants and products are in thermodynamic equilibrium.

The thermodynamic constraints can be expressed as a linear program (1). The principle behind this linear program is to find a set of metabolite concentrations such that thermodynamics is respected throughout the metabolic network. The objective function will thus be expressed as the minimisation of 0, *i.e.* no attempt is made to minimise or maximise a function. Constraints on chemical reactions are tested for a set of active reactions in the metabolic network, support of a potential EFM e.

$$\begin{cases} \min & 0 \\ \text{s.t} & \Delta_R G_r \leq 0, \quad \forall r \in supp(e) \\ & \Delta_R G_r = \sum_{m=1}^{M} S_{m,r} \Delta_F G'_m \\ & \Delta_F G'_m = \Delta_F G'^{\circ}_m + RT \ln\left(C_m / C_0\right), \quad C_0 = 1M \\ & \ln\left(C_m^{\min}/C_0\right) \leq \ln\left(C_m/C_0\right) \leq \ln\left(C_m^{\max}/C_0\right) \end{cases} \tag{1}$$

Finally, the linear program verifies that all reactions contributing to an EFM (present in the support), are simultaneously feasible and consistent with a metabolome with given error bounds C_m^{min} and C_m^{max}.

The integration of Gibbs free energy into EFM calculations has already been implemented as an extension to *efmtool* [17] in the thermodynamics EFM analysis (tEFMA) [7,8]. *tEFMA* proposes to add the same thermodynamic constraints in the same LP form. *tEFMA* uses the double description algorithm [12], which is an incremental method and does not enumerate solutions one-by-one. Moreover, it can integrate only support monotone constraints during the computation. Filtering EFMs with multiple constraints during enumeration is promising for finding EFMs of interest in large networks. In this article, we integrate this linear program into *aspefm*, which allows additional constraints to be combined simultaneously and can enumerate solutions one-by-one. The *aspefm* tool defines a logic program in ASP to be solved using a modified version of the *clingo*[LP] solver, making use of high-performance state-of-the-art SAT-solving techniques, as well as expressive IBM CPLEX linear constraints [10].

2.2 *DeltaGChecker* Extension to *aspefm*

Thermodynamic constraints require additional data, including the standard Gibbs free energies of formation $\Delta_F G'^{\circ}_m$ and the allowable concentration ranges for each metabolite, and can be found in the database as Equilibrator [5]. The minimum and maximum concentrations for each metabolite are user-defined. Users can base their choices on their own experimental data or from relevant databases. To retrieve the thermodynamic data, we developed a script using the well-documented Equilibrator API. In particular, we used the following standard conditions recommended by Equilibrator: a pH of 7.4, an ionic strength of 0.25M and a temperature of 298.15K [2]. More details on input data requirements, API scripts, file transformation utilities, and code for *aspefm* and the new *DeltaGChecker* extension are available in our GitLab repository: https://gitlab.inria.fr/erable/aspefm.

The *DeltaGChecker* extension is directly integrated into the EFMs calculation of *aspefm*. In the SAT resolution algorithm of *aspefm*, EFMs are constructed reaction by reaction, as partial supports. A partial support is defined here as an unfinished sketch of an EFM support in the constraint propagation. By evaluating the thermodynamic constraints on partial supports, the *DeltaGChecker* extension ensures that any added reactions in the EFM retain their its feasibility. The linear program solver CPLEX (1) is called for each partial support constructed and the feasibility of the $\Delta_R G_r$ energies for the calculated metabolite concentration variables constrained by the min/max concentration bounds is checked [4]. If the linear program has no solutions, then the thermodynamics for that partial set of active reactions is considered not respected. The invalidation of thermodynamics leads to the addition of 'nogoods' in the SAT resolution, which cut off the search for certain EFM search branches. In the case of *aspefm* and *DeltaGChecker*, nogoods are added for sets of reactions that are incompatible with each other. For rest of the resolution, the branches explored will not consider the forbidden sets of reactions together.

3 Application to Central Carbon Metabolism of *E.coli*

The aim, with a simple model as *E.coli* from the published model [3], is to show that the extension encoding thermodynamic constraints can restrict us to only relevant EFMs by eliminating those for which it is impossible to find a metabolite concentration solution. When validating the thermodynamic extension, the idea is to specify for each metabolite an interval around the reference value associated with the metabolite (to be found in the literature) and to compare the number of associated EFMs. In other words, if the percentage is x, then each metabolite will be assumed to have a concentration within a range of values between:

$$\begin{cases} 0 \leq x \leq 100, & \text{for all } x \quad \text{in percentage} \\ C_{\text{ref}} \times (1 - \frac{x}{100}) \leq C_{\text{ref}} \leq C_{\text{ref}} \times (1 + \frac{x}{100}) \end{cases} \tag{2}$$

Thus, the higher the percentage, the less the metabolites are constrained in their concentration as they have a wider range of values, *i.e.* the solver has more freedom in choosing the optimal concentration set to find. Increasing the percentage range directly reduces the constraints on thermodynamics. In the rest of the study, we will compare the EFMs found in *aspefm* with and without the use of thermodynamics, to show how the new extension reduces the number of relevant EFMs. The metabolic network of *E.coli* consists of 55 reactions, 46 metabolites and 1658 EFMs without any constraint.

3.1 Selection of EFMs of Thermodynamic Interest in *E.coli*

In this *E. coli* model, some metabolite concentrations were unknown. We first set their reference values to 10^{-5} M. Relevant EFMs (satisfying thermodynamic constraints) were found only from 50% onwards, increasing to 99% where we reached 1658 EFMs, matching the number without thermodynamic constraints (Fig. 1).

Instead of setting a reference concentration for the metabolite that does not have one, we then set the predefined interval from 10^{-7} to 1 M as in *tEFMA*. It can be seen that with the interval set for metabolite concentrations, more EFMs are obtained than with the use of a reference concentration, *e.g.* for 75%, 100 times more EFMs are obtained with the imposed interval. This is because the fixed interval is large, so less thermodynamic stress is applied.

Even though about 15% of the metabolite concentrations were unknown, we were successfully able to limit the computation to thermodynamically consistent EFMs by setting default reference concentrations or default bounds. Therefore, we highlight that knowing the concentrations for all the metabolites in the metabolic model might not always be necessary. In summary, the *DeltaGChecker* extension efficiently reduces the amount to only relevant thermodynamics EFMs by incorporating thermodynamic constraints. However, for these constraints to be effective, the concentration intervals provided must be sufficiently stringent.

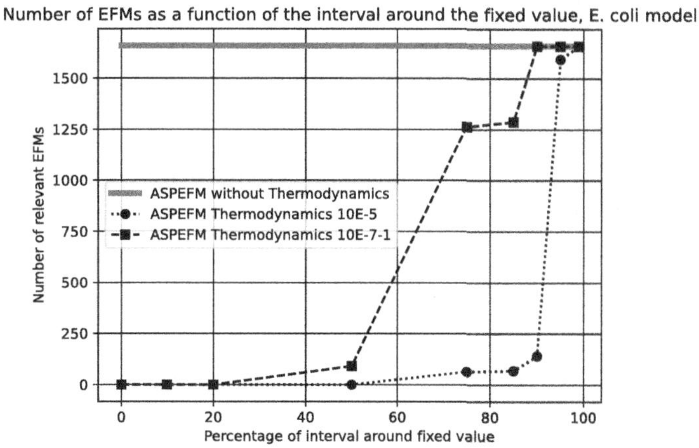

Fig. 1. Evolution of the number of EFMs selected as a function of metabolite concentration constraints, in *E.coli.* Abscissa are percentages of interval around the reference concentration values.

4 Conclusion

The *DeltaGChecker* extension gives *aspefm* a strong advantage in finding relevant EFMs. The addition of thermodynamic constraints significantly reduces the number of EFMs in most cases, provided the system has enough constraints. Our *DeltaGChecker* extension determines, for a potential EFM support, whether there exists a set of metabolite concentrations such that the system satisfies the thermodynamic constraints. The extension has been tested on the *E. coli* model, and we have shown that it reduces the number of EFMs, by an order of magnitude, allowing faster post-processing analysis.

It should be noted that at the scale of the applications studied, the aim is mainly to reduce the number of EFMs, rather than to reduce the computation time at all costs. For larger networks, where the combinatorial explosion is strong, it will be important to know whether the extension actually multiplies computation time or reduces it. For this reason, we have performed additional analysis on the computational time (Appendix A.1). We have concluded that the extension has a certain cost in computation time, but this time could probably be worthwhile on larger networks. In the case of large networks, where in some cases the aim is simply to enumerate EFMs of interest, the idea of using thermodynamic constraints to avoid traversing certain paths, and thus reduce computation, seems promising. Further extensions will be developed in the future, in particular to incorporate kinetic and enzyme concentration constraints.

A Appendix

A.1 Computation Times on *E.coli*

The thermodynamic extension makes it possible to reduce the number of EFMs of interest, and therefore the post-processing search, but it's interesting to look at and compare what the extension can bring to processing. The extension reduces the number of paths used to search for EFMs, and therefore saves time on paths that we know to be thermodynamically infeasible. On the other hand, the extension will waste time on path traversal, because for each EFM support tested, it will be necessary to check whether the support has a concentration set solution that validates the thermodynamics, which implies running the CPLEX solver. So, for the extension to be cost-effective in terms of computation time, the absence of some non-thermodynamically feasible EFM paths would have to be beneficial in terms of the computation time added to test whether each partial support has a thermodynamically feasible solution.

This figure shows the same percentages used in the previous study (Fig. 2). It was seen that the lower the percentage, the more the extension restricted the number of EFMs of interest, so that the extension must have had an impact

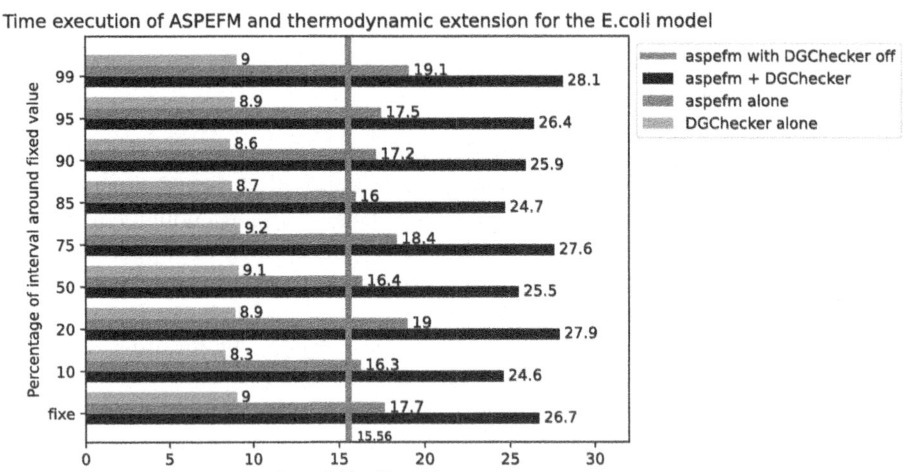

Fig. 2. Evaluation of *aspefm* computation times for *E. coli* depending on percentages around the selected intervals; 1) *aspefm* with DGChecker off: *aspefm* times for computing *E.coli* model EFMs without using extension *DeltaGChecker*; 2) *aspefm* + *DGChecker*: *aspefm* times for computing *E. coli* model EFMs using the *DeltaGChecker* extension, for each percentage around the fixed concentration value; 3) *aspefm* alone: for each percentage, portion of time spent computing *aspefm* solutions, without checking thermodynamics; 4) *DGChecker* alone: for each percentage, portion of time spent checking thermodynamics in the *DeltaGChecker* extension.

on the paths taken by the EFMs. As the percentage increased, the number of EFMs of interest increased, approaching the number found by *aspefm* without the addition of the extension, due to the fact that the thermodynamic constraints were not strong enough. We note that in the case of the *E.coli* model, computation times of *aspefm* oscillates between 25.6 and 28.4 s while not increasing with the percentage as would be expected; time within and outside the extension remains approximately constant. Surprisingly, filtering out more solutions had no effect on the computation time. This result might be explained by the fact that the network is not large enough, i.e. the time saved by not traversing certain paths is not visible at this scale. Overall, when removing the times taken by the extension, we find that the differences in time are almost all attributable to the extension, with a time difference of around ~ 2 s (on average 17.51 against 15.56).

This is a promising result, and we hope that computation times for our filtering procedure will be able to scale to larger networks.

References

1. Atkins, P.: The Laws of Thermodynamics: A Very Short Introduction. OUP, Oxford (2010)
2. Beber, M.E., et al.: eQuilibrator 3.0: a database solution for thermodynamic constant estimation. Nucleic Acids Res. **50**, D603–D609 (2022)
3. Carlson, R., Srienc, F.: Fundamental Escherichia coli biochemical pathways for biomass and energy production: identification of reactions. Biotechnol. Bioeng. **85**, 1–19 (2004)
4. CPLEX Users Manual: IBM ILOG CPLEX optimization studio. Version. 12, 1 (1987)
5. Flamholz, A., Noor, E., Bar-Even, A., Milo, R.: eQuilibrator-the biochemical thermodynamics calculator. Nucleic Acids Res. **40**, D770–D775 (2012)
6. Gebser, M., Kaminski, R., Kaufmann, B., Ostrowski, M., Schaub, T., Wanko, P.: Theory solving made easy with Clingo 5. In: Technical Communications of the 32nd International Conference on Logic Programming (ICLP 2016). Schloss-Dagstuhl - Leibniz Zentrum für Informatik (2016)
7. Gerstl, M.P., Jungreuthmayer, C., Zanghellini, J.: tEFMA: computing thermodynamically feasible elementary flux modes in metabolic networks. Bioinformatics **31**, 2232–2234 (2015)
8. Gerstl, M.P., Ruckerbauer, D.E., Mattanovich, D., Jungreuthmayer, C., Zanghellini, J.: Metabolomics integrated elementary flux mode analysis in large metabolic networks. Sci. Rep. **5**, 8930 (2015)
9. Jungreuthmayer, C., Ruckerbauer, D.E., Zanghellini, J.: regEfmtool: speeding up elementary flux mode calculation using transcriptional regulatory rules in the form of three-state logic. Biosystems **113**, 37–39 (2013)
10. Mahout, M., Carlson, R.P., Peres, S.: Answer set programming for computing constraints-based elementary flux modes: application to escherichia coli core metabolism. Processes **8**, 1649 (2020)
11. Morterol, M., Dague, P., Peres, S., Simon, L.: Minimality of metabolic flux modes under boolean regulation constraints. In: Workshop on Constraint-Based Methods for Bioinformatics (WCB) (2016)

12. Motzkin, T.S., Raiffa, H., Thompson, G.L., Thrall, R.M.: The double description method. In: Kuhn, H.W., Tucker, A.W. (eds.) Contributions to Theory of Games, vol. 2. Princeton University Press, Princeton (1953)
13. Peres, S., Morterol, M., Simon, L.: SAT-based metabolics pathways analysis without compilation. In: Mendes, P., Dada, J.O., Smallbone, K. (eds.) CMSB 2014. LNCS, vol. 8859, pp. 20–31. Springer, Cham (2014). https://doi.org/10.1007/978-3-319-12982-2_2
14. Peres, S., Schuster, S., Dague, P.: Thermodynamic constraints for identifying the elementary flux modes. Biochem. Soc. Trans. **46**, 641–647 (2018)
15. Peres, S., Jolicœur, M., Moulin, C., Dague, P., Schuster, S.: How important is thermodynamics for identifying elementary flux modes? PLoS ONE **12**, 1–20 (2017)
16. Schuster, S., Hilgetag, C.: On elementary flux modes in biochemical reaction systems at steady state. J. Biol. Syst. **2**, 165–182 (1994)
17. Terzer, M., Stelling, J.: Large-scale computation of elementary flux modes with bit pattern trees. Bioinformatics **24**, 2229–2235 (2008)

Batch Effect Correction in a Confounded Scenario: a Case Study on Gene Expression of Chornobyl Tree Frogs

Elen Goujon[1,3], Olivier Armant[2], Clément Car[2], Jean-Marc Bonzom[2], Arthur Tenenhaus[3], and Imène Garali[1](✉)

[1] French Institute for Radiation Protection and Nuclear Safety (IRSN), PSE-SANTE/SESANE/LRTOX, 92260 Fontenay-aux-Roses, France
imene.garalizineddine@irsn.fr
[2] IRSN, PSE-ENV/SERPEN/LECO, 13115 Saint-Paul-Lez-Durance, France
[3] Laboratoire des Signaux et Systèmes, Université Paris-Saclay, CNRS, CentraleSupélec, 91190 Gif-sur-Yvette, France

Abstract. When large omics datasets present unwanted latent variability, a critical analysis step is to control these so-called batch effects properly. However, most batch effect-correction algorithms (BECAs) face limitations when the source of unwanted variation and the variable of interest are confounded. In this paper, we use RNA-seq data to study the effects of radiation contamination on tree frogs (*Hyla orientalis*) collected in the Chornobyl Exclusion Zone. We identify the site of collection of the frogs as a confounding factor in the transcriptomics analysis. We present our strategy to correct this confounding effect using the following BECAs: ComBat-seq, linear residualization, and Surrogate Variable Analysis. We show that the severe confounding between the site and radiocontamination level makes the correction step challenging. Instead, we investigate the site-to-site variability and successfully deconvolute the batch variable from the radiation level by adjusting for the population genetic structure. Our strategy allowed us to reveal the effects of low-dose radiation on the gene expression of Chornobyl tree frogs and appropriately preprocess the RNA-seq dataset for future multimodal integrative analyses.

Keywords: Confounding factors · Batch effect-correction algorithms · Transcriptomics · Chornobyl tree frogs · Low-dose radiation

1 Introduction

In the current age of research in biology, high-throughput data such as omics measurements represent an immense wealth of information, assuming that they are processed appropriately. The analysis of such datasets often aims at identifying differential responses between classes of study –for example, groups of treatment, disease, or levels of exposure to a pollutant. However, unwanted variation can arise in datasets, stemming from technical or biological origins, and be

R. Gori et al. (Eds.): CMSB 2024, LNBI 14971, pp. 89–107, 2024.
https://doi.org/10.1007/978-3-031-71671-3_8

detrimental to the analysis by leading to misguided interpretations [12,13,21]. The term "batch effects" describes this underlying variability, and their handling has become a crucial step in data analysis.

Batch effects can be tackled by one of two strategies. Either include the batch in the model as a covariate in a one-step approach or correct for the batch in a pre-processing step using batch effect-correction algorithms (BECAs) before any downstream analysis [24,29]. In RNA-Seq data analysis, various differential expression methods allow the integration of batch effects and covariates in the model design [9,25,30]. While one-step methods can be preferred because of their simplicity, it is not always possible to integrate covariates, depending on the model type. Two-step batch adjustment allows more flexibility, and the myriad of bioinformatics tools dedicated to batch-effect correction should allow one to choose a context-appropriate BECA [41].

Still, difficulties emerge in scenarios where the batch and the study class or variable of interest have a simultaneous influence on the data. If batch and class are confounded, then adjusting the dataset to differentiate the effects of one from the other becomes more challenging [12,29,39,41]. BECA comparison studies have produced conflicted results in the case of batch-class design imbalance, with tools such as ComBat [16], Surrogate Variable Analysis (SVA) [22], batch mean-centering (BMC) [32], and ratio-based methods performing differently from one example to another [39,41].

In this study, we use a published dataset on the effects of chronic exposure to low-dose radiation on wildlife in the aftermath of the nuclear accident in Chornobyl[1] [3,8]. In particular, 87 tree frogs from species *Hyla orientalis* were sampled in 8 sites inside and outside of the Chornobyl Exclusion Zone (CEZ) in 2018 [8]. Previous research efforts have shown the presence of recent evolutionary processes at play in the CEZ [7]. Also, altered transcriptomic and metabolic pathways were identified, as well as changes in physiological traits for the frogs living in the most contaminated sites [8].

Different data modalities have been obtained, and we plan to conduct a multi-omics data integration study using transcriptomics, proteomics, and genomics data. In particular, for RNA-Seq expression data, we have found that this dataset is impacted by the site of collection of the frogs in a batch effect-like manner. Given the distribution of the geographical sites in the Chornobyl area, the batch is highly correlated with the level of exposure (μGy/h), which is our variable of interest. Before integrating data in a multi-omics study, we want to identify an appropriate approach for handling this batch effect.

In this paper, we present three different strategies to correct for the effect of the site. First, we adjusted the count matrix for the site effect using classical techniques such as ComBat-seq [40] and linear residualization [11], with and without integrating exposure levels in the correction step. In a second strategy,

[1] We use the official spelling Chornobyl, in accordance with the romanization of Ukrainian geographical names recommended by the 10th United Nations Conference on the Standardization of Geographical Names (see https://mfa.gov.ua/en/correctua).

we dissected the site variable to decouple the site effects from the radiation dose rate. We provide a new approach by integrating another data source: the genetic distance between individuals, which partly explains the site variability, is treated as a batch effect and corrected. Lastly, we use an exploratory approach to estimate the directions of systematic expression variation not explained by radiation exposure, thanks to SVA [22].

We structure the remaining parts as follows. In Sect. 2, we present the data handled in this study. Section 3 introduces the methods used for batch-effect correction, batch-effect assessment, and performance evaluation. The results of the different correction strategies are shown in Sect. 4, along with statistical and biological comparisons. We further discuss our results in Sect. 5 and conclude in Sect. 6.

2 Tree Frog Sampling and Dataset Description

The 87 individuals studied here are male Eastern tree frogs (*Hyla orientalis*) collected in Ukraine in 2018 across a gradient of radioactive contamination. Population sampling was done in 8 different wetlands in the Chornobyl Exclusion Zone (sites A18, B18, C18, D18, E18, and F18) and a neighboring non-contaminated area near Slavutych (sites G18, and H18) [7,8]. The individual total dose rate (ITDR, in μGy/h) absorbed by each tree frog was estimated by assessing internal and external exposure levels, as previously detailed [5,7]. The ITDR reflects the energy deposited into the frog's body per unit of time. Given that radiation sources are external (the soil, for example) and internal (by ingestion, respiration, ...), the intensity of this deposit is a function of radiation energy, as well as of the organisms' shape, composition, and lifestyle [4,5,7]. Missing values for ITDR of 3 individuals were imputed using the site median.

2.1 Transcriptomic Data

The RNA-Seq expression data is published in [3]. For each frog, total RNA was extracted from tibia muscle, and RNA-Seq analysis was performed as previously explained [8]. *De novo* assembly of transcriptome was performed with Trinity using 3 additional tree frogs captured in non-contaminated sites [8,14]. Estimations of transcript abundance were mapped against the transcriptome with Bowtie2, and their quantification was done with RSEM [19,23]. R packages `tximport` and `DESeq2` were used to import the quantification data and assemble the RNA-Seq pseudo count matrix [25,33]. Filtering was applied to remove genes with an expression quantified to 0 for all samples. Given the characteristic mean-variance dependence in RNA-Seq count data, a variance-stabilizing transformation (VST) was applied to the count matrix, producing an approximately homoskedastic matrix with normalized and log-transformed counts [1].

2.2 Genetic Distance

Interindividual genetic distances were obtained by computing the Euclidean distance between SNP genotypes with R package `vcfR` [17]. Genetic distances were summarized into a 3-level categorical variable by performing hierarchical clustering with a complete linkage method.

3 Methods

In this section, we first present the models we used for the correction of batch effects. In the context of our tree frog study, we then describe the exploratory analyses that led to identifying the sources of unwanted variation, the application of BECAs, and the qualitative and quantitative assessment of their results. Computations were performed using R.

3.1 Batch Effect-Correction Algorithms

In the presence of confounding factors, BECAs can be needed to remove the unwanted sources of variation before performing further analyses. Here, we describe three popular batch effect-removal strategies. Linear regression on the batch variable followed by extraction of residuals and ComBat-seq assume that batch factors are known and correct their effect. SVA is used to estimate the unwanted sources of expression heterogeneity.

Linear Residualization. Probably the most straightforward strategy when dealing with batch effects is to regress out (or residualize) the batch effects in a pre-processing step. This is done by adjusting a simple (or multiple) linear regression model on each feature, using one or more batch variables as regressors, and extracting the residuals [11]. This approach allows the inclusion of several confounders of varied types in the correction process, assuming that batch factors have additive effects.

In the classic scenario where there is only one categorical batch factor, removing the batch effect-associated variability using a linear model amounts to performing batch mean-centering (BMC) [32]. BMC adjustment leads to the average value of each variable being zero within each batch.

When batch effects and variable of interest have an unbalanced distribution, removing batch effects can lead to the suppression of actual between-group variation [29]. To avoid over-correction and loss of meaningful biological signal, it is possible to model the effects of the outcome y in the correction step to preserve its influence on the data, as is done by other tools [16,30,36,40]. Let $z^{(1)},\dots,z^{(b)}$ denote b factors, either categorical batch effects or other continuous covariates. Under the assumption that outcome-associated effects are additive to batch effects, the expression of gene g for sample $i \in \{1,\dots,n\}$ is modelled as such:

$$\begin{aligned} X_{ig} &= \hat{\beta}_{g0} + \hat{\beta}_{g1} z_i^{(1)} + \cdots + \hat{\beta}_{gb} z_i^{(b)} + \hat{\beta}_{g(b+1)} y_i + \varepsilon_{ig} \\ &= \hat{\beta}_g^\top z_i + \varepsilon_{ig} \end{aligned} \quad (1)$$

with $z_i = (1, z_i^{(1)}, \ldots, z_i^{(b)}, y_i)^\top$ containing the batch variables and the variable of interest y, used as regressors, $\hat{\beta}_g = (\hat{\beta}_{g0}, \hat{\beta}_{g1}, \ldots, \hat{\beta}_{gb}, \hat{\beta}_{g(b+1)})^\top$ the vector of coefficients estimated by Ordinary Least Squares, and ε_{ig} the residual part. The expression profile is then corrected for the batch effects:

$$\begin{aligned} X_{ig}^\star &= X_{ig} - (\hat{\beta}_{g0} + \hat{\beta}_{g1} z_i^{(1)} + \cdots + \hat{\beta}_{gb} z_i^{(b)}) \\ &= \varepsilon_{ig} + \hat{\beta}_{g(b+1)} y_i \end{aligned} \tag{2}$$

This is also the spirit of the function `removeBatchEffects` from R package `limma` which allows removing the effects of batches and other continuous covariates, while preserving the effect of a condition variable on the log-transformed counts [30].

ComBat-Seq. ComBat-seq is designed for RNA-Seq count data and is a descendant of ComBat, a widely popular tool for correcting batch effects from known sources [16,40]. ComBat-seq uses a gene-wise negative binomial regression model to estimate batch effects on the mean and dispersion parameters. After regression parameters are estimated using the raw count matrix, a new batch-corrected distribution is computed and used to adjust the count matrix by quantile mapping. The biological condition of interest can be included in the modeling to protect between-group differences. We used ComBat-seq both with and without providing ITDR in the covariate model matrix. ComBat-seq is part of the R package `sva` [20].

Surrogate Variable Analysis. Unknown sources of variability present in the datasets can be uncovered using SVA [22]. Unobserved variation sources are identified and estimated as follows: after the expression dataset is residualized on the primary variable of interest, the remaining sources of expression heterogeneity are extracted by eigendecomposition and translated into surrogate variables (SVs). These estimated batches can later be used as covariates in subsequent analyses, such as differential expression modeling, or can be adjusted for by residualization prior to visualization algorithms such as Principal Component Analysis (PCA). SVA is implemented in the namesake R package [20]. Function `num.sv` allows the estimation of the number of SVs to be calculated with `sva`.

3.2 Application to the Tree Frog RNA-Seq Study

In the context of the transcriptomic data analysis of Chornobyl tree frogs, we considered the batch effects stemming from the site of collection and the different genetic backgrounds. We also computed and studied surrogate batch variables.

Batch Effect Assessment. We performed PCA on the gene expression dataset to extract components that reflect the largest sources of variability in the data. Data exploration of the first components allowed visualizing the effect of the confounders. Indeed, we expect that principal components will correlate strongly with factors that have the most impact on gene expression. PCA was computed on the VST-transformed count matrix using R function `prcomp`.

Following published batch effect-assessment methodology [21], we tested the association between the 5 first principal components and the qualitative covariates by analysis of variance (ANOVA). We show the rank of the principal component presenting the strongest correlation with the batch using the R-squared value, denoted R^2. R^2 is interpreted as the percentage of variation in the principal component explained by the batch effects.

Similarly, the strength of confounding between categorical batch variables and total dose rate was assessed by generalized R^2 [21]. A correlation of 0 indicates independence between batch and radiocontamination, i.e., a low confounding level, while a value close to 1 indicates severe to complete confounding.

Batch Effect Correction

Targeted Variables. The collection site was primarily considered as the batch variable to target. Further investigation led to the creation of groups based on genetic distance, which were also used as a batch factor. Lastly, a Surrogate Variable Analysis was carried out to estimate 3 SVs, as advised by `num.sv`.

Correction Techniques. For all candidate variables, we applied batch-adjustment algorithms, which returned batch-adjusted matrices. Linear residualization was performed on the VST-transformed count matrix. We combined this method with our three approaches: correction aimed at the site, the genetic group, or the SVs. ComBat-seq was applied to the pseudo count matrix, using either the site or the genetic group, and the adjusted matrices were then transformed via VST. ComBat-seq could not be used to remove the effects associated with the 3 SVs, as it can only accommodate a single batch variable.

For each experiment, we examined the impact of incorporating the variable of interest into the correction step. In the residualization method, this was accomplished by introducing the dose rate as a regressor. With ComBat-seq, the dose rate was included in the covariate model matrix. Table 1 recapitulates the batch-adjustment approaches used.

Performance Evaluation

Visualization and Summary Statistics. We conducted PCA on the batch-adjusted matrices to evaluate the performance of BECAs. We used visual comparisons by plotting individuals in the first factorial plane. The association of principal components with the radiation level was tested using the absolute value of Pearson's correlation to check the preservation of relevant biological information.

Table 1. Batch effect-correction strategies

Strategy	Targeted batch	BECA
1	Site	Linear residualization
2	Site	ComBat-seq
3	Genetic group	Linear residualization
4	Genetic group	ComBat-seq
5	Surrogate variables	SVA & Linear residualization

Functional Enrichment Analysis. We performed a biologically informed comparison of batch-adjustment strategies by studying the functional classes of genes that were differentially regulated among individuals [26]. Using sparse principal component analysis (sPCA), we selected features (genes) that capture a maximum amount of variance across the radiocontamination gradient for all batch-adjusted datasets. Indeed, with the addition of an L1-penalty, sPCA allows the parametrization of the number of non-zero weights assigned to the variables to form principal components [37]. Using function `SPC` from R package `PMA`, we applied sPCA on the variance-stabilized counts and the batch-adjusted matrices. For each dataset, the degree of sparsity in the weight vector was adjusted to yield around 400 non-zero weighted genes in the first component.

A list of stable genes was identified by replicating sPCA on 1000 bootstrap samples. The most stable genes correspond to those selected in more than 70% of cases in either the first or second component, or in both. Bootstrap-iterating the computing of sPCA reveals the robustness of feature selection with respect to resampling. We focused on the first two components to capture the most important directions of variation in the expression. Stable gene names were then mapped to UniProt IDs for a GO term enrichment analysis performed with R packages `clusterProfiler` [38] and `enrichplot` [38]. This enrichment analysis allowed us to study the biological processes over-represented in the list of deregulated genes.

4 Results

We examine the confounding effect of the collection site and the impact of its correction using ComBat-seq and residualization. We then present our second approach, which involves the correction of genetic stratification of the populations. Additionally, we examine the results of SVA-based correction, which we use to identify potential sources of unwanted variation that may have been overlooked. Finally, batch effect-mitigation strategies targeting the genetic group and SVs are further compared through functional enrichment analysis.

4.1 Confounding Effect of the Collection Site

The tree frogs were collected in 8 geographical sites in order to gather samples across a continuum of radiation-pollution levels. The distribution of the sites, both inside the CEZ and outside, is shown in Fig. 1a. Tree frogs living in different sites exhibit distinct expression patterns, which is illustrated by individuals clustering by site in the PCA of Fig. 1c. Given the distribution of radiation pollution in the Chornobyl area, individuals captured in one site have likely been exposed with a similar intensity, as shown in Fig. 1b. This is reflected by the high level of confounding between the site and the variable of interest. However, Fig. 1c illustrates that the variation in expression captured by the first PCs poorly reflects the radiocontamination variation, which raises the need to distinguish between site- and radiation exposure-related effects.

To address this confounding effect, we applied batch-correction techniques targeting the site, and we obtained batch-corrected matrices. We handled the site effect using linear residualization on the site and ComBat-seq, both with or without including radiation levels in the correction step. In Fig. 1d, we correlate principal components obtained on the raw and site-adjusted matrices. Adjusting the dataset without modeling the radiation-associated effect appears to have failed to preserve the biological information of interest since no principal components are correlated with ITDR after correction. On the other hand, batch-mitigation models that included the radiation level seem to have increased the radiation-correlated variance. This evidence suggests that removing the site-associated variation leads to the suppression of relevant biological information if the variable of interest is not protected in the correction step. However, several studies have raised the issue of the unreliability of BECAs when adjusting for a highly confounded batch while preserving class differences [24,29,41]. The concern is that artificial variation in the direction of dose rate variation could be introduced into the "corrected" dataset, resulting in exaggerated significance in downstream analyses [29]. Without ground-truth measurements to test for this issue, we sought a way to deconvolute the batch effects from the variable of interest.

4.2 Genetic Diversity Treated as Batch Effects

In a second approach, we wondered which parameters, other than radionuclide pollution, could drive the expression heterogeneity among populations. Firstly, all the frogs were captured in wetlands and presumably had the same living conditions in terms of food availability, temperature, or habitat type. Secondly, tree frog populations were sampled site by site, visiting one or two sites per day over a 9-day period. Therefore, even with a carefully thought-out protocol, we cannot exclude that the effect of the site may have a technical cause related to the capture and handling of the frogs. Lastly, the geographical distance between sites (80 km between the two most distant) and the limited dispersal radius of this species could translate into genetic background dissimilarities, and in turn have an impact on transcription levels.

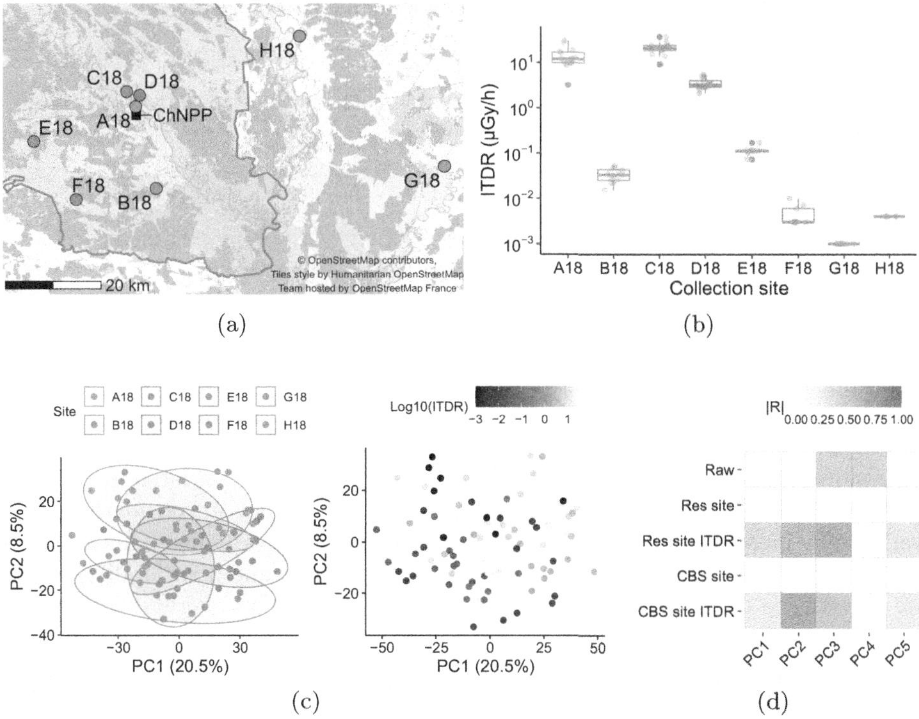

Fig. 1. Confounding effect of the site of collection: effect visualization and batch correction **a.** Geographical location of tree frog-collection sites inside the CEZ (purple line) and in the Slavutych region (ChNPP = Chornobyl nuclear power plant) **b.** Radiation dose rate distribution among sites **c.** PCA of raw RNA-seq count matrix with individuals colored by collection site (left) or dose rate (right) **d.** Absolute value of the correlation between radiation level and PCs of site-corrected count matrices (CBS = ComBat-seq, Res = Residualization, ITDR = ITDR was included in the batch-correction step to preserve its effects)

Interindividual genetic distances were used to integrate this aspect into the analysis and were translated into a categorical batch variable by hierarchical clustering, as represented by Fig. 2a. The obtained batch factor is less confounded with ITDR than the site: the site exhibited a correlation of 80% with ITDR, compared with 8% between genetic group and ITDR (Table 2). Also, the range of dose rate values per group is greater within genetic groups than within sites, as we can observe in Fig. 2b. Table 2 shows that the genetic factor is still strongly correlated with the third principal component. We note that the principal component that carries the most variance in the direction of radiation contamination is of a lesser rank.

We performed genetic group adjustment with the techniques mentioned previously, and we present in Fig. 2 the results of PCA conducted on the adjusted matrices. Figure 2c shows that the first two PCs, and especially the second,

Table 2. Confounding effects of site and genetic distance

	Batch variables		Variable of interest
	Site of collection	Genetic distance group	Radiocontamination level (ITDR)
Confounding with ITDR, R^2	0.80	0.08	×
Rank of PC with highest correlation (R^2)	3 (0.72)	3 (0.35)	4 (0.42)

capture expression heterogeneity along the contamination gradient. Individuals still appear to cluster by site, but this is explicable by the dose rate disparity between sites. Regarding the correlation of the PCs with the variable of interest, in Fig. 2d, it appears that correcting the effect of genetic structure protects the radiation-associated variation even without integrating the variable in the correction step. With this approach, we identified a previously unconsidered batch variable by incorporating genetic distances into the pre-processing. This led us to consider the possibility of other unknown sources of unwanted variability, which we investigate in the next step.

4.3 Search of Other Possible Confounders with Surrogate Variable Analysis

In order to reveal potential targets for batch correction, we applied SVA to diagnose the sources of variation in the VST-transformed expression dataset. SVA removed the linear relationships between expression and radiation exposure, and then identified 3 surrogate variables (SVs) as significant sources of systemic variation. In Fig. 3a, we compare the SVs to the known and measured covariates by computing the adjusted R^2 of the linear regression of SVs on covariates. The constructed SVs correlate with known potential covariates, such as the mass and length (SV1) and the site and genetic group for SVs 2 and 3. None of the estimated SVs capture age-related expression patterns.

As we expected from the primary exploration of the dataset, the collection site appears to have the strongest influence among known covariates. SVs capture expression trends that correlate with the site factor, even after radiation effects were residualized by SVA, which confirms that radionuclide pollution is not the only driver of differences between sites. We applied linear residualization to adjust for the estimated SVs, both with and without modeling ITDR effects, and present the results of PCA on the SVA-corrected matrix in Figs. 3b–3c. After correction, the first principal component correlates significantly with radiocontamination levels and allows us to distinguish well between the highest and lowest dose rate groups.

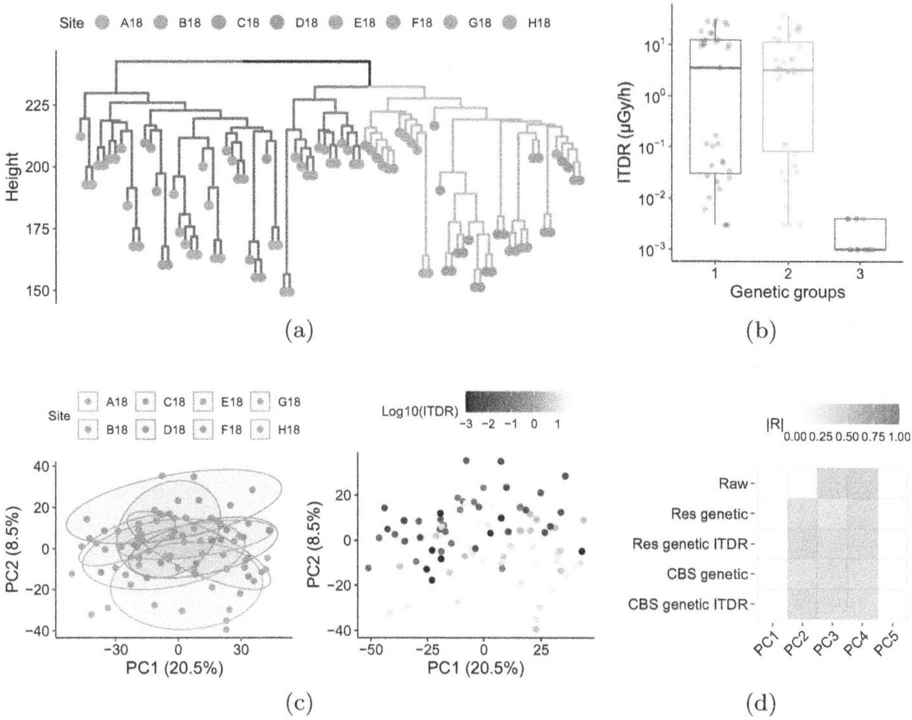

Fig. 2. Construction of genetic population groups and impact of genetic batch-correction **a.** Hierarchical clustering on genetic distances **b.** Radiation dose rate distribution among genetic groups **c.** PCA after correction of genetic group with ComBat-seq without including ITDR in the correction **d.** Absolute value of the correlation between radiation level and PCs of genetic group-corrected count matrices (CBS = ComBat-seq, Res = Residualization, ITDR = ITDR was included in the batch-correction step to preserve its effects)

4.4 Biological Interpretation

Correction methods targeting either the genetic structure or the surrogate variables both appear to have removed confounding effects. We now aim to compare the two approaches through functional enrichment analysis. Using the raw and adjusted datasets, we performed sPCA coupled to bootstrap to identify stably selected genes that carry the most variance in each dataset. Enrichment analysis then searches for the biological pathways in which the selected genes participate, thus indicating the deregulation of expression in the associated biological functions.

Table 3 presents the number of stably selected genes for each dataset and the number of significantly enriched Gene Ontology (GO) terms. We note that highly similar numbers of genes and biological processes were recovered in the raw and ComBat-seq-corrected datasets. This good performance of the uncor-

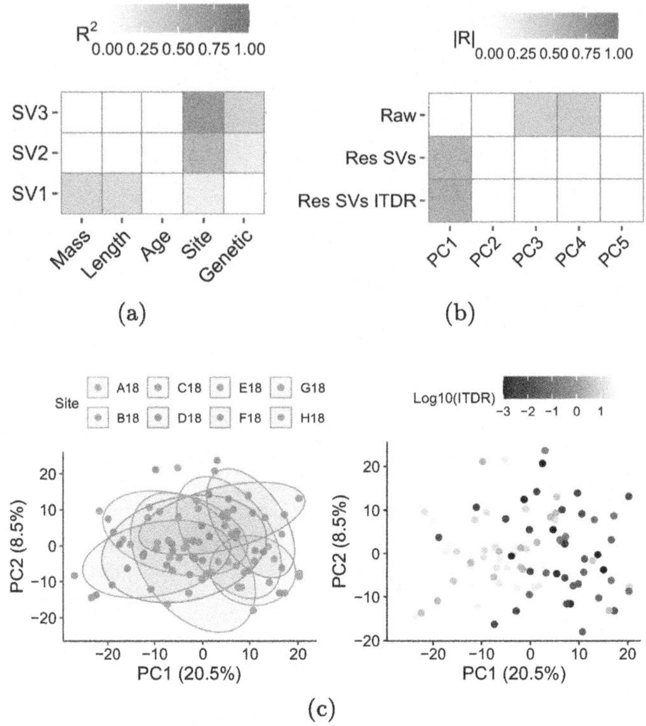

Fig. 3. Surrogate variables: connection to known covariates and correction **a.** Correlation between estimated SVs and covariates **b.** Absolute value of the correlation between radiation level and PCs of SVs-corrected count matrices (Res = Residualization, ITDR = ITDR was included in the batch-correction step to preserve its effects) **c.** PCA after correction of SVs-corrected by residualization without including ITDR in the correction

rected matrix could be explained by the robustness of sparse PCA compared to standard PCA, allowing the true biological signal to be distinguished from noise even without batch correction (see Appendix A). ComBat-seq-adjusted matrices yielded approximately as many stable genes as the raw matrix, while residualized matrices returned considerably less. Residualization coupled with SVA led to the selection of very few genes, which suggests overcorrection and removal of relevant information. Also, for the SVA method, we note that including the radiation levels in the correction step doubles the number of GO terms identified, whereas it does not impact the count for the other methods. This could indicate residual confounding between the SVs and the variable of interest.

In Fig. 4, we present the most significant deregulated pathways identified after the different correction methods. The nodes, which represent biological processes, are linked according to the sharing of genes involved in the processes. The various pre-processing methods lead to the selection of pathways that do not perfectly overlap. In particular, GO terms connected to cellular respiration

Table 3. Feature selection and Gene Ontology terms enrichment before and after batch correction

Correction method	Number of stably selected genes	Number of enriched GO terms
Raw	642	77
Residualization on genetic group	363	56
Residualization on genetic group with preservation of ITDR effects	441	58
ComBat-seq on genetic group	660	80
ComBat-seq on genetic group with preservation of ITDR effects	653	77
Residualization on SVs	33	24
Residualization on SVs with preservation of ITDR effects	95	48

or mitochondrial activity are enriched in datasets corrected for genetic structure and the raw dataset but disappear after SVA correction. Conversely, deregulated pathways identified consecutively to SVA correction are heavily biased toward muscle processes. We find that correction methods targeting the genetic groups preserved more relevant biological information than the residualization of SVs.

Overall, the study of deregulated biological processes highlights effects on energy metabolism (GO terms such as "ATP metabolic process" or "energy derivation by oxidation of organic compounds") and muscle processes (GO terms "muscle system process" or "muscle tissue development"). Perturbation in these pathways is consistent with impacts of exposure to low-dose radiation reported in studies of other organisms [18,27,28].

5 Discussion

In a first approach, classical BECAs were used to correct for the confounding effect of the site. However, this approach resulted in either the suppression of valuable biological variation or in the lack of reliability of the adjusted dataset when including the radiation level in the modeling. Instead, we successfully deconvoluted the batch from the variable of interest by including the genetic distance between individuals as a new variable to correct. The computed genetic batch reflects population structure and observed admixture between populations [8]. In the context of the CEZ, genetic distance between individuals could be a testimony of inherited mutations due to historic radiation pollution. We hypothesize that the removal of population structure effects, which we associate with historical exposure, has allowed for a better investigation of the impact of current exposure to low-dose radiation.

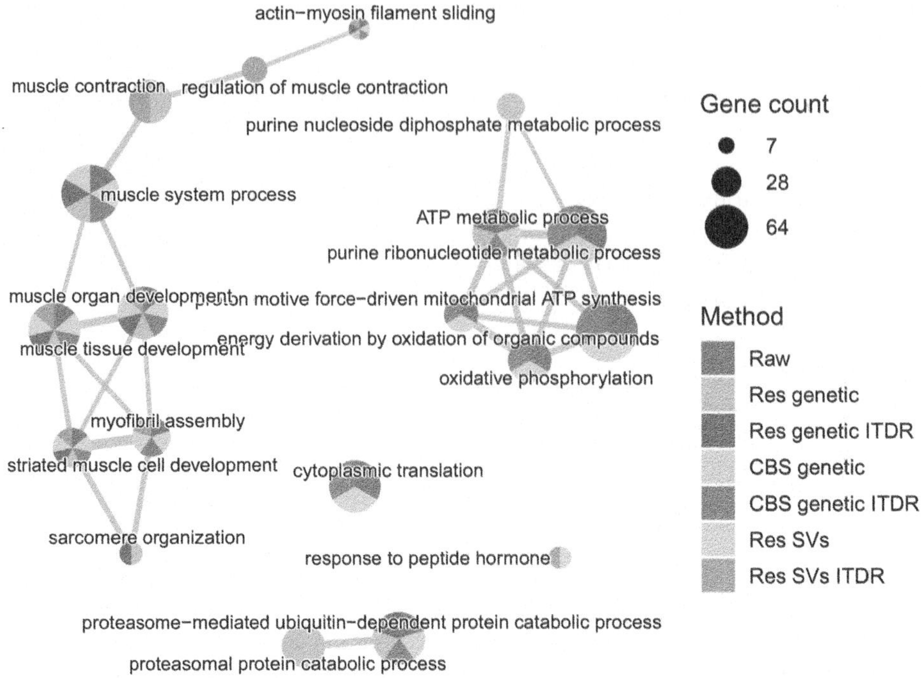

Fig. 4. Enriched Gene Ontology-terms network after correction strategies targeting genetic groups or SVs (CBS = ComBat-seq, Res = Residualization, ITDR = ITDR was included in the batch-correction step to preserve its effects)

When the origin of a confounder is unknown, a proxy variable is used to represent it. This variable encompasses the true source of variation. For example, in the case of technical batch effects, the day and group of processing can be the only accessible information [21]. Regarding the site effects, we assume that their origin is not technical, but we lack the information to assert it with certainty. Nevertheless, the consistency between the genetic structure of CEZ populations and their geographical distribution makes the genetic batch a suitable candidate. Additionally, it is not uncommon to correct for population structure, for instance, in genome-wide association studies [34]. We acknowledge that this approach is modality-dependent and may be less relevant when studying proteomics or metabolomics.

Our experiments illustrated the limitations of ComBat-seq and residualization in a batch-class imbalance design, as was acknowledged by several studies that suggest using BECAs with caution in confounded scenarios [12,24,29,41]. The genetic distance-adjusted datasets appear to be less affected by the integration or not of the variable of interest in the correction step. Feature selection on adjusted datasets followed by functional enrichment demonstrated that relevant biological information was preserved more effectively with ComBat-seq than with residualization. SVA recovered sources of unwanted variation, which represent an appealing strategy for disentangling the confounding variable from the level of radiation exposure. The residualization of estimated SVs yielded satisfactory results in the PCA visualization, but the enrichment analysis revealed a substantial loss of information relative to radiation exposure. Indeed, SVA authors caution against correcting SVs in a pre-processing step and advise that SVs be used as covariates in following analyses, such as differential expression [15].

When batch effects cannot be avoided, and happen to covary with a variable of biological interest, the recommended mitigation strategy is to take into account the batch factor as a covariate directly in the main analysis [29]. For gene expression analysis, this can be achieved using popular differential expression tools [9,25,30]. For all types of data, if the confounder is a categorical variable, multi-group approaches allow integrating batches in a joint factorial analysis [10,31,35,36]. However, in the multi-omics framework, researchers are advised to assess and address confounding effects in the different modalities prior to conducting joint analysis [39]. As the field of multi-omics studies continues to evolve, we hope that more tools will be developed that allow accounting for batch effects or other group structures, with already a few examples in the single cell context [2,6].

6 Conclusion

In the presence of a confounded batch effect, we were able to deconvolute the batch variable from the variable of interest through the integration of additional information. Our correction strategy based on genetic distance effectively mitigated the confounding effect of the frogs' collection site. With this approach, we identified distinctive changes in gene expression associated with chronic radiation exposure in Chornobyl tree frogs, by implementing this approach (see [8]). Future work will include the development of a new method for the incorporation of confounding factors into multimodal analyses based on the RGCCA framework.

Acknowledgments. EG and CC are supported by PhD grants funded by the French Institute for Radiation Protection and Nuclear Safety (IRSN). S. Gashchack, Y. Gulyaichenko, G. Orizaola, and P. Burraco helped in the field, and S. Gashchack also with measurements of radioactive contamination in tree frogs.

A Appendix: Close-Up on the Impact of Sparsity in Gene Selection

To assess how forcing sparsity in PCA influenced the selected gene list, we also ran a similar approach using standard PCA. We performed PCA on 1000 bootstrap samples of the non-corrected (raw) variance-stabilized matrix. In each model, genes were ranked by the absolute value of their weight and the top 400 genes were selected from components 1 and 2. Genes stably selected across bootstrap iterations in components 1, 2, or both were submitted to gene functional annotation, as mentioned previously.

Table 4 shows that the genes selected using sparse PCA were more stable across bootstrap samples than with standard PCA. This led to identifying a larger number of deregulated pathways in the uncorrected dataset with sparse PCA than with PCA. In Fig. 5, we notice that the alteration of biological processes related to energy metabolism (GO terms "oxidative phosphorylation" or "energy derivation by oxidation of organic compounds") was recovered with sPCA and not with PCA. The identification of pathways typically linked with low-dose radiation, despite the presence of batch effects, suggests that sparsity in the PCA weight vectors mitigated the influence of noise.

Table 4. Feature selection approaches and Gene Ontology terms enrichment

Feature selection method	Number of stably selected genes	Number of enriched GO terms
Standard PCA (Raw)	384	52
Sparse PCA (Raw)	642	77

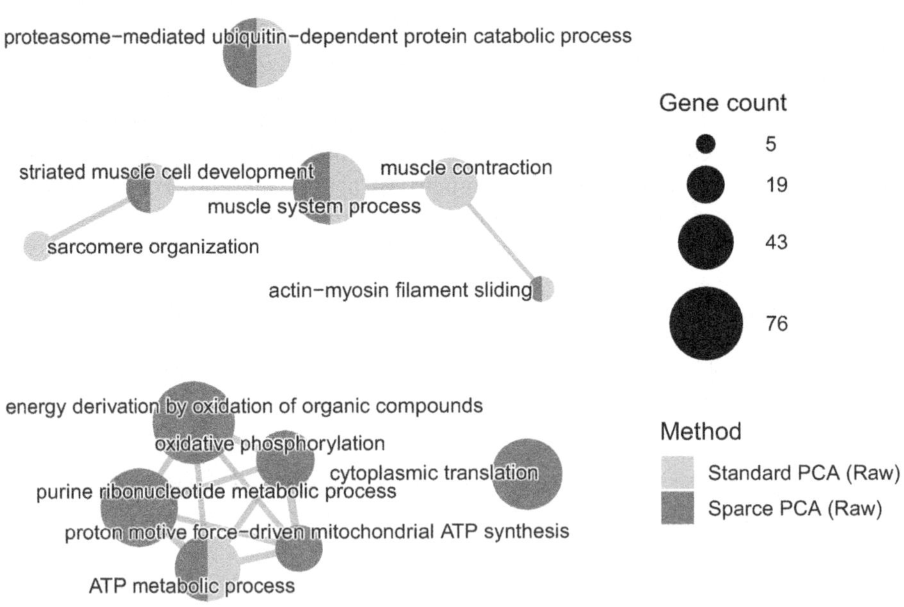

Fig. 5. Enriched Gene Ontology-terms network after feature selection by PCA or sparse PCA on the uncorrected count matrices

References

1. Anders, S., Huber, W.: Differential expression analysis for sequence count data. Genome Biol. **11**(10), R106 (2010). https://doi.org/10.1186/gb-2010-11-10-r106
2. Argelaguet, R., et al.: MOFA+: a statistical framework for comprehensive integration of multi-modal single-cell data. Genome Biol. **21**(1), 111 (2020). https://doi.org/10.1186/s13059-020-02015-1
3. Armant, O., Car, C., Frelon, S., Camoin, L.: Population transcriptogenomics highlights impaired metabolism and small population sizes in tree frogs living in the Chernobyl Exclusion Zone (2023). https://www.ncbi.nlm.nih.gov/geo/query/acc.cgi?acc=GSE211060
4. Beaugelin-Seiller, K., Jasserand, F., Garnier-Laplace, J., Gariel, J.C.: Modeling radiological dose in non-human species: principles, computerization, and application. Health Phys. **90**(5), 485–493 (2006). https://doi.org/10.1097/01.HP.0000182192.91169.ed
5. Burraco, P., Car, C., Bonzom, J.M., Orizaola, G.: Assessment of exposure to ionizing radiation in Chernobyl tree frogs (Hyla orientalis). Sci. Rep. **11**, 20509 (2021). https://doi.org/10.1038/s41598-021-00125-9
6. Cao, Z.J., Gao, G.: Multi-omics single-cell data integration and regulatory inference with graph-linked embedding. Nat. Biotechnol. **40**(10), 1458–1466 (2022). https://doi.org/10.1038/s41587-022-01284-4
7. Car, C., et al.: Unusual evolution of tree frog populations in the Chernobyl exclusion zone. Evol. Appl. **15**(2), 203–219 (2022). https://doi.org/10.1111/eva.13282
8. Car, C., et al.: Population transcriptogenomics highlights impaired metabolism and small population sizes in tree frogs living in the Chernobyl Exclusion Zone. BMC Biol. **21**(1), 164 (2023). https://doi.org/10.1186/s12915-023-01659-2
9. Chen, Y., Chen, L., Lun, A.T.L., Baldoni, P.L., Smyth, G.K.: edgeR 4.0: powerful differential analysis of sequencing data with expanded functionality and improved support for small counts and larger datasets. bioRxiv (2024). https://doi.org/10.1101/2024.01.21.576131
10. Eslami, A., Qannari, E.M., Kohler, A., Bougeard, S.: Algorithms for multi-group PLS. J. Chemom. **28**(3), 192–201 (2014). https://doi.org/10.1002/cem.2593
11. García, C.B., Salmerón, R., García, C., García, J.: Residualization: justification, properties and application. J. Appl. Stat. **47**(11), 1990–2010 (2020). https://doi.org/10.1080/02664763.2019.1701638
12. Goh, W.W.B., Wang, W., Wong, L.: Why batch effects matter in omics data, and how to avoid them. Trends Biotechnol. **35**(6), 498–507 (2017). https://doi.org/10.1016/j.tibtech.2017.02.012
13. Goh, W.W.B., Yong, C.H., Wong, L.: Are batch effects still relevant in the age of big data? Trends Biotechnol. **40**(9), 1029–1040 (2022). https://doi.org/10.1016/j.tibtech.2022.02.005
14. Grabherr, M.G., et al.: Trinity: reconstructing a full-length transcriptome without a genome from RNA-Seq data. Nat. Biotechnol. **29**(7), 644–652 (2011). https://doi.org/10.1038/nbt.1883
15. Jaffe, A.E., et al.: Practical impacts of genomic data "cleaning" on biological discovery using surrogate variable analysis. BMC Bioinform. **16**(1), 372 (2015). https://doi.org/10.1186/s12859-015-0808-5
16. Johnson, W.E., Li, C., Rabinovic, A.: Adjusting batch effects in microarray expression data using empirical Bayes methods. Biostatistics **8**(1), 118–127 (2007). https://doi.org/10.1093/biostatistics/kxj037

17. Knaus, B.J., Grünwald, N.J.: VCFR: a package to manipulate and visualize variant call format data in R. Mol. Ecol. Resour. **17**(1), 44–53 (2017). https://doi.org/10.1111/1755-0998.12549
18. Kostyuk, S.V., et al.: Effect of low-dose ionizing radiation on the expression of mitochondria-related genes in human mesenchymal stem cells. Int. J. Mol. Sci. **23**(1), 261 (2021). https://doi.org/10.3390/ijms23010261
19. Langmead, B., Salzberg, S.L.: Fast gapped-read alignment with Bowtie 2. Nat. Methods **9**(4), 357–359 (2012). https://doi.org/10.1038/nmeth.1923
20. Leek, J.T., Johnson, W.E., Parker, H.S., Jaffe, A.E., Storey, J.D.: The SVA package for removing batch effects and other unwanted variation in high-throughput experiments. Bioinformatics **28**(6), 882–883 (2012). https://doi.org/10.1093/bioinformatics/bts034
21. Leek, J.T., et al.: Tackling the widespread and critical impact of batch effects in high-throughput data. Nat. Rev. Genet. **11**(10), 733–739 (2010). https://doi.org/10.1038/nrg2825
22. Leek, J.T., Storey, J.D.: Capturing heterogeneity in gene expression studies by surrogate variable analysis. PLoS Genet. **3**(9), e161 (2007). https://doi.org/10.1371/journal.pgen.0030161
23. Li, B., Dewey, C.N.: RSEM: accurate transcript quantification from RNA-Seq data with or without a reference genome. BMC Bioinform. **12**(1), 323 (2011). https://doi.org/10.1186/1471-2105-12-323
24. Li, T., Zhang, Y., Patil, P., Johnson, W.E.: Overcoming the impacts of two-step batch effect correction on gene expression estimation and inference. Biostatistics **24**(3), 635–652 (2023). https://doi.org/10.1093/biostatistics/kxab039
25. Love, M.I., Huber, W., Anders, S.: Moderated estimation of fold change and dispersion for RNA-seq data with DESeq2. Genome Biol. **15**(12), 550 (2014). https://doi.org/10.1186/s13059-014-0550-8
26. Martinelli, F., et al.: Gene regulatory networks elucidating huanglongbing disease mechanisms. PLoS ONE **8**, e74256 (2013). https://doi.org/10.1371/journal.pone.0074256
27. Murat El Houdigui, S., Adam-Guillermin, C., Armant, O.: Ionising radiation induces promoter DNA hypomethylation and perturbs transcriptional activity of genes involved in morphogenesis during gastrulation in zebrafish. Int. J. Mol. Sci. **21**(11), 4014 (2020). https://doi.org/10.3390/ijms21114014
28. Murat El Houdigui, S., et al.: A systems biology approach reveals neuronal and muscle developmental defects after chronic exposure to ionising radiation in zebrafish. Sci. Rep. **9**(1), 20241 (2019). https://doi.org/10.1038/s41598-019-56590-w
29. Nygaard, V., Rødland, E.A., Hovig, E.: Methods that remove batch effects while retaining group differences may lead to exaggerated confidence in downstream analyses. Biostatistics **17**(1), 29–39 (2016). https://doi.org/10.1093/biostatistics/kxv027
30. Ritchie, M.E., et al.: limma powers differential expression analyses for RNA-sequencing and microarray studies. Nucleic Acids Res. **43**(7), e47 (2015). https://doi.org/10.1093/nar/gkv007
31. Rohart, F., Eslami, A., Matigian, N., Bougeard, S., Lê Cao, K.A.: MINT: a multivariate integrative method to identify reproducible molecular signatures across independent experiments and platforms. BMC Bioinform. **18**(1), 128 (2017). https://doi.org/10.1186/s12859-017-1553-8

32. Sims, A.H., et al.: The removal of multiplicative, systematic bias allows integration of breast cancer gene expression datasets - improving meta-analysis and prediction of prognosis. BMC Med. Genomics **1**(1), 42 (2008). https://doi.org/10.1186/1755-8794-1-42
33. Soneson, C., Love, M.I., Robinson, M.D.: Differential analyses for RNA-seq: transcript-level estimates improve gene-level inferences. F1000Research **4**, 1521 (2016). https://doi.org/10.12688/f1000research.7563.2
34. Sul, J.H., Martin, L.S., Eskin, E.: Population structure in genetic studies: confounding factors and mixed models. PLoS Genet. **14**(12), e1007309 (2018). https://doi.org/10.1371/journal.pgen.1007309
35. Tenenhaus, A., Tenenhaus, M.: Regularized generalized canonical correlation analysis for multiblock or multigroup data analysis. Eur. J. Oper. Res. **238**(2), 391–403 (2014). https://doi.org/10.1016/j.ejor.2014.01.008
36. Wang, Y., Lê Cao, K.A.: PLSDA-batch: a multivariate framework to correct for batch effects in microbiome data. Briefings Bioinform. **24**(2), bbac622 (2023). https://doi.org/10.1093/bib/bbac622
37. Witten, D.M., Tibshirani, R., Hastie, T.: A penalized matrix decomposition, with applications to sparse principal components and canonical correlation analysis. Biostatistics **10**(3), 515–534 (2009). https://doi.org/10.1093/biostatistics/kxp008
38. Wu, T., et al.: clusterProfiler 4.0: a universal enrichment tool for interpreting omics data. Innovation **2**(3), 100141 (2021). https://doi.org/10.1016/j.xinn.2021.100141
39. Yu, Y., et al.: Correcting batch effects in large-scale multiomics studies using a reference-material-based ratio method. Genome Biol. **24**(1), 201 (2023). https://doi.org/10.1186/s13059-023-03047-z
40. Zhang, Y., Parmigiani, G., Johnson, W.E.: ComBat-seq: batch effect adjustment for RNA-seq count data. NAR Genomics Bioinform. **2**(3), lqaa078 (2020). https://doi.org/10.1093/nargab/lqaa078
41. Zhou, L., Chi-Hau Sue, A., Bin Goh, W.W.: Examining the practical limits of batch effect-correction algorithms: when should you care about batch effects? J. Genet. Genomics **46**(9), 433–443 (2019). https://doi.org/10.1016/j.jgg.2019.08.002

Approximate Reductions of Rational Dynamical Systems in CLUE

Antonio Jiménez-Pastor[1], Alexander Leguizamon-Robayo[1(✉)], Max Tschaikowski[1], and Andrea Vandin[2,3]

[1] Aalborg University, Aalborg, Denmark
alexanderlr@cs.aau.dk
[2] Sant'Anna School of Advanced Studies, Pisa, Italy
[3] DTU Technical University of Denmark, Kongens Lyngby, Denmark

Abstract. In life sciences, deriving insights from dynamical systems can be challenging due to the large number of state variables involved. To address this, model reduction techniques can be used to project the system onto a lower-dimensional state space. CLUE is a tool that computes exact reductions for rational systems of ordinary differential equations. In this paper, we present an extension of CLUE to include approximate reductions which allow for larger aggregating power at the expense of a bounded error. Additionally, our extension includes new functionalities such as an interface to the model database ODEBase repository and simulation techniques for exploratory analyses.

Keywords: Approximate reduction · Dynamical systems · Constrained lumping

1 Introduction

Dynamical systems are used to model biochemical systems [9,29], performance models [12,31,32,34], electrical circuits [7] and even neural networks [14]. Model reduction is a family of techniques that that provides a modeler with a lower-dimensional dynamical system, while preserving properties of interest of the original system. In biology, e.g., it is key to preserve physical interpretability as these models validate mechanistic insights [2,6,22,26,28].

Lumping is a set of model reduction techniques for different mathematical formalisms [1,4,6,10,11], including ODEs. Given an original model, lumping produces a self-consistent system of ODEs involving macro-variables, each given in terms of combinations of the original ones [9,13,23,27]. In linear lumping, the reduction is a linear transformation of the original state variables. To avoid a complete loss of physical intelligibility, *constrained lumping* restricts the reduced state space so that previously defined linear combinations of state variables are preserved in the reduction [20].

R. Gori et al. (Eds.): CMSB 2024, LNBI 14971, pp. 108–116, 2024.
https://doi.org/10.1007/978-3-031-71671-3_9

The tool CLUE[1] was introduced along with a theoretical framework to efficiently compute *the smallest* constrained linear lumping for polynomial ODEs (i.e., ODEs with polynomial derivatives) that preserves the evolution of an arbitrary combination of state variables given by the user [24]. The efficiency of this framework with respect to previous work [19,20] stems from the fact that it does not rely on the symbolic computation of eigenvalues of a non-constant matrix. By leveraging this, the first version of CLUE (`v1.0`) was used to analyse models with several thousands of equations on standard hardware [24]. While it is possible to apply the polynomial theory to rational systems by means of symbolic computations, this approach quickly becomes computationally unfeasible [16]. This problem can be avoided by efficiently creating a representation of the dynamics by sampling it at different points using automatic differentiation [16], as implemented in the current version of CLUE (`v1.5`).

In this paper, we introduce a new version of CLUE (`v1.7`), extending [16, 24]. We add support for approximate reductions from [18] which have greater aggregation power at the expense of the exactness of the reduction. Moreover, we provide importing support from the public online repository ODEBase [21], as well as analysis capabilities by means of model simulation for exploratory purposes. Last but not least, we present the updated architecture of the new version of CLUE, and illustrate how the tool can be applied on a concrete model from the literature. The presented version of CLUE is publicly available and can be installed via `pip` with the following command

```
$ pip install git+https://github.com/clue-developers/CLUE@v1.7
```

2 Preliminaries

Constrained lumping is a reduction method that allows to reduce an ODE system in a way that it preserves a given linear combination of variables of interest. Suppose we are interested in the evolution of $x_{obs} = x_2 + 2x_3$ for the following system of ODEs

$$\dot{x}_1 = \frac{x_2^2 + 4x_2x_3 + 4x_3^2}{x_1^2 + 1}, \qquad \dot{x}_2 = \frac{2x_1 - 4x_3}{x_2 + 2x_3 + 1}, \qquad \dot{x}_3 = \frac{-x_1 - x_2}{x_2 + 2x_3 + 1}. \tag{1}$$

The matrix $L = (1\ 0\ 0, 0\ 1\ 2)^T$ is an *exact constrained lumping* of dimension 2, since it allows the construction a smaller self-consistent system in terms of the two *macro-variables* $y_1 = x_1$ and $y_2 = x_2 + 2x_3$, given by $(\dot{y}_1, \dot{y}_2)^T = (y_2^2/(y_1^2 + 1), -2y_2/(y_2 + 1))^T$.

The matrix L is a *constrained* lumping since the evolution of x_{obs} can be directly recovered from the evolution of y_2. It is *exact*, as the evolution of x_{obs} can be recovered exactly without any errors by using the lower dimensional system. In general, given a system of ODEs $\dot{x} = f(x)$ and an exact lumping L, a self-consistent reduced system is given by $y = Lf(\bar{L}y)$, where $y = Lx$, and $\bar{L}$ is a right-pseudo inverse of L. In this case, $\bar{L} = (1\ 0, 0\ 0.2, 0\ 0.4)^T$.

[1] https://github.com/clue-developers/CLUE.

Now, suppose that the evolution of x_{obs} is described by a slightly modified system with $4.05x_2x_3$ in the numerator of the first equation:

$$\dot{x}_1 = \frac{x_2^2 + 4.05x_2x_3 + 4x_3^2}{x_1^2 + 1}, \ \dot{x}_2 = \frac{2x_1 - 4x_3}{x_2 + 2x_3 + 1}, \ \dot{x}_3 = \frac{-x_1 - x_2}{x_2 + 2x_3 + 1}. \quad (2)$$

In this case, it is not possible to construct an exact approximate constrained lumping. However, by relaxing the conditions for lumping, the matrix $L = (1\ 0\ 0, 0\ 1\ 2)^T$ can be used to compute a smaller system that preserves x_{obs} up to an error, given by $(\dot{y}_1, \dot{y}_2)^T = (1.004y_2^2/(y_1^2+1), -2y_2/(y_2+1))^T$.

This means that the evolution of x_{obs} will be recovered up to an error. Figure 1 shows the evolution of x_{obs} computed via the original and reduced system for initial conditions $x = (1, 1, 1)^T$. Note that, by using an approximate reduction, it is possible to reduce a system that was not exactly reducible while still obtaining a simulation with low error. The detailed theory of approximate constrained lumping for polynomial systems, including the algorithm to compute them and how to bound the introduced errors, can be found in [18].

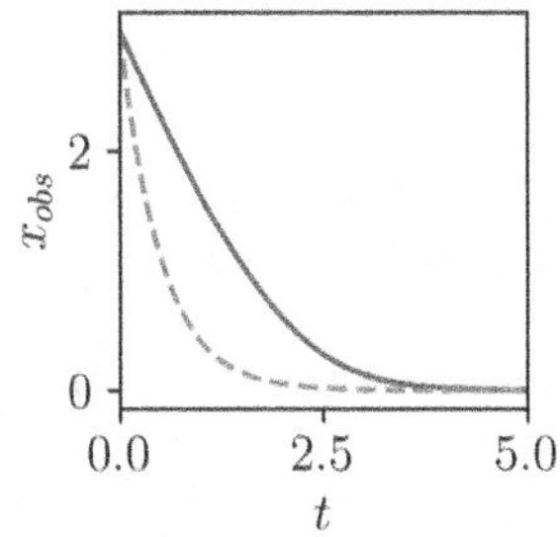

Fig. 1. x_{obs} in original (blue) and reduced (orange) model. (Color figure online)

3 Implementation

CLUE is designed as an open source python library to boost model analysis and exploration. Its architecture, summarized in Fig. 2, has been planned for efficient interaction and data handling through three main components: *model input*, *core functionalities*, and *outputs*. The *model input* component handles the steps of model acquisition and processing, while the *core functionalities* component, instead, contains all the logic to apply model reduction or approximation onto the models. The *outputs* component conducts simulations and allows for preliminary data analysis. We now proceed to explain each component in detail.

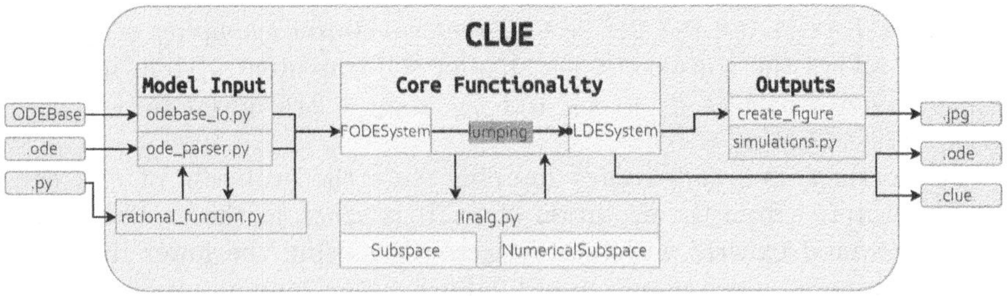

Fig. 2. Architecture of CLUE (red). The arrows indicate data flow. The main components of CLUE (blue) are composed of modules (green) of which the main classes and methods (e.g., `lumping`) are reported (orange) .External files and sources are displayed in purple. (Color figure online)

Model Input. The main goal of this component is to construct instances of the `FODEsystem` class from the supported sources: ODEBase [21], whose support has been added in this version of the tool, and ERODE [8] `.ode` files, and systems of ODEs symbolically written using `sympy`. The modules realizing the import functionalities of aforementioned formats are, respectively, `odebase_io.py`, `ode_parser.py` and `rational_function.py`. Regardless of the format, the creation of `FODESystem` instances relies on the efficient representation of polynomials and rational functions provided in the module `rational_function.py`.

Core Functionalities. The core functionalities of CLUE are available in the module `clue.py`. In this module, the main class is `FODEsystem` which contains all necessary information to represent a model, e.g., equations, observables, parameters and initial conditions. Simulations can be computed via the `simulate` method. Additionally, `FODESystem` offers export functionality to `.ode` files and to serialized `.clue` files.

Given an observable, a constrained lumping is computed by finding the smallest invariant subspace from which the evolution of the observable can be recovered. Following the theory presented in [16,18], lumpings are computed using the `lumping` and `app_lumping` methods which find exact and approximate lumpings, respectively. The outputs of these methods are instances of `LDESystem`. This class inherits from `FODEsystem`, while including lumping information, e.g., the lumping subspace and mappings from the original to the lumped variables.

Exact lumping subspaces are instances of the `Subspace` class, which stores them as matrices in row-echelon form. In contrast, approximate lumping subspaces are instances of the `NumericalSubspace` class, which stores them as orthonormal matrices. Both these classes are described in the module `linalg.py`. It should be noted that all exact computations are carried out using the rational numbers implementation provided by `sympy`. For both numerical and exact computations, we use our own implementation of matrix arithmetic. Matrices are stored as hash tables where the keys are the nonzero rows and the values are `SparseVectors` representing the actual row. Instances of `SparseVector` are hash tables storing the number of the nonzero value as keys and the actual vector values as values.

Outputs. Utilities to handle simulations are provided in the `simulations.py` module. Basic manipulation of simulations and data is supported. This corresponds to merging, comparing and applying matrices to simulation results. Similarly, basic plot functionality is supported, namely, exporting plots (e.g., Fig. 1), and simulation data (as CSV files).

4 Illustration of Model Workflow Using CLUE

In this section, we show an example workflow displaying the main features of CLUE on a simple, yet informative, model from the literature. A full listing of this section is available as a Jupyter notebook[2]. Suppose we want to study the *signaling mechanism in apoptosis*, as described by the kinetic model in [17].

Step 1: Model Retrieval. To access the required model from ODEBase, we use the following line

```
>>> model = ode_scrapper(name="BIOMD0000000102")
>>> model = model.remove_parameters_ic()
```

In general, CLUE treats parameters as variables of the differential system. We can replace each known parameter by its value using the method `remove_parameters_ic`.

Step 2: Model Validation. Having retrieved the model, we can examine its attributes:

```
>>> print(model.name)        # Outputs the name of the model
>>> print(model.size)        # Outputs the size of the model
>>> print(model.equations)   # Shows the model equations
```

In particular, we infer that the original model size is 13.

Step 3: Analyses and Lumping. Following the original paper presenting the model [17], we are interested in studying the evolution of the observable C3, formally given by the variable `x7` in the model. As a first step, we compute the respective exact lumping which has `x7` as observable:

```
>>> exact_lump = model.lumping(['x7'])
```

This command raises a warning that the system could not be reduced.

```
[lumping] Warning: lumped size (13)
and original size (13) are the same.
```

This means that it was impossible to find an exact reduction.

We can use the approximate lumping functionality of CLUE to increase the aggregation power with the following command:

```
>>> app_lump_1 = model.app_lumping(['x7'])
```

This command relaxes the conditions of exact lumping until it finds a reduction that is smaller than the exact one. In this case, the resulting reduction is of size 12. To find more compact reductions, it is possible to add a size limit as follows:

```
>>> app_lump_2 = model.app_lumping(['x7'], max_size=10)
```

The output will be the largest reduction possible having at most 10 lumped variables (species), following the approach of [18]. The resulting lumped model in this case is of size 9.

[2] https://github.com/clue-developers/CLUE/tree/v1.7/notebooks/ODEBase_example.ipynb.

Step 4: Simulating and Comparing Models. To evaluate the effectiveness of the lumpings, we can simulate the original model and its reduction with the initial conditions and the time horizon from the original paper [17].

```
>>> exact_sim = model.simulate(0, 2500, view=["x7"])
>>> app_sim_1 = app_lump_1.simulate(0, 2500,
                   view=app_lump_1.observe(["x7"]))
>>> app_sim_2 = app_lump_2.simulate(0, 2500,
                   view=app_lump_2.observe(["x7"])
```

It is possible to choose the observable of interest with the third argument in the `simulate` method. The method `observe` computes the value of a given observable based on the lumped variables. We now merge all simulations into one.

```
>>> merged = merge_simulations(exact_sim,app_sim_1,app_sim_2)
```

Step 5: Visualizing the Results. To visualize the merged results (Fig. 3), we use the following line.

```
>>> create_figure(merged,names=['Exact','App_12','App_9'])
```

Figure 3 shows that it was possible to find further reductions beyond the exact one. We have two exemplary situations: on one hand, the reduction with 12 species provides a small error. On the other hand, the one with 9 is too aggressive, being close to the original only for a few units of time. This example shows that approximate reductions provide a trade-off between the aggressiveness of the reduction and the error of the reduced model. Depending on the modeller's goals, any of the reduced models can be further analyzed in Python or ERODE [8].

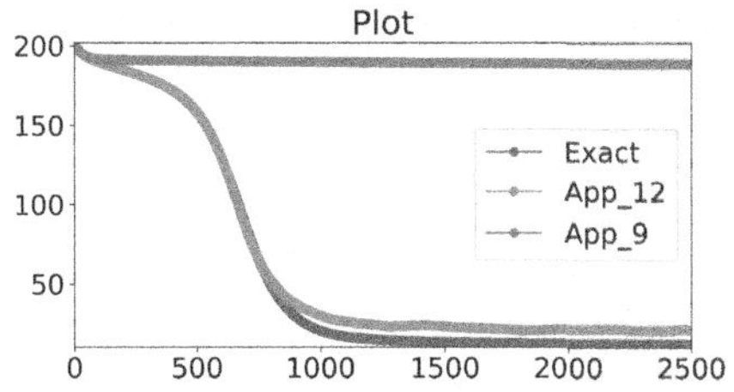

Fig. 3. Time evolution of C3 using approximate constrained lumping.

5 Conclusion

We presented an extension of the tool CLUE [16,24] to support approximate reductions. The underlying theory increases the aggregation power of exact lumping approaches [3,15,30,33] by relaxing the exactness criteria of aggregations. The extension also included simulation functionalities which allow for the exploration of different model reductions, and importing capabilities for popular online model repositories. Future work will extend the functionalities of the CLUE tool to include analytic functions and to support importing and exporting other formats such as SBML [25] and BioNetGen [5].

Acknowledgement. This work was partially supported by Poul Due Jensen Foundation grant no. 883901, the Villum Investigator Grant S4OS, the Fsc regional Tuscan project AUTOXAI2 J53D21003810008, the project SERICS (PE00000014) and the Tuscany Health Ecosystem (THE) project with CUP B83C22003920001, under the MUR National Recovery and Resilience Plan funded by the European Union - NextGenerationEU.

References

1. Abate, A., Andriushchenko, R., Ceska, M., Kwiatkowska, M.: Adaptive formal approximations of Markov chains. Perf. Eval. **148**, 102207 (2021)
2. Apri, M., de Gee, M., Molenaar, J.: Complexity reduction preserving dynamical behavior of biochemical networks. J. Theor. Biol. **304**, 16–26 (2012)
3. Bacci, G., Bacci, G., Larsen, K.G., Mardare, R.: On-the-fly exact computation of bisimilarity distances. In: Piterman, N., Smolka, S.A. (eds.) TACAS 2013. LNCS, vol. 7795, pp. 1–15. Springer, Heidelberg (2013). https://doi.org/10.1007/978-3-642-36742-7_1
4. Beica, A., Feret, J., Petrov, T.: Tropical abstraction of biochemical reaction networks with guarantees. Electron. Notes Theor. Comput. Sci. **350**, 3–32 (2020)
5. Blinov, M.L., Faeder, J.R., Goldstein, B., Hlavacek, W.S.: BioNetGen: software for rule-based modeling of signal transduction based on the interactions of molecular domains. Bioinformatics **20**(17), 3289–3291 (2004)
6. Cardelli, L., Perez-Verona, I.C., Tribastone, M., Tschaikowski, M., Vandin, A., Waizmann, T.: Exact maximal reduction of stochastic reaction networks by species lumping. Bioinformatics **37**(15), 2175–2182 (2021)
7. Cardelli, L., Tribastone, M., Tschaikowski, M.: From electric circuits to chemical networks. Nat. Comput. **19**(1), 237–248 (2020)
8. Cardelli, L., Tribastone, M., Tschaikowski, M., Vandin, A.: ERODE: a tool for the evaluation and reduction of ordinary differential equations. In: Legay, A., Margaria, T. (eds.) TACAS 2017. LNCS, vol. 10206, pp. 310–328. Springer, Heidelberg (2017). https://doi.org/10.1007/978-3-662-54580-5_19
9. Cardelli, L., Tribastone, M., Tschaikowski, M., Vandin, A.: Maximal aggregation of polynomial dynamical systems. PNAS **114**(38), 10029–10034 (2017)
10. Cardelli, L., Tribastone, M., Tschaikowski, M., Vandin, A.: Syntactic Markovian bisimulation for chemical reaction networks (2017)
11. Feret, J., Danos, V., Krivine, J., Harmer, R., Fontana, W.: Internal coarse-graining of molecular systems. Proc. Natl. Acad. Sci. **106**(16), 6453–6458 (2009)
12. Gast, N., Bortolussi, L., Tribastone, M.: Size expansions of mean field approximation: transient and steady-state analysis. ACM SIGMETRICS Perf. Eval. Rev. **46**(3), 25–26 (2019)
13. Großmann, G., Kyriakopoulos, C., Bortolussi, L., Wolf, V.: Lumping the approximate master equation for multistate processes on complex networks. In: McIver, A., Horváth, A. (eds.) QEST, vol. 11024, pp. 157–172 (2018)

14. Hasani, R., et al.: Closed-form continuous-depth models. arXiv preprint arXiv: 2106.13898 (2021)
15. Hillston, J., Tribastone, M., Gilmore, S.: Stochastic process algebras: from individuals to populations. Comput. J. **55**(7), 866–881 (2011)
16. Antonio, J.-P., Jacob, J.P., Pogudin, G.: Exact linear reduction for rational dynamical systems. In: Petre, I., Paun, A. (eds.) CMSB 2022. LNCS, vol. 13447, pp. 198–216. Springer, Heidelberg (2022). https://doi.org/10.1007/978-3-031-15034-0_10
17. Legewie, S., Blüthgen, N., Herzel, H.: Mathematical modeling identifies inhibitors of apoptosis as mediators of positive feedback and bistability. PLoS Comput. Biol. **2**(9), e120 (2006)
18. Leguizamon-Robayo, A., Jiménez-Pastor, A., Tribastone, M., Tschaikowski, M., Vandin, A.: Approximate constrained lumping of polynomial differential equations. In: CMSB, pp. 106–123 (2023)
19. Li, G., Rabitz, H.: A general analysis of exact lumping in chemical kinetics. Chem. Eng. Sci. **44**(6), 1413–1430 (1989)
20. Li, G., Rabitz, H.: New approaches to determination of constrained lumping schemes for a reaction system in the whole composition space. Chem. Eng. Sci. **46**(1), 95–111 (1991)
21. Lüders, C., Sturm, T., Radulescu, O.: ODEbase: a repository of ODE systems for systems biology. Bioinf. Adv. **2**(1), vbac027 (2022)
22. Maus, C., Rybacki, S., Uhrmacher, A.M.: Rule-based multi-level modeling of cell biological systems. BMC Syst. Biol. **5**, 1–20 (2011)
23. Okino, M., Mavrovouniotis, M.: Simplification of mathematical models of chemical reaction systems. Chem. Rev. **2**(98), 391–408 (1998)
24. Ovchinnikov, A., Pérez Verona, I., Pogudin, G., Tribastone, M.: CLUE: exact maximal reduction of kinetic models by constrained lumping of differential equations. Bioinformatics **37**(12), 1732–1738 (2021)
25. Pérez-Verona, I.C., Tribastone, M., Vandin, A.: A large-scale assessment of exact model reduction in the biomodels repository. In: Bortolussi, L., Sanguinetti, G. (eds.) CMSB 2019. LNCS, vol. 11773, pp. 248–265. Springer, Cham (2019). https://doi.org/10.1007/978-3-030-31304-3_13
26. Schmidt, H., Madsen, M., Danø, S., Cedersund, G.: Complexity reduction of biochemical rate expressions. Bioinformatics **24**(6), 848–854 (2008)
27. Snowden, T., van der Graaf, P., Tindall, M.: Methods of model reduction for large-scale biological systems: a survey of current methods and trends. Bull. Math. Biol. **79**(7), 1449–1486 (2017)
28. Sunnaker, M., Cedersund, G., Jirstrand, M.: A method for zooming of nonlinear models of biochemical systems. BMC Syst. Biol. **5**(1), 140 (2011)
29. Tognazzi, S., Tribastone, M., Tschaikowski, M., Vandin, A.: EGAC: a genetic algorithm to compare chemical reaction networks. In: Genetic and Evolutionary Computation Conference, GECCO, pp. 833–840 (2017)
30. Tribastone, M.: Behavioral relations in a process algebra for variants. In: Gnesi, S., Fantechi, A., Heymans, P., Rubin, J., Czarnecki, K., Dhungana, D. (eds.) SPLC, pp. 82–91. ACM (2014)
31. Tribastone, M., Mayer, P., Wirsing, M.: Performance prediction of service-oriented systems with layered queueing networks. In: International Symposium on Leveraging Applications of Formal Methods, Verification and Validation (ISOLA), pp. 51–65 (2010)
32. Tschaikowski, M., Tribastone, M.: Tackling continuous state-space explosion in a Markovian process algebra. Theoret. Comput. Sci. **517**, 1–33 (2014)

33. Tschaikowski, M., Tribastone, M.: Spatial fluid limits for stochastic mobile networks. Perf. Eval. **109**, 52–76 (2017)
34. Wirsing, M., et al.: Sensoria patterns: augmenting service engineering with formal analysis, transformation and dynamicity. In: Margaria, T., Steffen, B. (eds.) Leveraging Applications of Formal Methods, Verification and Validation, pp. 170–190 (2008)

Discovering Biochemical Reaction Models by Evolving Libraries

Justin N. Kreikemeyer[1(✉)], Kevin Burrage[2,3], and Adelinde M. Uhrmacher[1]

[1] University of Rostock, Rostock, Germany
{justin.kreikemeyer,adelinde.uhrmacher}@uni-rostock.de
[2] Queensland University of Technology (QUT), Brisbane, Australia
kevin.burrage@qut.edu.au
[3] Department of Computer Science, University of Oxford, Oxford, UK

Abstract. In a time of data abundance, automatic methods increasingly support manual modeling. To this end, the Sparse Identification of Non-linear Dynamics (SINDy) provides a solid foundation for identifying non-linear dynamical systems in the form of differential equations. In biochemistry, reaction networks imply coupled differential equations. It has recently been demonstrated how this intrinsic coupling can be achieved within the SINDy framework, providing a straightforward interpretation of the learned equations as reaction systems with mass-action kinetics. However, this extension inherits from SINDy the requirement to enumerate all candidate reactions in a library, resulting in ill-posed optimization problems and long model descriptions, limiting its utility for identifying models with many species. Here, we elaborate on the recent advances in bringing SINDy to the biochemical domain by considering the subsampling of reaction libraries as part of an evolutionary optimization scheme. This enables the generation of parsimonious models, as well as the inclusion of model-level constraints, and allows the consideration of large numbers of candidate reactions. We evaluate the approach on two smaller case studies and the recovery of a large Wnt signaling model.

Keywords: sparse regression · genetic programming · automatic model generation · machine learning · reaction systems

1 Introduction

In *regression*, models are used to infer the relationships among variables from a series of measurement points to, e.g., predict future measurements. This is done by choosing from various available models, from simple linear functions to deep neural networks, and fitting methods. However, the most accurate deep models quickly become opaque to the scientist and end-user, especially when learning complex and non-linear relationships. Thus, *symbolic regression* seeks to identify expressive yet human-interpretable expressions for the relationships inside a system. Instead of choosing a specific model structure, these approaches learn the

R. Gori et al. (Eds.): CMSB 2024, LNBI 14971, pp. 117–136, 2024.
https://doi.org/10.1007/978-3-031-71671-3_10

structure alongside suitable parameters. For example, the Sparse Identification of Non-linear Dynamics (SINDy) [7] can often discover parsimonious differential equations that govern the observed evolution of dynamical systems. Remarkably, it achieves the learning of non-linear dynamics by solving a *linear* equation system, relying on a comprehensive set of time series for all variables of interest and a library of possible function evaluations on those variables (cf. Fig. 1, top).

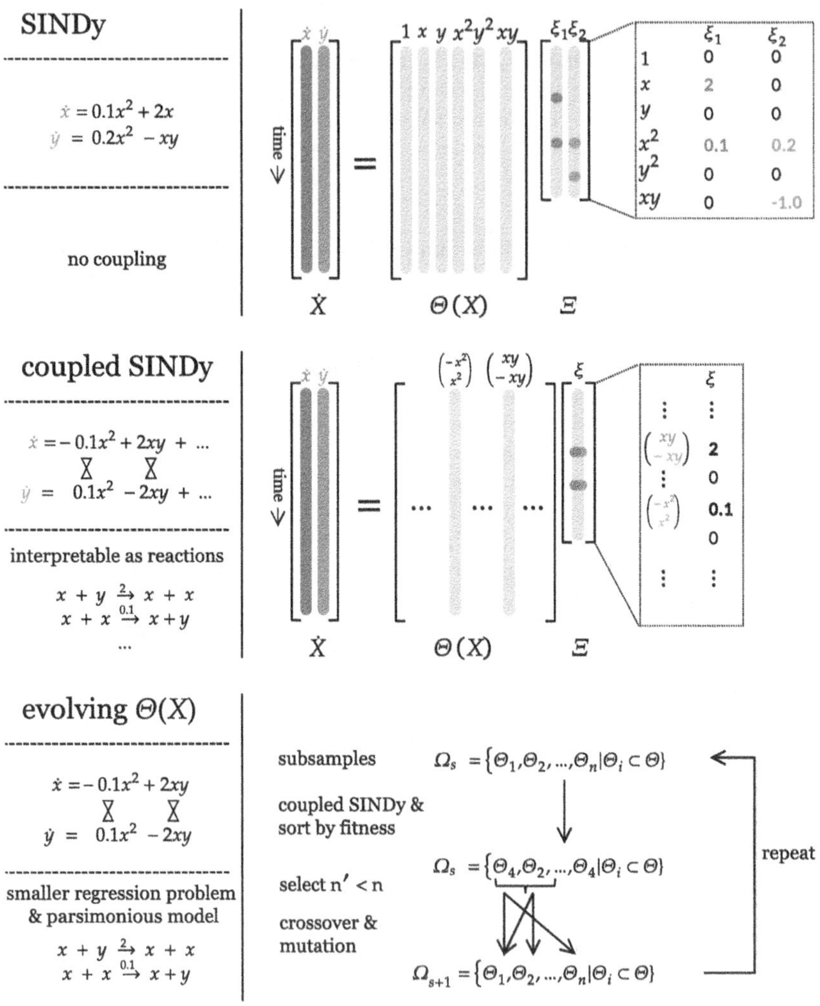

Fig. 1. Schematic overview of SINDy ([7], top), coupled SINDy ([8], center) and evolving libraries (this paper, bottom). For simplicity, we neglect the 1/2 factor for x^2 and y^2 resulting from mass-action kinetics. Graphic inspired by [7].

For many engineering domains, ordinary differential equations (ODEs) are a suitable modeling formalism and, thus, a target for symbolic regression. In systems biology, reaction or rule-based modeling approaches prevail [11], also reflected in systems biology standards [16]. Although reactions are frequently mapped to ODEs for simulation, the opposite mapping is not as straightforward [32]. Particularly, if ODEs are learned from data, they might easily miss an important feature of the modeled biochemical reactions that is intrinsically captured by reaction-based modeling approaches: reactions introduce a *coupling* between ODEs. For example, a single reaction $A + B \xrightarrow{k} C$, translates into three coupled ODEs $dA/dt = dB/dt = -dC/dt = -k \cdot A \cdot B$. Note how terms in the differential equation for one variable constrain the terms that may occur in the other variables' derivatives (cf. Fig. 1, center). Based on this observation, in [8], a coupled approach to SINDy is proposed. By using a library of possible reactions instead of possible functions, reaction-based models can be learned from data.

One problem with introducing such a coupling is the combinatorial nature of the possible couplings. Compared to standard SINDy, its coupled counterpart exhibits a much wider design matrix, which can lead to problems with linear solvers whose ability to find good solutions depends on a well-conditioned matrix (cf. Sect. 5.1). This motivates a subsampling approach, where only a well-chosen subset of the reaction candidates is used to fit a model. To this end, we here embed coupled SINDy inside a genetic algorithm (cf. Fig. 1, bottom) to find an optimal subset and thus model. This allows constraining the model size to a minimum, ensuring human interpretability, as well as handling large libraries that would otherwise result in extremely large regression problems. We evaluate this idea on three case studies (cf. Sect. 5): the discovery of disease spread, predator-prey dynamics, as well as the complex dynamics of a Wnt pathway [34] involving 19 variables, which results in a library of around 37 000 reactions. To the best of our knowledge, the latter is one of the largest search spaces tackled so far for the learning of reaction models. For example, the large case study in [17] searches only around 6000 possible reactions. Before introducing our method in greater detail in Sect. 4, we provide some background on biochemical reaction models and (sparse) symbolic regression (cf. Sect. 2), including related work (cf. Sect. 3). We conclude with a discussion of the results and an outlook in Sects. 6 and 7.

2 Background

This section reviews the formalism of reaction systems, which is widely used to develop models in biochemistry. Typically, these models are hand-crafted, making extensive use of domain knowledge. The reaction systems' parameters are often calibrated based on measurement data. The second part discusses how to additionally identify a model structure from (time-series) data, which is the main topic of this paper and further discussed in Sect. 3.

2.1 Biochemical Reaction Models

In systems biology, reaction networks are important in describing system dynamics. Their simple syntax allows a natural specification of simulation models in terms of reactants, products, and reaction rate constants:

$$R_j\colon\ l_{0,j}S_0 + \cdots + l_{i,j}S_i \xrightarrow{c_j} r_{0,j}S_0 + \cdots + r_{i,j}S_i \tag{1}$$

where S is a vector of species, $l_{\cdot,j}$ and $r_{\cdot,j}$ are vectors of coefficients determining the stoichiometry and change $\nu_j = r_{\cdot,j} - l_{\cdot,j}$ of reaction j, and c_j determines the rate at which the reactants are consumed to form products. It is common to restrict $\sum_i l_{i,j} = 2$ for all reactions j, i.e., only consider *binary reactions*, as a collision of more than two reactants is unlikely. Starting from an initial vector of amounts for each species, their quantities evolve over time t, which we denote as $S(t)$. Many systems obey the *law of mass action*, such that the effective rate $a_j(S(t))$ (also called propensity) of a reaction depends on the reactant's amounts at the current time. For example, for deterministic systems $a_j(S(t)) = A(t)B(t)c_j$ in the case of binary reactions with reactants A and B ($A \neq B$) and $a_j = 0.5A^2(t)c_j$ ($A = B$).

Assuming a *homogeneous mixture* of species, the semantics of this representation are defined by a system of ODEs called the chemical master equation (CME) [35]. However, as this system is intractable for many relevant cases, simulation is often applied in practice. The CME can, for example, be understood as a continuous-time Markov chain (CTMC) over the evolution of the amounts of species $S(t)$, accounting for the discreteness and stochasticity of small amounts. Using the stochastic simulation algorithm, sample trajectories of the CTMC can be obtained, approaching the exact distribution in the limit of samples. Here, we focus on approximating the CME under the assumption of a deterministic system so that reactions can be directly translated to (coupled) ODEs over time. An example reaction system and its translation to a system of differential equations are depicted in Fig. 1 (bottom left).

Many models use rule-based formalisms, which offer even more expressiveness by introducing the concepts of attributes and compartments [4,11,15]. In cases where attributes take on only a finite set of values and compartments are not dynamic, it is possible to derive the underlying "flattened" reaction network. We use this approach in Sect. 5 when considering a Wnt signaling model.

2.2 Symbolic Regression for Reaction Systems

The field of symbolic regression develops methods to derive symbolic expressions fitting a given set of measurement points. When the fitting is successful, these are powerful alternatives to black-box regression models, such as neural networks. Symbolic regression methods have also been applied to reaction systems. Early approaches relying on genetic programming [18,26], start with a random population of reaction systems and converge to increasingly fit solutions by applying evolutionary operators.

Whereas genetic programming can, in principle, work with any symbolic expression including imperative programs, the sparse identification of non-linear dynamics (SINDy) [7] is tailored toward differential equations describing the behavior of a system over time, as are common in computational science. This is made possible by constructing a linear equation system over derivative measurements and possible functions involved in them and solving a least-squares problem. The latter is defined over (1) N measured variables and M discrete measurements of their time derivatives $Y' \in \mathbb{R}^{M \times N}$ (which may be determined numerically from Y) and (2) a matrix of candidate derivatives (design matrix) $\Theta(Y) \in \mathbb{R}^{M \times K}$, determined by the application of a vector of candidate functions $\Theta \in \mathbb{R}^{K}$ with $\theta \in \Theta \colon \mathbb{R}^{N} \rightarrow \mathbb{R}$ to each measurement in Y:

$$\min_{\Xi} ||Y' - \Theta(Y)\Xi||_2 \tag{2}$$

The weights or parameters $\Xi \in \mathbb{R}^{K \times N}$ then provide a factor with which each function from the *library* Θ contributes to each of the N ODEs. A schematic depiction is given in Fig. 1 (top). This problem formulation can be further extended in many ways, for example, by using regularization [36], a two-step procedure [17], or ensembling [12].

By requiring a specific coupling of the variables, SINDy can also be used to discover reaction networks [8]. The key idea is that Θ now consists of *vectors* of functions, reflecting the terms that will appear in each of the species' derivatives when a particular reaction is included in the model (cf. Fig. 1, center). In this formulation, the library Θ consists of the corresponding derivatives of all possible candidate reactions and there is only one coefficient ξ per reaction indicating its rate constant.

3 Related Work

As the automatic discovery of mechanistic models from data promises enormous benefits not only for predicting dynamics but also for furthering their understanding, it has a long-standing history under many names [19]. Nevertheless, there are still many unsolved challenges and the amount of available data steadily increases, making it an active research area. Here, we present some recent work closely related to our approach.

Discovering ODEs from data, with a focus on biochemical systems, is discussed in [10]. The authors automatically discover parsimonious models by systematically fitting models of increasing complexity. In [7], the seminal Sparse Identification of Nonlinear Dynamics (SINDy) is proposed as already introduced in the previous section. [12] presents an extension to SINDy in the form of E-SINDy, which allows for uncertainty quantification by subsampling either by time (bagging E-SINDy) or library terms (library bagging E-SINDy). As opposed to uncertainty quantification, a random subsampling of the library is not enough in our case (cf. Section 4), hence we perform a goal-driven evolution of our library of reactions. The authors of [8] introduce a method to generate reaction models with SINDy by considering the coupling of species. They test their method by building a reaction system surrogate for an agent-based model. We here extend this approach with a genetic algorithm and evaluate its ability to recover a ground truth reaction model from synthetic measurement data.

The latter is just one example of a method for the data-driven discovery of reaction systems, and the number of proposed methods is growing, particularly in the last decade. Recently, a deep learning architecture based on a variational autoencoder has been demonstrated to discover stochastic reaction models [3]. Preliminary results show the successful recovery of a single reaction with two species from data. This work is part of ongoing efforts in the field of relational inference using variational autoencoders to learn interaction graphs from data, such as [13]. Closely related to (coupled) SINDy, the "ReactioNet Lasso" [17] is a method to fit *stochastic* population models to heteroscedastic time series distribution snapshots. It works by considering the moment equations, an ODE system describing the evolution of stochastic moments derived from the chemical master equation (cf. Sect. 2). Another recent method is Reactmine [24], which builds models one reaction at a time by stochastically searching a model tree. By limiting the depth of the search tree, the discovery of parsimonious models is enforced. Here, we evaluate a library-based approach and limit model size using an evolutionary algorithm.

Combining sparse regression with genetic programming approaches is not a new idea. Most recently, [2] uses a combination of SINDy and genetic programming to allow the inclusion of rational functions in the library for the recovery of multibody physical systems. Similarly, [22] presents a method (not relying on genetic programming) to handle rational functions occurring in biological networks in the SINDy framework, deriving parsimonious models. A similar idea to our approach is presented in [23], where SINDy is integrated with symbolic regression via genetic programming in a framework called "Deep Symbolic Regression". The authors also find that, in their applications to orbital mechanics, genetic programming enables a limitation of the learned models' complexity. Here, we are interested in the special case of reaction systems and coupled SINDy.

4 Evolving Libraries

As introduced in Sect. 2, SINDy requires a library of all possible functions to be applied to the measured time series. The number of possible interactions and, thus, library terms increases superlinearly with the number of variables. When requiring a coupling of reactions as with coupled SINDy, this problem becomes significantly more pronounced (cf. Fig. 1), which can lead to a design matrix that poses major challenges for linear solvers in terms of problem size and collinearities (cf. Sect. 5.1).

To circumvent this issue, a straightforward idea would be to reduce the library size by subsampling, as also proposed in [12]. In the former, library subsampling is used to estimate the algorithm's uncertainty over the learned coefficients Ξ. However, for this to work, the samples must be large enough to ensure a high probability of including many "helpful" reactions. Typically only one or two library terms are omitted. Taking much smaller subsamples would result in a random search over models. Hence, this approach does not readily solve the problems that arise from large libraries. Thus, we take the idea one step further.

4.1 Searching the Sub-library Space Using a Genetic Algorithm

Instead of taking purely random subsamples, we perform a genetic search over possible subsamples of the library, assigning a higher fitness to libraries that lead to a better fit. Whereas coupled SINDy alone has to consider all possible models "at once", an evolutionary strategy can consider a small subset at a time. Limiting the number of reactions in the library also enforces a certain degree of parsimony, as observed in [23] for symbolic expressions. Further, combinations of reactions that lead to low regression accuracy are discouraged. A schematic overview can be found in Fig. 1 (bottom).

Our genetic algorithm (GA) follows a standard procedure [20] and starts with a randomly sampled initial population of n reaction sublibraries, each containing a fixed number of reactions. The samples are taken from an enumeration of all possible reaction structures up to configurable bounds for the number of reactants and products and a given number of species. Duplicate reactions are removed from a subsampled library.

In each step, coupled SINDy is applied to all library subsamples, yielding estimates of the rate constants. The R^2 score of each sublibrary, parametrized with the corresponding constants, is used as its *fitness value*. After sorting the population members by their fitness, the first $n' < n$ libraries are *selected* for reproduction, i.e., as the basis for generating the next generation's population. In our experiments, we typically use a relatively small value for n' between one and ten percent of n. From the n' selected libraries, n new libraries are obtained by (1) picking a library and randomly replacing one reaction with a newly sampled one (*mutation*) and (2) picking two libraries and exchanging several reactions among them (*crossover*). While mutation allows the exploration of unknown parts of the optimization space, crossover exploits good solutions under the assumption that, often, the combination of two fit solutions achieves a higher fitness than the original solutions alone. The latter is intuitive in the case of evolving libraries, as the presence of reactions explaining a certain behavior correlates with a high fitness of the sublibrary. In other words, the fitness of a sublibrary is expected to be correlated with the presence of certain "helpful" reactions. The above procedure is repeated over a given number of steps (generations) or until the current fittest sublibrary reaches a threshold for the R^2 score. Between steps, one elite population member (the currently fittest sublibrary) is preserved as-is for the next generation.

The coupled SINDy approach used to determine the rate constants and fitness values follows the description in [8] and was, together with all experiments, implemented[1] in the Python programming language. In preliminary experiments, we found that the non-negative least squares (nnls) algorithm described in [5] and, in our case, implemented in the `SciPy` Python module [31] produced the best results. Application of STLSQ [7] or the LASSO [36] often resulted in worse solutions in our coupled case. For numerical differentiation, a central differences method as implemented in `NumPy`'s [27] `gradient` function is used, as we found that forward differences performed consistently worse in our applications.

[1] https://zenodo.org/doi/10.5281/zenodo.11654439.

4.2 Including Background Knowledge

In many cases, biochemical models are not built from the ground up but based on existing models and knowledge [1,34]. Considering the data-driven learning of models, this calls for a method to include such background knowledge.

The simplest form of this is the removal of certain reactions from the library. For example, if there is information about a compartment structure of the biochemical system, such as a division into cytosol and nucleus, one may want to exclude reactions between species located in different compartments. These can be removed from the reaction enumeration that underlies the initial population generation, as well as the mutation operator.

When some reactions are known, i.e., a model is extended, the coefficients of known reactions can be constrained to values greater than zero using a suitable regression algorithm [17]. Alternatively, if $\Theta = \Theta_1 \cup \Theta_2$ and a subset of reactions Θ_1 is known a priori along with their rate constants Ξ_1, the effects of these can be subtracted from the observations, and the problem from Eq. 2 becomes:

$$\min_{\Xi_2} ||(Y' - \Theta_1(Y)\Xi_1) - \Theta_2(Y)\Xi_2||_2$$

This idea is also employed in [24] as part of a tree-based search over models.

An advantage of the GA is the possibility of including arbitrary (numerical) terms in the fitness function. This can be used to introduce an inductive bias, which encourages sublibraries with certain properties. In contrast to the "library-level" constraints already possible with SINDy, this allows the definition of constraints on the "model level", e.g., on the existence of a reaction that produces/consumes a certain species. For example, in the case of a compartment structure discussed above, this can be used to encourage the existence of reactions for shuttling between parts of the cell (cf. Section 5.3).

5 Case Studies

In this section, we apply the concept of evolving libraries in three benchmarks. First, we aim to recover a compartmental model of epidemic dynamics (SIR) [25] and a well-known model of predator-prey dynamics based on the works of Lotka and Volterra [28]. These models contain a rather small number of reactions and species (≤ 3). However, as noted before, even for these, the search space is quite large (around 30 to 100 candidate reactions or 2^{30} to 2^{100} models when allowing models of all sizes). As a final benchmark, we also study the recovery of a Wnt pathway model as presented in [34]. After flattening (cf. Sect. 2) and some reformulations, this model consists of 43 reactions over 19 species. For this model, we consider a search space of around 37 000 candidate reactions.

As is commonly done, we focus on a synthetic environment for model learning to highlight the raw capabilities of our approach. For each benchmark model, we generate time-series measurements by simulating the ground truth model over a certain time period with the LSODA integrator [29]. We found this variable step-size method to consistently yield the best results among the methods we tested (explicit Runge-Kutta methods of orders three, five, eight; Backward Differentiation Formulas). The likely reason is its automatic adaptation to the observed stiffness of the system, which can vary widely between learned models. Measurements are taken at fixed intervals or at each (variable-size) integration step. The latter can yield better results than fixed steps, as regions with larger derivatives, higher frequencies, and turning points are sampled more densely. Generally, measurement data is rarely well-formed like this but may be obtained by an appropriate experiment design. Even then, it will exhibit noise depending on the measurement technique, which we disregard here for simplicity.

We compare the results of coupled SINDy (c-SINDy) to the results obtained with the evolving libraries (evolib) approach. Additionally, we compare a random search that follows the same procedure as the GA but samples a new population at each step, reporting the fitness of the best-found solution until the current step. For each problem, the "complete" library consists of all reactions between the respective number of species with up to two reactants and up to three (SIR, Predator-Prey) or two (Wnt) products. In every case we limit the maximum change $\nu_{i,j}$ per species i (cf. Sect. 2) to one, excluding some cases where the underlying derivatives are collinear, e.g., $A \xrightarrow{1} 3A$ ($dA/dt = 2A$) and $A \xrightarrow{1} 2A$ ($dA/dt = A$) where only the latter will be part of the library. As the results of evolib and random search are stochastic, we perform ten replications of the complete optimization, additionally reporting the mean. An overview of the hyperparameters used in our experiments can be found in Appendix A.

5.1 Susceptible-Infected-Recovered Model

The susceptible-infected-recovered (SIR) model describes the dynamics of an infectious disease spreading in a population. Susceptible individuals are infected upon interaction with an infected individual ($S + I \rightarrow 2I$). Infected individuals recover at a certain rate ($I \rightarrow R$). Here, the rate constants of the ground truth model are set to 0.02 for the infection and 5 for the recovery reaction; the initial population consists of 1980 susceptible, 20 infected, and 0 recovered individuals. The data comprises 100 points measured at equal distances over a period of one time unit. The GA worked with 100 sublibraries, i.e., each generation of the GA contains 100 individuals. Each sublibrary contains 2 reactions, and the total number of candidate reactions to select from is 97.

The results shown in Fig. 2 indicate that, on this dataset, the solver used with c-SINDy and the GA in evolib converge to a solution yielding a good fit to the data. In this rather small search space of choosing two out of 97 reactions, the random search also eventually converges to the same solution as evolib, but at a much slower rate. Compared to c-SINDy, the evolib approach results in a more parsimonious model, in this case, recovering the ground truth model. The resulting model of the c-SINDy approach requires many more reactions with lower rate constants to achieve its fit, learning "wrong" mechanisms. This disadvantage can be attributed to the design matrix's properties. Whereas evolib's smaller matrices often have full rank, the complete matrix representing 97 reactions only has rank 18, posing a rather ill-conditioned problem.

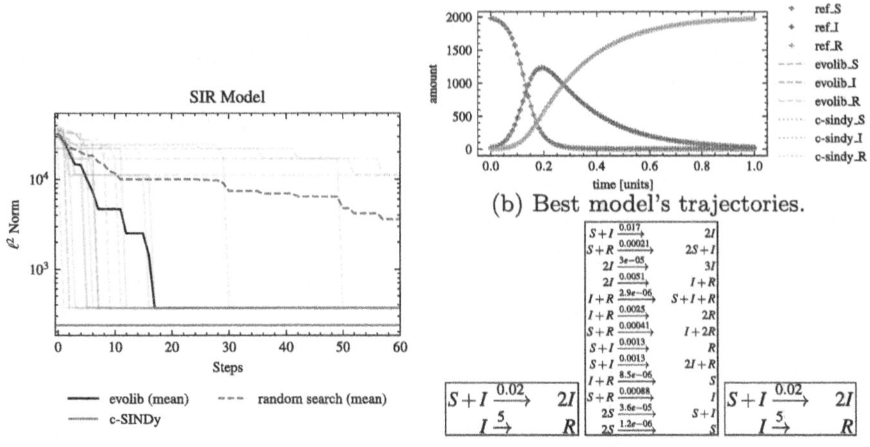

(a) Convergence of optimization.

(b) Best model's trajectories.

(c) Learned models (ground truth, coupled SINDy, evolving libraries).

Fig. 2. Results for recovering the SIR model from synthetic measurement data. Even though the solver used with coupled SINDy is an iterative procedure, the steps are not comparable to the GA steps and we only indicate the final achieved loss for coupled SINDy by a straight line in all figures.

5.2 Lotka-Volterra Model

Our second example is a model of predator-prey dynamics. The predator (here W/wolf) interacts with the prey (here S/sheep), increasing the predator population ($S + W \rightarrow 2W$). The predator population decays over time ($W \rightarrow \emptyset$), whereas the prey population multiplies ($S \rightarrow 2S$). The rate constants and initial population for the ground truth model were set to 0.01, 8, 10.0 for predation, predator decay, and prey multiplication, respectively, and the initial population is given by $(S, W) = (1000, 20)$. The data comprises 100 equally-spaced measurements over two time units. Analogous to the SIR benchmark, the GA worked with 100 sublibraries of size 3 (total number of candidate reactions 28).

We see that the results shown in Fig. 3 are very similar to the observations when learning the SIR model. In contrast to these results, for this ground truth parametrization featuring an oscillation between the predator and prey species, we found that 100 measurements were not enough for an accurate fit, although very similar dynamics producing an oscillation of different frequencies were still discovered. Notably, c-SINDy achieves a lower Loss, but the model learned by evolib provides a better fit to the reference trajectory. This is because the solution quality is measured by the goodness of fit to the *numerical derivatives* (cf. Y' in Eq. 2 in Sect. 2), which in this case are not reliable as there is an insufficient number of samples at the population peaks. Thus, whereas c-SINDy can overfit the (misleading) derivatives, evolib is forced to generalize, abiding by the configured maximum model size. Using the LASSO with c-SINDy and different regularization weights also yielded similar results.

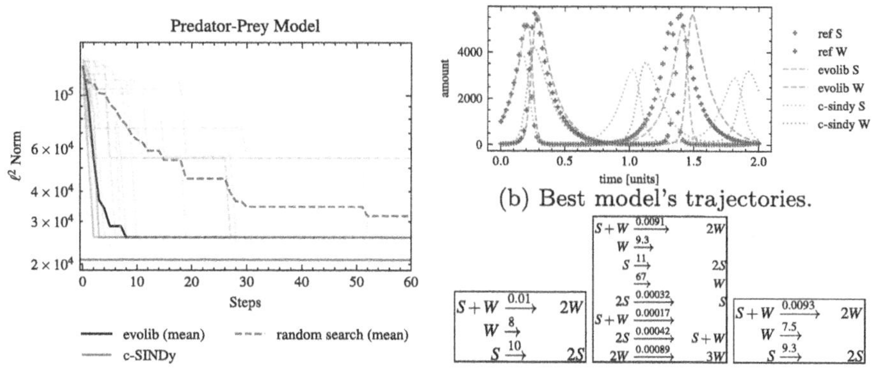

(a) Convergence of optimization.

(b) Best model's trajectories.

(c) Learned models (ground truth, coupled SINDy, evolving libraries).

Fig. 3. Results for recovering the Predator-Prey model from synthetic measurement data.

5.3 Wnt Pathway

Finally, we also study a model of the Wnt signaling pathway as a very large application with significant relevance to systems biology. We use the model given in [34], an extension of both, the seminal Wnt model by Lee et al. [21], and an adaptation of Haack et al. [14]. The goal of the study [34] was to explain the cellular response of osteoblasts to increased oxidative stress induced by placing them on a micropillar structure found on a titanium implant. The original model, written in ML-Rules [15], uses attributes and compartments. For this study, we transformed the model into a flattened version so it presents a simple reaction system (cf. Sect. 2). Further, without changing the model behavior, the Sox17 species was replaced by respectively increasing the degradation rate of the TCF/β-catenin complex (R14). The Ros synthesis (R1), for which the rate depends on a counting species P (H1) was rewritten with mass-action kinetics. The final model consists of 43 reactions between 19 species (cf. Appendix C), resulting in a highly non-trivial learning task.

Hence, we make some additional assumptions. First, as we are mostly limited to the capabilities of SINDy, we can only readily consider the case where all species are measured. This is rarely the case in systems biology, despite continuous advances in measurement techniques, such as single-cell mass-cytometry [33]. In [34], solely measurements of Axin, β-catenin, Sox17, and ICAT at a small number of time points were available in addition to the basis models. Secondly, we here use the steps proposed by the numeric integration with LSODA as measurement points as described at the beginning of this section. We additionally truncated the time series to start at the 16-min mark and end at 500 min, as we found that the extremely high derivatives at the beginning, which are a result of a long and fast chain of reactions, hindered a fast convergence of the optimization. In fact, the ground truth model was fitted with the primary goal of reproducing the steady-state behavior observed in the wet lab experiments, disregarding the initial transient response phase. Note, however, that this may generally discard important information on the model dynamics, and we leave an in-depth analysis of alternative ways to deal with this situation inside the SINDy framework as future work.

Similar to [17], we consider two cases: (1) learning of the complete model "from scratch" and (2) extending a base model (representing the known reactions from [14,21]) with the six rules that have been added and fitted in [34] (amounting to nine reactions when flattened). For both cases, we also tested the inclusion of background knowledge about the compartment structure of the system. This entails reducing the library only to include reactions where species within the same compartment (nucleus or cytosol) interact and further the adjustment of the fitness function to penalize libraries that do not include shuttling of species, encouraging the existence of at least one migration into/out of the nucleus, respectively. We implement this by subtracting 1 from the R^2 score fitness if there is no shuttling into the nucleus (similarly for shuttling out of the nucleus). Given that the R^2 score quickly approaches its maximum value of one during optimization, this is a large enough penalty to downrank sublibraries when sorting them by their fitness. As it enables unwanted constructs, we also penalize the use of the counting species P with 0.01. We choose a low penalty as two reactions in the ground truth model still depend on it.

With this in mind, the complete library size for c-SINDy is 37 018. When extending the model, the 35 fixed reactions are removed from this library. In the case where we constrain reactions to occur only between reactants within the same compartment, the library size is significantly reduced to 10 462 reactions.

Figure 4 shows the convergence of the four scenarios described above. As a first result, we see that, as expected, the evolutionary algorithm significantly outperforms the random search, which we only performed for the unconstrained cases. In all cases, evolib yields a higher loss than c-SINDy. Looking at the resulting models reveals that c-SINDy's solutions hinge on a huge amount of around 500 reactions, some of which exhibit enormous rates. These cannot reasonably be depicted here. The models learned with evolib are shown in Appendix C. Thus, we conclude that, similar to our smaller case studies, evolib trades off a higher loss value for a more comprehensible model.

In contrast to the smaller case studies, we observed that there is no significant overlap between the ground truth model and the learned models. Some ground truth reactions are recovered, but often their effects are scattered over multiple different reactions in the learned model so that a single reaction is expressed by multiple ones.

For example, the model learned by evolib from scratch shown in Appendix C Fig. 6a includes the Axin-induced degradation of β-catenin, which is a central part of the Wnt pathway. In the constrained case Fig. 7a, the shuttling of β-catenin is recovered. The small overlap with the ground truth model is a hint that our chosen constraints are not enough to enforce biologically viable models, which underlines the problem of model identifiability (cf. Sect. 6). This is further underpinned by the fit to the reference trajectory shown in Appendix B. In the extension cases, a visually accurate fit is achieved when simulating the learned models, despite them being different from the ground truth. Judging from Fig. 4, the (excessive) models learned by c-SINDy seem to produce much more accurate fits but were not amenable to integration with LSODA, which resulted in errors regarding numerical precision (cf. Sect. 6).

Note, that at this time, the implementation in Python is not optimized for performance, so we can only reliably estimate theoretically the performance benefits of evolib (cf. Sect. 6). With our current setup and hyperparameters, we found that performing ten replications of evolib in parallel, also parallelizing the fitness evaluation, took around five times as long as a single run of c-SINDy for the Wnt experiments (1h vs. 5h), but this result depends heavily on the population size and number of steps performed. For example, increasing the population size allows for decreasing the number of steps, where the computations on the former can be fully parallelized, leading to much fewer sequential steps and faster convergence. Note also that we limited the number of iterations for non-negative least squares to 10^8 for both approaches.

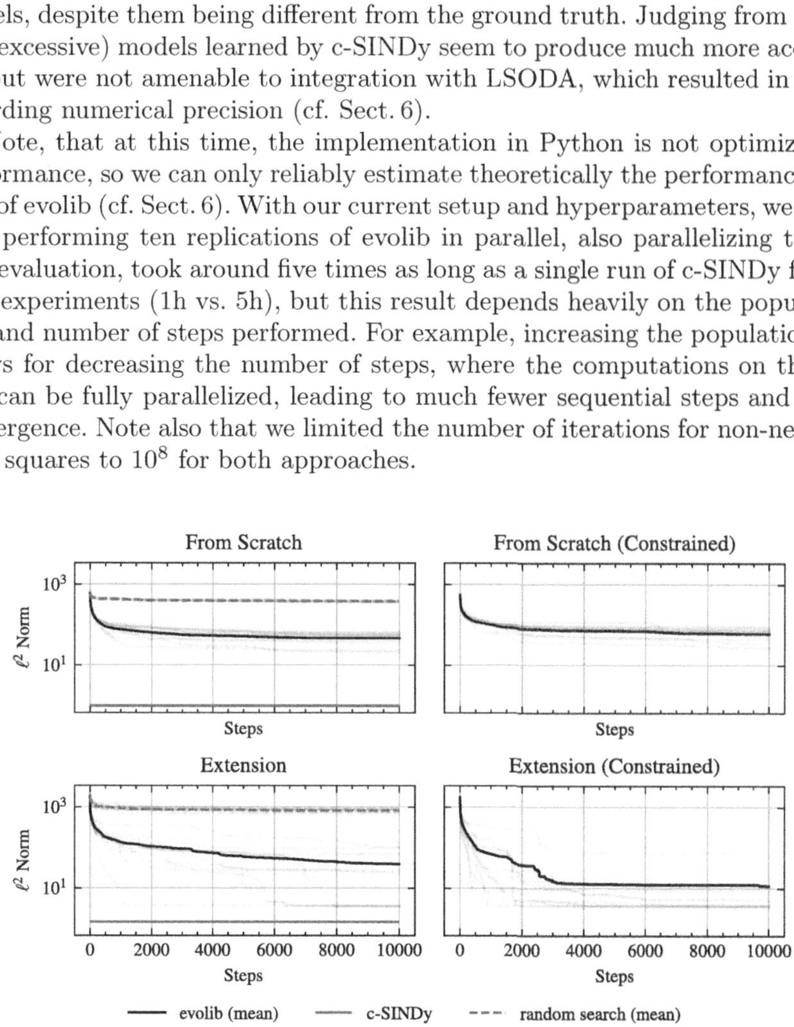

Fig. 4. Convergence of evolving libraries for learning the Wnt pathway and loss of the solution found by coupled SINDy. In the constrained case, coupled SINDy is not applicable and we deemed a random search uninformative.

6 Discussion

Our results demonstrate that the genetic evolution of libraries can provide benefits in terms of model parsimony and fit. In the smaller case studies, evolib enables the recovery of the ground truth model, which was not possible with coupled SINDy alone. However, evolib was not able to recover a biologically viable structure for the complex Wnt pathway, even though achieving a visually accurate fit to the reference trajectory in some scenarios. Partially, the missing ground truth structures can be explained by our truncation of the time-series data to a smaller interval. Note also that we here set the hyperparameters of the genetic search, e.g., population size, to values we determined manually (cf. Appendix A). We leave a systematic (and costly) exploration of hyperparameters, which may be able to achieve slightly better results, as future work. Generally, however, we attribute the mismatch between the learned and the ground truth model to the fact that, in most cases, there is no single model that exclusively explains a given time series. Rather, it is known that reaction systems are not *identifiable* [9], i.e., multiple models fit the same data, necessitating an informed selection. This is why background knowledge, e.g., about the possible interactions, known reactions, and model-level constraints, plays an important role in ensuring that learned models are grounded in the (known) laws of science, and are coherent with the current knowledge about mechanisms. To this end, evolib enables the inclusion of model-level constraints, in addition to constraints on the reaction level possible with (c-)SINDy. Here, we explored how both constraints can be combined to limit reactions to occur within compartments while encouraging the inclusion of shuttling reactions between them. Future work may adjust the fitness function to encourage libraries where the species/reactions match an ontology or are aligned with information gathered from biological databases. In fact, some methods, such as [1], approach the problem from the perspective of knowledge rather than data by automatically constructing models from facts found in literature databases.

Whereas a regression considering all possible reactions at once often leads to very long model descriptions, the GA produced much more sparse solutions, as it allows enforcing a certain maximum number of reactions. This leads to a tradeoff between model complexity and goodness of fit. Limiting model complexity results in humanly comprehensible models, but typically less accurate predictions, which, however, may be beneficial in cases with low-quality data (cf. the predator-prey case study in Sect. 5). Further, in contrast to the much shorter model descriptions learned by evolib in our Wnt case study, the models learned by c-SINDy could not be readily integrated with LSODA, yielding errors regarding numerical precision. These may be attributed to the presence of singularities, extreme oscillations or due to the fact that LSODA is not switching effectively between stiff and non-stiff regimes.

In our evaluation, we used relatively ideal conditions to test the principal capabilities of the approach. However, in a real-world scenario, measurement data would be available only for a subset of species and be subject to noise. Handling these scenarios is a major limitation of SINDy and, thus, also a limitation of our approach. Due to its relevance, there have been attempts to also include unmeasured variables, e.g., by using dynamic mode decomposition (DMD) to identify those variables before employing SINDy [6] or implicitly in a black-box manner by training neural networks to act as correction terms [30].

In terms of execution time, we did not observe a speedup using the GA on smaller libraries as opposed to c-SINDy on a very large library. Theoretically, assuming coupled SINDy takes $f(n_a)$ steps (n_a is the library size), its wrapping in a GA results in $cf(n_b)$ steps for some $n_b << n_a$ (n_b is the sublibrary size). The constant c is determined by the number of generations and individuals in each generation. Note that the fitness calculations are easily parallelizable, reducing the factor c. While not changing the complexity class, evolib may still provide a speedup in practice when n_b and c can be chosen small enough. Here, we work with a first prototypical implementation and leave a comprehensive empirical evaluation of the execution time performance as future work.

7 Conclusion

We demonstrated the use of a genetic algorithm together with an extension of SINDy to the case of coupled differential equations, which allows learning compact biochemical reaction models (with many species) from time-series data. Our results show the benefits of the approach for two small and one very large reaction system to be learned. In the smaller cases, the approach can recover the ground truth model. For the case of identifying a very large Wnt model, where coupled SINDy failed to learn models amenable to numerical integration, it learned well-fitting parsimonious models, but they lack biological meaning. Thus, we believe that including background knowledge (inductive bias) is essential to learning a model that can serve as a useful theory to explain the mechanisms at work. Our approach offers the ability to include background knowledge at the level of the model as well as on the level of individual reactions. Although our first and rather simple restrictions that enforced a compartment structure and encouraged shuttling reactions between compartments did not suffice to learn a biologically meaningful Wnt signaling model, we will further pursue this line of research as particularly promising. In addition, we plan to evaluate the method in comparison to other methods for learning reactions, such as the Reactionet LASSO [17] (e.g., by only considering the mean) or Reactmine [24]. For that purpose, identifying a set of suitable benchmark models/data sets to test different capabilities will be essential. In addition, to make these approaches widely applicable in systems biology, the possibility of inferring (time series for) unmeasured species needs further research. This may, for example, be done by including a pre-processing step [6].

Acknowledgments. JNK and AU acknowledge the funding of the DFG Project GrEASE (grant number 320435134). KB acknowledges the funding of the ARC Centre of Excellence for Plant Success in Nature and Agriculture CE 200100015. The authors thank Fiete Haack for many helpful discussions, particularly over his model of the Wnt pathway, and his feedback on the evaluation.

A Complete List of Experiment (Hyper-)parameters

(See Table 1).

Table 1. Hyperparameters used for evolib. The random search and c-SINDy use the same parameters where applicable. In particular, for the non-negative least squares, the maximum number of iterations was limited to 10^8. Abbreviations: X extended, cstr. constrained.

parameter↓/model→	SIR	Predator-Prey	Wnt	Wnt-X	Wnt (cstr.)	Wnt-X (cstr.)
replications	10					
measurement points	100		78			
measurement interval	0.0–1.0	0.0–2.0	16.0–500.0			
measurement steps	equidistant		determined by LSODA			
sublibrary size	2	3	60	20	60	20
population size	100		200			
max steps	100		10000			
crossover probability	0.8					
mutation probability	0.2					
crossover points	1		5			
num parents	10					
max. number of reactants	2					
max. number of products	3		2			

B Learned Model's Trajectories for the Wnt Pathway

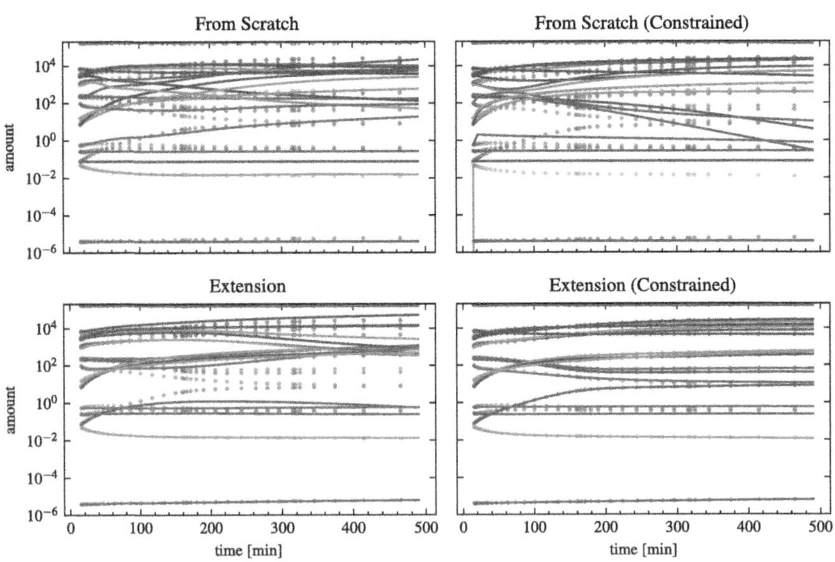

Fig. 5. Results from simulating the models learned with evolib for the Wnt pathway. The + symbols mark measurement points, and the lines are the trajectories simulated for the 19 species. Note that, for clarity, we omit the labeling of species, and only every second measurement point is shown. The integration of the (large) models produced by c-SINDy resulted in errors due to numerical problems.

C Learned Models for the Wnt Pathway

(See Figs. 6 and 7).

(a) Evolib from scratch (b) Evolib extension (c) Ground truth

Fig. 6. The models inferred in the Wnt case study (unconstrained) compared to the ground truth model. For the extension, the reactions above the line are fixed. Only reactions with a rate above 10^{-6} are shown, and if applicable the number of excluded reactions is shown in the lower left. Bolded reactions indicate an overlap with the ground truth reactions. Note in particular how in (a) the Axin-induced degradation of β-catenin and in (b) the synthesis of Ros was recovered, which are both central components of the ground truth model of the Wnt pathway.

(a) Evolib from scratch (b) Evolib extension (c) Ground truth

Fig. 7. The models inferred in the Wnt case study (constrained) compared to the ground truth model. For the extension, the reactions above the line are fixed. Only reactions with a rate above 10^{-6} are shown, and if applicable, the number of excluded reactions is shown in the lower left. Bolded reactions indicate an overlap with the ground truth reactions. Note in particular how in (a) the shuttling of β-catenin in and out of the nucleus and in (b) the production of TCF was recovered.

References

1. Ahmed, Y., Telmer, C., Miskov-Zivanov, N.: Accordion: Clustering and selecting relevant data for guided network extension and query answering. arXiv preprint arXiv:2002.05748 (2020). https://doi.org/10.48550/arXiv.2002.05748
2. Askari, E., Crevecoeur, G.: Evolutionary sparse data-driven discovery of multibody system dynamics. Multibody Syst. Dyn. **58**, 197–226 (2023). https://doi.org/10.1007/s11044-023-09901-z
3. Bortolussi, L., Cairoli, F., Klein, J., Petrov, T.: Data-driven inference of chemical reaction networks via graph-based variational autoencoders, pp. 143–147. Springer, Heidelberg (2023). https://doi.org/10.1007/978-3-031-43835-6_10

4. Boutillier, P.: The kappa platform for rule-based modeling. Bioinformatics **34**(13), i583–i592 (2018). https://doi.org/10.1093/bioinformatics/bty272
5. Bro, R., De Jong, S.: A fast non-negativity-constrained least squares algorithm. J. Chemom. **11**(5), 393–401 (1997). https://doi.org/10.1002/(SICI)1099-128X(199709/10)11:5<393::AID-CEM483>3.0.CO;2-L
6. Brummer, A.B., et al.: Data driven model discovery and interpretation for car t-cell killing using sparse identification and latent variables. Front. Immunol. **14** (2023). https://doi.org/10.3389/fimmu.2023.1115536
7. Brunton, S.L., Proctor, J.L., Kutz, J.N.: Discovering governing equations from data by sparse identification of nonlinear dynamical systems. Proc. Natl. Acad. Sci. **113**, 3932–3937 (2016). https://doi.org/10.1073/pnas.1517384113
8. Burrage, P.M., Weerasinghe, H.N., Burrage, K.: Using a library of chemical reactions to fit systems of ordinary differential equations to agent-based models: a machine learning approach. Numer. Algor. (2024). https://doi.org/10.1007/s11075-023-01737-0
9. Craciun, G., Pantea, C.: Identifiability of chemical reaction networks. J. Math. Chem. **44**(1), 244–259 (2008). https://doi.org/10.1007/s10910-007-9307-x
10. Daniels, B.C., Nemenman, I.: Automated adaptive inference of phenomenological dynamical models. Nat. Commun. **6** (2015). https://doi.org/10.1038/ncomms9133
11. Faeder, J.R., Blinov, M.L., Hlavacek, W.S.: Rule-based modeling of biochemical systems with bionetgen. In: Systems Biology, pp. 113–167. Springer, Heidelberg (2009). https://doi.org/10.1007/978-1-59745-525-1_5
12. Fasel, U., Kutz, J.N., Brunton, B.W., Brunton, S.L.: Ensemble-sindy: robust sparse model discovery in the low-data, high-noise limit, with active learning and control. Proc. Royal Soc. A: Math. Phys. Eng. Sci. **478** (2022). https://doi.org/10.1098/rspa.2021.0904
13. Großmann, G., Zimmerlin, J., Backenköhler, M., Wolf, V.: Unsupervised relational inference using masked reconstruction. Appl. Netw. Sci. **8**(1), 18 (2023). https://doi.org/10.1007/s41109-023-00542-x
14. Haack, F., Lemcke, H., Ewald, R., Rharass, T., Uhrmacher, A.M.: Spatio-temporal model of endogenous ros and raft-dependent wnt/beta-catenin signaling driving cell fate commitment in human neural progenitor cells. PLoS Comput. Biol. **11**(3), 1–28 (2015). https://doi.org/10.1371/journal.pcbi.1004106
15. Helms, T., Warnke, T., Maus, C., Uhrmacher, A.M.: Semantics and efficient simulation algorithms of an expressive multilevel modeling language. ACM Trans. Model. Comput. Simul. (TOMACS) **27**(2), 1–25 (2017). https://doi.org/10.1145/2998499
16. Keating, S.M., et al.: Sbml level 3: an extensible format for the exchange and reuse of biological models. Molec. Syst. Biol. 16(8), e9110 (2020). https://doi.org/10.15252/msb.20199110
17. Klimovskaia, A., Ganscha, S., Claassen, M.: Sparse regression based structure learning of stochastic reaction networks from single cell snapshot time series. PLoS Comput. Biol. **12**, e1005234 (2016). https://doi.org/10.1371/journal.pcbi.1005234
18. Koza, J.R., Mydlowec, W., Lanza, G., Yu, J., Keane, M.A.: Reverse Engineering of Metabolic Pathways From Observed Data Using Genetic Programming, pp. 434–445. World Scientific (2000). https://doi.org/10.1142/9789814447362_0043
19. Kozin, F., Natke, H.: System identification techniques. Struct. Saf. **3**(3–4), 269–316 (1986). https://doi.org/10.1016/0167-4730(86)90006-8
20. Kramer, O.: Genetic Algorithms. In: Genetic Algorithm Essentials. SCI, vol. 679, pp. 11–19. Springer, Cham (2017). https://doi.org/10.1007/978-3-319-52156-5_2

21. Lee, E., Salic, A., Krüger, R., Heinrich, R., Kirschner, M.W.: The roles of apc and axin derived from experimental and theoretical analysis of the wnt pathway. PLoS Biol. **1**(1), e10 (2003). https://doi.org/10.1371/journal.pbio.0000010
22. Mangan, N.M., Brunton, S.L., Proctor, J.L., Kutz, J.N.: Inferring biological networks by sparse identification of nonlinear dynamics. IEEE Trans. Molec. Biol. Multi-Scale Commun. **2**, 52–63 (2016). https://doi.org/10.1109/tmbmc.2016.2633265
23. Manzi, M., Vasile, M.: Orbital anomaly reconstruction using deep symbolic regression. In: 71st International Astronautical Congress, IAC 2020 (2020)
24. Martinelli, J., Grignard, J., Soliman, S., Ballesta, A., Fages, F.: Reactmine: a statistical search algorithm for inferring chemical reactions from time series data. arXiv preprint arXiv:2209.03185v2 (2022)
25. Milgroom, M.G.: Epidemiology and SIR Models, pp. 253–268. Springer, Cham (2023). https://doi.org/10.1007/978-3-031-38941-2_16
26. Nobile, M.S., Besozzi, D., Cazzaniga, P., Pescini, D., Mauri, G.: Reverse engineering of kinetic reaction networks by means of cartesian genetic programming and particle swarm optimization. In: 2013 IEEE Congress on Evolutionary Computation (CEC). IEEE (2013).https://doi.org/10.1109/cec.2013.6557752
27. NumPy team an contributors: Numpy. Version 1.24.3 (2023). https://numpy.org/
28. Parker, M., Kamenev, A.: Extinction in the lotka-volterra model. Phys. Rev. E **80**, 021129 (2009). https://doi.org/10.1103/PhysRevE.80.021129
29. Petzold, L.: Automatic selection of methods for solving stiff and nonstiff systems of ordinary differential equations. SIAM J. Sci. Stat. Comput. **4**(1), 136–148 (1983). https://doi.org/10.1137/0904010
30. Rackauckas, C., et al.: Universal differential equations for scientific machine learning. arXiv preprint arXiv:2001.04385v4 (2020). https://doi.org/10.48550/arXiv.2001.04385
31. SciPy team and contributors: Scipy. Verison 1.10.1 (2023). https://scipy.org/
32. Soliman, S., Heiner, M.: A unique transformation from ordinary differential equations to reaction networks. PLoS ONE **5**(12), e14284 (2010). https://doi.org/10.1371/journal.pone.0014284
33. Spitzer, M.H., Nolan, G.P.: Mass cytometry: single cells, many features. Cell **165**(4), 780–791 (2016). https://doi.org/10.1016/j.cell.2016.04.019
34. Staehlke, S., et al.: Ros dependent wnt/β-catenin pathway and its regulation on defined micro-pillars-a combined in vitro and in silico study. Cells **9**(8) (2020). https://doi.org/10.3390/cells9081784
35. Székely, T., Burrage, K.: Stochastic simulation in systems biology. Comput. Struct. Biotechnol. J. **12**(20), 14–25 (2014). https://doi.org/10.1016/j.csbj.2014.10.003
36. Tibshirani, R.: Regression shrinkage and selection via the lasso. J. Roy. Stat. Soc. Series B (Methodological) **58**(1), 267–288 (1996). http://www.jstor.org/stable/2346178

Flexible Nets to Improve GEM Cell Factories by Combining Kinetic and Proteomics Data

Jorge Lázaro[1(✉)], Jorge Júlvez[2], and Jürgen Zanghellini[3]

[1] Department of Computer Science and Systems Engineering, University of Zaragoza, Zaragoza, Spain
jorgelazaro@unizar.es

[2] Aragón Institute for Engineering Research, University of Zaragoza, Zaragoza, Spain
julvez@unizar.es

[3] Department of Analytical Chemistry, University of Vienna, Vienna, Austria
juergen.zanghellini@univie.ac.at

Abstract. Alzheimer's disease is expected to reach a prevalence of 152 million people worldwide caused by the aggregation of amyloid β-proteins leading to apoptosis of neurons and loss of cognitive function. Although there is no effective treatment for this disease, molecules such as *scyllo*-inositol have been shown to be promising. *Bacillus subtilis* has been proposed as a suitable organism for the production of *scyllo*-inositol. Metabolic computational models have proven useful in the prediction of the production of a metabolite. However, most genome-scale metabolic models lack detailed parameters and tend to overestimate the production of a metabolite with respect to the consumption of medium resources. In order to reduce the solution space and, hence, obtain a more realistic model, additional constraints from experimental data can be added to the model. This work exploits the modeling capabilities of Flexible Nets to model the production of *scyllo*-inositol in a genome-scale metabolic model of *Bacillus subtilis* that has been previously enriched with proteomic and enzymatic data. We assess how these constraints limit the *scyllo*-inositol production to more realistic levels. Moreover, the integration of different types of constraints allowed us to uncover which one of them limits the production of *scyllo*-inositol for a given growth rate.

Keywords: Metabolic modeling · Flexible Nets · genome-scale models · *scyllo*-inositol · proteomic constraints · enzymatic constraints

1 Introduction

Genome-scale metabolic models (GEMs) are mathematical representations of the metabolism of an organism that account for the stoichiometry and flux bounds of the metabolic reactions (Gu et al. [2019]). GEMs have various practical applications such as strain development for producing bio-based chemicals and materials (Senger [2010], designing bio-processes Gotsmy et al. [2023], Lázaro et al. [2022]), drug targeting pathogens (Kim et al. [2010]), predicting enzyme functions (Orth et al. [2011]), modeling interactions among multiple cells or organisms (Zorrilla et al. [2021]), and understanding human diseases (O'brien et al. [2013]). The most widely-used method

R. Gori et al. (Eds.): CMSB 2024, LNBI 14971, pp. 137–154, 2024.
https://doi.org/10.1007/978-3-031-71671-3_11

for analyzing GEMs is Flux Balance Analysis (FBA) (Orth et al. [2010]), which consists of solving a linear programming problem.

Although GEMs have proven useful in many cases, they are still far from reliably predicting growth rates or product efficiency because they are exclusively based on stoichiometric and mass balance constraints. Consequently, the FBA optimal solution still tends to be relatively large and usually fails to correctly approximate reality (Durot et al. [2008], Shaw and Cheung [2021], Mao et al. [2015]). The implementation of additional constraints to GEMs is gaining more and more attention, since they can reduce the solution space and, hence, result in a more realistic prediction of the metabolic fluxes. It must be noticed that, in general, the bottleneck that limits the rate at which a reaction occurs is the enzyme abundance and its kinetic parameters (Massaiu et al. [2019]).

This work focuses on the implementation of enzymatic and proteomic constraints. The metabolic flux of each reaction strongly depends on the availability of the enzyme that catalyzes that reaction. These properties can be integrated into the GEM in the form of parameters such as protein abundance and catalytic constants. Here, we show how to exploit the modeling formalism of Flexible Nets, an extension of Petri nets, to integrate enzymatic constraints into a GEM of *Bacillus subtilis*.

B. subtilis, which is a model organism for Gram-positive bacteria, stands out for its extensively mapped genome and remarkable adaptability to diverse environments and substrates. *B. subtilis* efficiently secretes products without genetic engineering. In particular, its ability to grow in minimal medium and produce *scyllo*-inositol, a promising therapeutic for Alzheimer's disease which inhibits β-amyloid aggregation, has been well-characterized. Recently, a method was devised by Michon et al. [2020] to synthesize *scyllo*-inositol from glucose, making *B. subtilis* a competitive producer compared to *E. coli*.

The remainder of this paper is organized as follows: Sect. 2 introduces Flexible Nets and shows their capabilities to model complex dynamics in biological systems. Sections 3, 4 and 5 introduce two methods to implement kinetic and proteomic constraints as well as their alternative using FNs. Sections 6 and 7 introduce the case study and discuss how we enriched the model and assessed its performance. The main conclusions are drawn in Sect. 8.

2 Flexible Nets to Model Biological Systems

2.1 Flexible Nets

Flexible Nets (FNs) (Júlvez et al. [2018]) is a modeling formalism for dynamic systems inspired by Petri Nets (Murata [1989], Silva [Chapman & Hall, London [1993]). An FN models the interaction between the state and the processes of a dynamic system through two interconnected nets: the event net and the intensity net. While the event net describes how processes alter the state variables, the intensity net captures how these variables determine the process rates.

In contrast to Petri Nets, both the event and intensity nets are tripartite graphs whose vertices are places, transitions, and handlers. In the event net, event handlers trigger changes in metabolite concentration resulting from reactions, while in the intensity net, intensity handlers model how metabolite concentrations affect reaction rates. Places (depicted as circles) are linked with metabolites; transitions (depicted as rectangles) are linked with reactions; event handlers (depicted as dots) monitor metabolite concentration changes; and intensity handlers (also dots) model the impact of metabolite concentrations on reaction rates.

If only the stoichiometry of a metabolic network is taken into account, i.e., no reaction rate is specified, then, an event net is enough to model the behavior of the network. Figure 1 shows a simple event net modeling a system consisting of 4 metabolites (places $\{A, B, C, D\}$), 2 reactions or transitions ($\{R_1, R_2\}$), and 2 event handlers ($\{v_1, v_2\}$) that connect places and transitions. In this simple system, A and B react (reaction R_1) to produce 2 molecules of C; and B is also substrate of a reaction (R_2) that transforms this metabolite into D. The reaction network modeled by the net is:

$$R_1{:}A + B \rightarrow 2C \tag{1}$$
$$R_2{:}B \rightarrow D \tag{2}$$

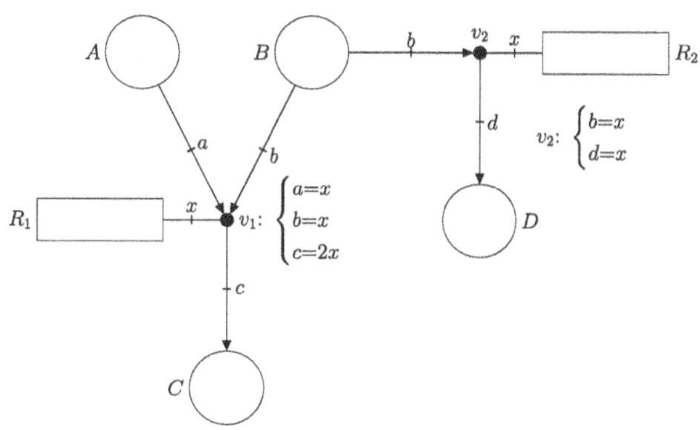

Fig. 1. Event net that only takes into account the stoichiometry of the reactions $R_1 : A{+}B \rightarrow 2C$ and $R_2 : B \rightarrow D$.

The equalities in v_1, $a{=}x$; $b{=}x$; and $c{=}2x$ imply that the occurrence of reaction R_1 triggers the consumption of 1 unit of metabolite A and 1 unit of metabolite B to produce 2 units of metabolite C. Similarly, the equation of v_2 models that when reaction R_2 happens, 1 unit of metabolite B is converted into 1 unit of metabolite D.

An appealing feature of FNs is that its intensity net can model complex dynamics in an intuitive way. For instance, assume that reaction R_2 is enzymatically catalyzed by an enzyme E that needs to incorporate a ligand, Lig, to be active (Fig. 2). This phenomenon is quite common in biology as an enzymatic regulation process. The reaction formulation would look like:

$$R_1{:}A + B \rightarrow 2C \tag{3}$$
$$R_2{:}B + E + L \rightarrow D + E + Lig \tag{4}$$

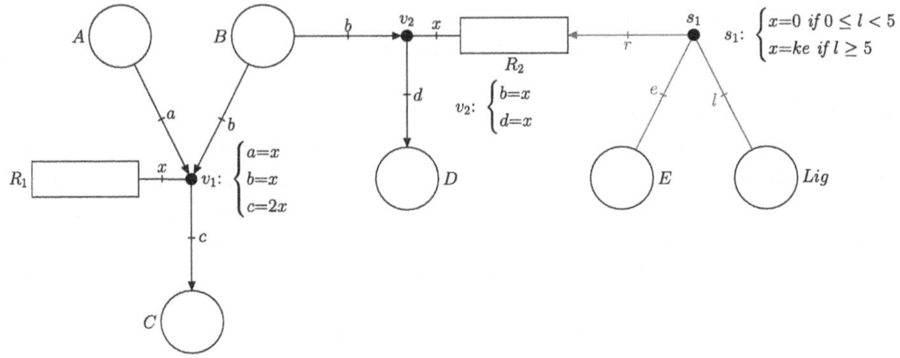

Fig. 2. Flexible Net modeling the reactions $R_1 : A + B \rightarrow 2C$ and $R_2 : B + E + L \rightarrow D + E + Lig$. The event net is represented by black arcs and edges. In contrast, blue arcs and edges depict the intensity net.

The regulation of R_2 is captured by the intensity net which, for clarity, is colored in blue. The enzyme, depicted by E catalyzes the reaction R_2 which will be activated depending on the ligand concentration, modeled by l. This relationship is determined by the inequalities that define the intensity handler s_1. The first condition, $r = 0$ if $0 \leq l < 5$, indicates that the reaction rate, r, of R_2 is null when the concentration of Lig, l, is less than a threshold set to 5 units. If the concentration of Lig is greater than or equal to 5 units, then the rate of R_2 is equal to a constant, k, multiplied by the enzyme abundance, e. This dependency is captured by the inequality $r = ke$ if $l \geq 5$.

In FNs, the rate of a reaction, R_i, is captured by the variable $\lambda[R_i]$. This variable is equal to $\lambda_0[R_i]$, which is the default intensity in a transition, plus the intensities associated with its input arcs minus the intensities of its output arcs.

If no $\lambda_0[R_i]$ is explicitly associated with a reaction R_i, then it is assumed that $\lambda_0[R_i] = 0$. Thus, in the presented example, the intensity of R_2 in Fig. 2 can be expressed as:

$$\lambda[R_2] = \lambda_0[R_2] + \Delta\lambda[(s_1, R_2)] = \Delta\lambda[(s_1, R_2)] \tag{5}$$

where, according to the equations associated with s_1, $\Delta\lambda[(s_1, R_2)]$ is equal to k times the enzyme abundance, e, if the concentration of Lig is greater than or equal to 5, and equal to 0 otherwise.

2.2 Modeling GEMs with FNs

GEMs are mathematically defined as a tuple $\{\mathcal{R}, \mathcal{M}, \mathcal{S}, L, U\}$, where $\mathcal{R}$ is the set of reactions, $\mathcal{M}$ is the set of metabolites, $\mathcal{S} \in \mathbb{R}^{|\mathcal{M}| \times |\mathcal{R}|}$ refers to the stoichiometric matrix whose rows(columns) are associated with the metabolites(reactions) of the model, and $L, U \in \mathbb{R}^{|\mathcal{R}|}$ are the lower and upper flux bounds that constrain the fluxes of the reactions. Since GEMs mainly account for mass balance and stoichiometric constraints, these flux bounds are usually extremely loose. A common assumption when analyzing GEMs is that the model is in steady state. Under this condition, all the metabolites have a constant concentration and, hence, its production rate is equal to its consumption rate. This is expressed by:

$$\mathcal{S} \cdot v = 0 \tag{6}$$

where v is a one-column vector with the fluxes of the reactions. The addition of the flux bounds, L and U, and an objective function to (6) results in the FBA linear programming problem:

$$\begin{aligned} \max\ c^T \cdot v \\ \mathcal{S} \cdot v = 0 \\ L \leq v \leq U \end{aligned} \tag{7}$$

where c is a vector of weights indicating the contribution of each reaction to the objective function. Notice that the FBA of GEMs can be mimicked by FNs by imposing constant concentrations of all metabolites, adding flux bounds and introducing an objective function (see Júlvez and Oliver [2020] for details on steady state analysis in general FNs).

Some methods, such as GECKO (Domenzain et al. [2022]) and sMOMENT (Bekiaris and Klamt [2020]), have been developed recently to restrict the metabolic fluxes of an organism with enzymatic and proteomic data. Nevertheless, they rely on the addition of various artificial reactions and metabolites that expand the original GEM stoichiometric matrix. In addition to modeling all the reactions in a GEM, FNs can capture complex enzymatic relationships that are difficult to accommodate with current methods.

3 GECKO Method

The GECKO method consists of constraining the metabolic flux of each catalyzed reaction for which there is information about the abundance of a specific enzyme. In particular, the flux of a given reaction R_i, v_i $(mmol\ g_{DW}^{-1}\ h^{-1})$, cannot exceed the product between the catalytic constant, $k_{cat,i}$ $(1/h)$ and the enzyme abundance, e_i $(mmol\ g_{DW}^{-1})$ (Equation (8)). This is applied to each metabolic enzyme, e_i, in the model:

$$v_i \leq k_{cat,i} \cdot e_i \tag{8}$$

GECKO constraints are integrated into the GEM as follows:

1. Add a metabolite, E_i, as a reactant with stoichiometric coefficient $\frac{1}{k_{cat,i}}$ to each metabolic enzyme-catalyzed reaction R_i. For instance, if R_i:$A \rightarrow B$, then the addition of E_i results in:

$$R_i'{:}A + \frac{1}{k_{cat,i}} E_i \rightarrow B$$

2. Add a reaction, R_i'' for each implemented enzyme abundance, e_i. For instance:

$$R_i''{:}\emptyset \rightarrow E_i$$

The upper bound of R_i'' is set to the enzyme abundance, i.e. $U[R_i''] = e_i$.

3.1 Modeling the GECKO Method with FNs

The GECKO method can be easily modeled by FNs. Consider a reaction R_1 catalyzed by an enzyme E that transforms metabolite A into B:

$$R_1{:}A + E \rightarrow B + E$$

The FN in Fig. 3 shows how the reaction rate of R_1 is constrained by enzymatic limitations according to Equation (8).

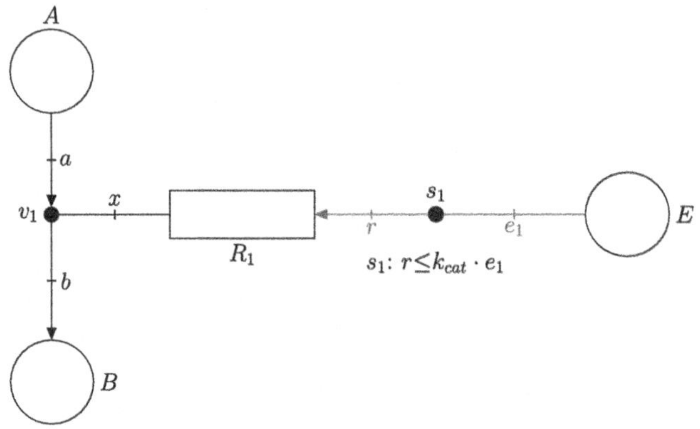

Fig. 3. Flexible Net that models the dynamics of a system where the reaction rate of R_1 is limited by a GECKO-like constraint. Since the intensity transferred from s_1 to R_1 satisfies that $r \leq k_{cat} \cdot e_1$, the intensity of R_1 will satisfy $\lambda[R_1] \leq k_{cat} \cdot e_1$.

In Fig. 3, the inequality presented in the intensity handler s_1 is equivalent to Equation (8). Thus, the flux of a reaction must be less or equal than the catalytic constant multiplied by the enzyme abundance. This strategy is applied to all the reactions for which information about their kinetic parameters and specific enzyme abundances is available.

4 sMOMENT Method

sMOMENT is a method to constrain metabolic fluxes of the reactions in a GEM when the k_{cat} and molecular weights of the enzymes that catalyze the reactions are known. As GECKO, sMOMENT also assumes steady state conditions, see Equation (6). Additionally, the specific constraint added in this method corresponds to the total concentration, P, of all the enzymes experimentally measured. A threshold is set so that the total abundance of enzymes cannot exceed this value, P (g/g_{DW}):

$$\sum_i e_i \cdot MW_i \leq P \qquad (9)$$

where MW_i is the molecular weight of an enzyme. Additionally, Equation (8) can be alternatively expressed as:

$$\frac{v_i}{k_{cat,i}} \leq e_i \qquad (10)$$

This leads to Equation (9):

$$\sum_i v_i \cdot \frac{MW_i}{k_{cat,i}} \leq P \tag{11}$$

which can be rewritten as:

$$v_{pool} - \sum_i v_i \frac{MW_i}{k_{cat,i}} = 0; v_{pool} \leq P \tag{12}$$

where v_{pool} is an auxiliary variable that corresponds to the total mass of metabolic enzymes expressed in g/g_{DW}.

This constraint is integrated into a GEM through the following steps:

1. Add an artificial metabolite called M_{pool} in all the reactions catalyzed by metabolic enzymes. For instance, if R_i:$A \rightarrow B$, then the addition of M_{pool} results in:

$$R_i''':A + \frac{MW_i}{k_{cat,i}} M_{pool} \rightarrow B$$

2. Add a new artificial reaction, R_{pool}:

$$R_{pool}:\emptyset \rightarrow M_{pool}$$

whose upper flux bound is equal to the sum of the abundances of all the metabolic enzymes, P.

This way the resulting model is constrained by the total abundance of metabolic enzymes.

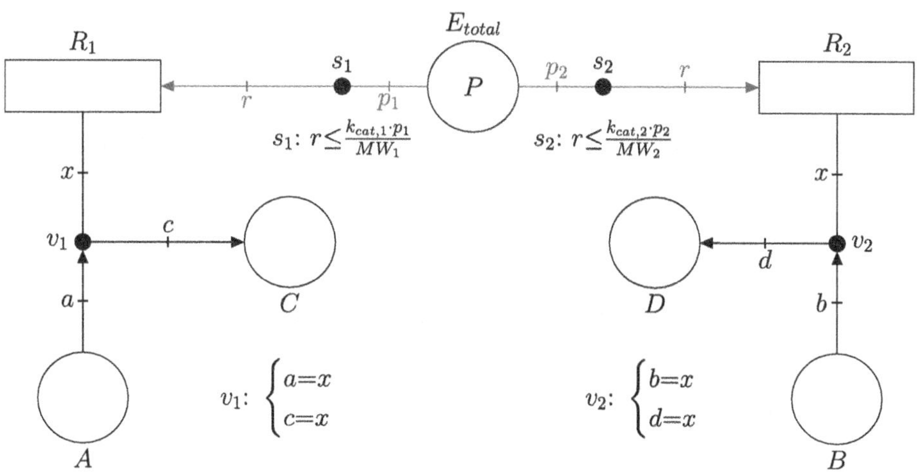

Fig. 4. FN that models the dynamics relationships of a system where the reaction rates of R_1 and R_2 are limited by an sMOMENT-like global constraint. The net determines that the intensity transferred from s_1 to R_1 satisfies that $r \leq \frac{k_{cat,1} \cdot p_1}{MW_1}$, thus, the intensity of R_1 will satisfy that $\lambda[R_1] \leq \frac{k_{cat,1} \cdot p_1}{MW_1}$. The same is applied to R_2.

4.1 Modeling the sMOMENT Method with FNs

In contrast to the sMOMENT method which requires the addition of artificial net elements, i.e. a metabolite and a transition, to account for the effect of the available enzymes, FNs can model the same behavior in a more natural and simple way by means of the intensity net. For instance, assume that we know the kinetic parameters of two reactions, R_1 and R_2, and the sum of the abundances of the enzymes that catalyze them, Y and Z, respectively. R_1 is catalyzed by an enzyme that transforms metabolite A into C and R_2 is catalyzed by a different enzyme converting B into D:

$$R_1{:}A + Y \rightarrow C + Y$$
$$R_2{:}B + Z \rightarrow D + Z$$

Both enzymes are part of a pool of enzymes E_{total} whose total amount is constrained by experimental data.

The global constraint referring to the total measured abundance of enzymes is implemented as in Fig. 4. The place E_{total} accounts for all the metabolic enzymes, two in this case, and it is assigned a marking, P, that represents the total enzyme concentration ($g\ g_{DW}^{-1}$) which is expressed in (12). The edges connected to E_{total} allow enzyme abundances to be distributed between reactions R_1 and R_2. The dynamics of the system are captured by the intensity handlers s_1 and s_2. These net elements, E_{total}, s_1 and s_2, model the following inequalities, which are derived from (12):

$$v_i \leq \frac{k_{cat,i} \cdot p_i}{MW_i} \tag{13}$$

$$\sum_{i=1}^{20} p_i \leq P \tag{14}$$

where v_i is the metabolic flux of reaction R_i, $k_{cat,i}$ is the catalytic constant of the enzyme that catalyzes reaction R_i, MW_i is its molecular weight, and p_i is the abundance of an individual metabolic enzyme from the pool directed to the reaction R_i.

5 Combining the sMOMENT and GECKO Methods

This section shows how FNs can accommodate simultaneously the global and individual enzymatic constraints (sMOMENT and GECKO methods) of a metabolic network by means of its intensity network.

Assume that we know the kinetic and proteomic information of two enzymes Y and Z that catalyze two reactions R_1 and R_2:

$$R_1{:}A + Y \rightarrow C + Y$$
$$R_2{:}B + Z \rightarrow D + Z$$

In the FN in Fig. 5, the intensity handlers s_1 and s_2 collect the enzymatic constraints that limit the fluxes of R_1 and R_2 respectively. This is extended to all the reactions for which enzymatic information is available.

The flux of the reaction R_1, r, is doubly constrained. On the one hand, it has to be less than or equal to the product of the k_{cat} ($k_{cat,1}$ in the figure) of the enzyme Y by its abundance from the pool of enzymes directed to the reaction that Y catalyzes, p_1, divided by its molecular weight, MW_1. This would be the implementation of the global constraint. On the other hand, the flux of R_1 must satisfy a second inequality, being r less or equal than the k_{cat} of Y enzyme multiplied by its specific abundance, e_1 (equivalent to GECKO). The same procedure is applied to the reaction R_2.

6 *Bacillus Subtilis* models

In this work, we enriched a *B. subtilis* GEM (iYO844 available in the BIGG Models database (Norsigian et al. [2019])) with enzymatic constraints using FNs to computationally simulate the work presented in Michon et al. [2020] where a mutant strain was genetically modified to produce *scyllo*-inositol from glucose.

Since a wild-type strain of *B. subtilis* cannot endogenously produce *myo*-inositol from glucose, Norsigian et al. [2019] introduced the gene *ino1* that encodes the inositol-1-phosphate (MI1P) synthase (MI1PS) from *Mycobacterium tuberculosis* to convert glucose-6-phosphate (G6P) into MI1P. Subsequently, the inositol monophosphatase (IMP) cleaves-off a phosphate from the MI1P originating a molecule of *myo*-inositol. This *myo*-inositol is the substrate of the inositol 2-dehydrogenase IolG encoded by the gene *iolG*. The last step of the *scyllo*-inositol production pathway is catalyzed by the IolW enzyme, the scyllo-inositol 2-dehydrogenase encoded by the gene *iolW* (Fig. 6). However, the metabolite produced by IolG, 2-inosose, can be degraded by a

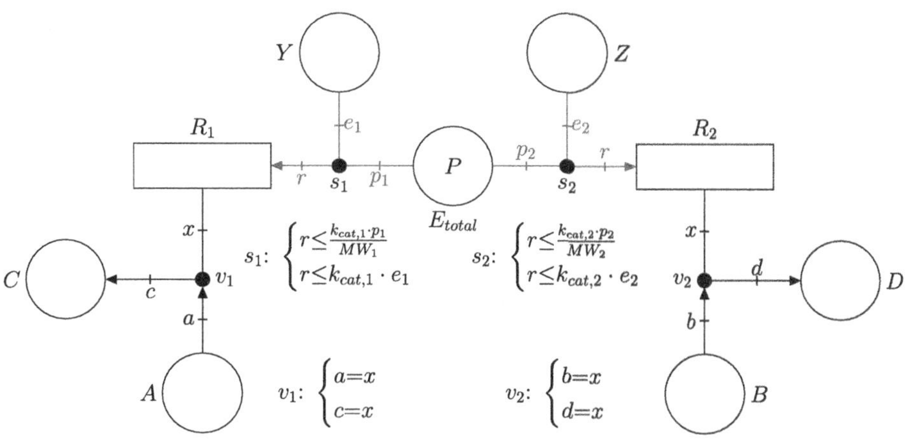

Fig. 5. Flexible Net that sets the dynamics relationships of a system where the reaction rates of R_1 and R_2 are limited by GECKO-like and an sMOMENT-like global constraints. Consider that the the intensity transferred from s_1 to R_1 satisfies the two inequalities associated to s_1. As a result, the intensity of R_1 will satisfy both conditions: $\lambda[R_1] \leq \dfrac{k_{cat,1} \cdot p_1}{MW_1}$ and $\lambda[R_1] \leq k_{cat,1} \cdot e_1$. The same is applied for s_2.

series of catabolic reactions. To avoid that the 2-inosose follows its catabolic pathway, researchers inhibited it, so all the 2-inosose goes through the *scyllo*-inositol pathway.

The above genetic modifications can be computationally integrated in the GEM of *B. subtilis* by means of the following steps:

1. Add the reaction catalyzed by the MI1PS. This way, the *B. subtilis* model can transform the G6P into MI1P (R_1 in Fig. 6).
2. Set a null flux of the reaction that degrades 2-inosose, *INSCR*:

$$\lambda[R_2] = 0 \tag{15}$$

3. Include the reaction that catalyzes the conversion of 2-inosose into *scyllo*-inositol (R_3 in Fig. 6).

For clarity, in Fig. 6, the transitions of the model that are not relevant for this case study have been omitted. The equations corresponding to the event handlers were also omitted because the stoichiometry of the represented metabolites is one to one in all the reactions of the *scyllo*-inositol production pathway, and hence, all the labels in the event handlers are equal (e.g. in v_1, $a = x$ and $b = x$). Moreover, the event handler v_{growth} models that the G6P can be consumed by a series of reactions to generate growth and $v_{exchange}$ models an exchange reaction that allows *scyllo*-inositol to leave the system in order to satisfy the steady state condition.

As a result, we obtained a *B. subtilis* model with the capacity to synthesize *scyllo*-inositol from glucose as carbon source (sc_iYO844). These modifications were included using the library COBRApy (Ebrahim et al. [2013]).

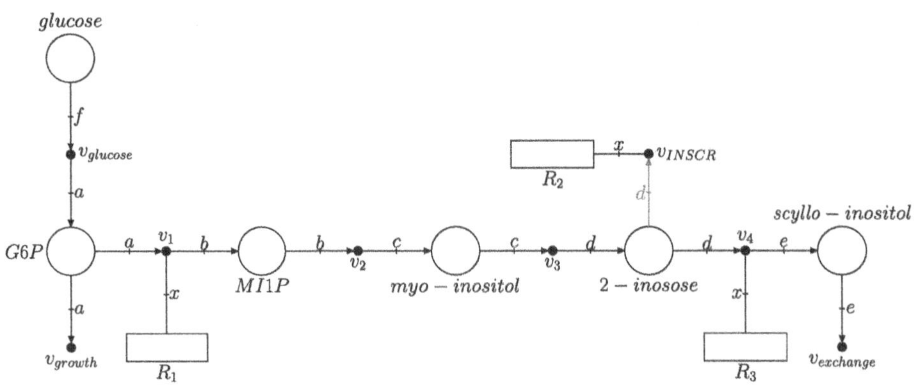

Fig. 6. Schematic representation of the analyzed pathways in the *B. subtilis* modified model. Glucose (*glc_c* in the scheme) can be either converted into MI1P, following the *scyllo*-inositol synthesis pathway or employed for the growth of the cell. The red arrow pointing to v_{INSCR} indicates the inhibited reaction in the model, *INSCR*, that simulates the experimental knock-out of the multiple reactions degrading the intermediate 2-inosose in Michon et al. [2020].

7 Enriching iYO844 with Kinetic and Proteomic Constraints Using FNs

7.1 Collection of Enzymatic Parameters

To obtain the enzymatic data for the GEM enrichment, we followed the process introduced in Massaiu et al. [2019]. This process consists in a manual search in several databases and literature. The kinetic parameters were obtained from the BRENDA (Chang et al. [2020]) and SABIO-RK (Wittig et al. [2017]) databases and the literature. The molecular weights (kDa) of the enzymes were obtained from the SubtiWiki (Pedreira et al. [2021]), a database that collects exclusively information about *B. subtilis* proteins and genes. The information for each enzyme is in Table 1 (see Annex).

Protein abundances expressed in $mmol\ g_{DW}^{-1}$ for each enzyme were quantified in Goelzer et al. [2015] at exponential phase. As stated in Massaiu et al. [2019], when the protein levels were under the detection limit, we took the minimum value among the protein level measurements in the same condition (6.8 x $10^{-8}\ mmol\ g_{DW}^{-1}$). Since we also integrated the sMOMENT global constraint, total protein abundance data were required. On the one hand, according to Milo [2013], the total amount of enzymes in an *E. coli* cell is 0.55 $g\ g_{DW}^{-1}$. On the other hand, the mass fraction of metabolic enzymes (the 20 enzymes taken into account) is 0.0194 (Massaiu et al. [2019]). This value was computed by summing the abundance expressed in parts per million of all the 20 enzymes retrieved in PaxDB (Huang et al. [2023]). Thus, the total enzyme abundance is $P = 0.0194 \cdot 0.55\ g\ g_{DW}^{-1} = 0.01067\ g\ g_{DW}^{-1}$.

7.2 Integrating Kinetic and Proteomic Constraints

Once all the necessary enzymatic data was collected, we applied the specific constraints on the enzyme-catalyzed reactions (17 in the glucose metabolism pathway and 3 that belong to the *scyllo*-inositol production pathway) and over the total concentration of metabolic enzymes.

For the GECKO-like constraints, we followed Equation (8) for the 17 enzyme-catalyzed reactions participating in glucose metabolism. Conversely, for the 3 enzymes participating in the *scyllo*-inositol pathway, we assumed the following: the flux of the *scyllo*-inositol synthesis reactions will satisfy Equation (16) because these enzymes are overexpressed and, consequently, their abundances must be strictly higher when *scyllo*-inositol production is maximized than when they were experimentally measured (when the growth rate is maximized).

$$v_i \geq k_{cat,i} \cdot e_i \tag{16}$$

For the constraint that accounts for the total amount of enzymes modeled by FNs as showed in Fig. 4, we applied the corresponding restriction, i.e. inequality (13), to each of the 20 enzyme-catalyzed reactions.

7.3 Assessing the Effect of the Integrated Constraints

The aim of this section is to assess the effect of kinetic, proteomic and glucose uptake rate constraints on the *scyllo*-inositol production and growth rate. Several models, which depend on the integrated constraints, were generated from the original GEM (iYO844):

- **sc_ec_iYO844**. This model implements the kinetic constraints for the 17 reactions of the carbon source metabolism (the GECKO-like constraints).
- **sc_ec_iYO844 + uptake constraint**. This model integrates the kinetic constraints for the 17 reactions and the constraint for the glucose uptake rate.
- **sc_ec_iYO844 +** v_{pool}. This model integrates the proteomic constraints that end up limiting the total amount of enzymes as a pool (sMOMENT-like constraint).
- **sc_ec_iYO844 +** v_{pool} **+ uptake constraint**. This model includes all the above presented constraints: kinetic, proteomic and glucose uptake rate constraints.

If the steady state condition is imposed to the sc_iYO844 FN and the synthesis of *scyllo*-inositol is maximized, then the obtained growth rate is null. Conversely, if the growth rate is maximized (0.624 h^{-1}), *scyllo*-inositol production is zero. This is biologically non-realistic and shows that, when there is no kinetic or proteomic constraints, the carbon source is either directed to the growth or to *scyllo*-inositol synthesis. In other words, there are pathways that compete for the carbon source, and the maximization of one implies zero flux in the other.

In further simulations, we ran the steady state condition applied in Eq. 7 on each of the presented models converted to FN, setting the *scyllo*-inositol synthesis as the objective function to be maximized. The maximization of *scyllo*-inositol synthesis was performed while fixing the growth rate to 20 different values, equally distributed in the interval between 0 and 0.624 h^{-1}. The obtained results are shown in Fig. 7.

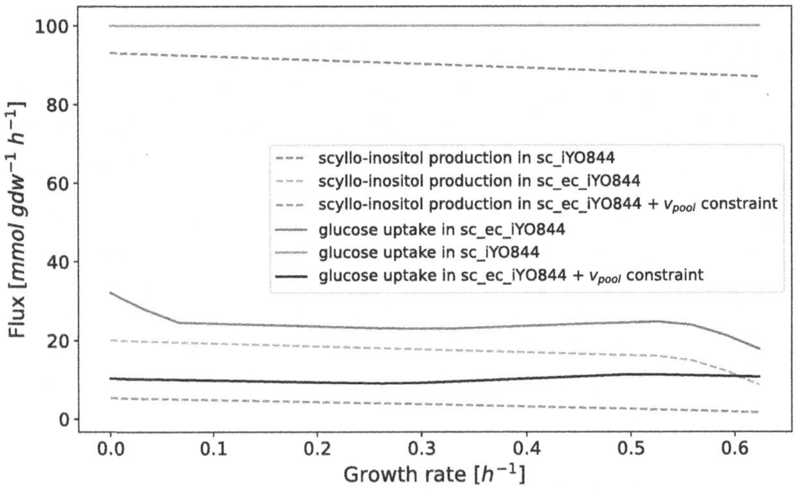

Fig. 7. Fluxes of *scyllo*-inositol synthesis (dashed lines) and glucose uptake (continuous lines) for growth rates in the interval $[0,\ 0.624]\ h^{-1}$.

The purple and blue lines refer to the glucose uptake and the *scyllo*-inositol production, respectively, in the model sc_iYO844. The red and orange lines correspond with the glucose uptake and *scyllo*-inositol production, respectively, in sc_ec_iYO844.

Finally, the black and green lines represent the glucose uptake and the *scyllo*-inositol production, respectively, when including all the constraints introduced in Sect. 7.2 (sc_ec_iYO844 + v_{pool} constraint). These results suggest that the inclusion of the kinetic and proteomic constraints results in a reduction of the solution space, and consequently predicts a more realistic production of *scyllo*-inositol. The initial production of *scyllo*-inositol is high (between 93.118 $mmol\ g_{DW}^{-1}\ h^{-1}$ and 87.034 $mmol\ g_{DW}^{-1}$ h^{-1} when the growth rates are minimum and maximum, respectively) due to the lack of tight constraints in sc_iYO844. Notice that the experimental uptake rate of glucose in *B. subtilis* is 7.71 $mmol\ g_{DW}^{-1}\ h^{-1}$ (Massaiu et al. [2019]) and that, according to the reaction network in Fig. 6, one molecule of glucose can produce at most one molecule of *scyllo*-inositol. Hence, 7.71 $mmol\ g_{DW}^{-1}\ h^{-1}$ is an upper bound for the *scyllo*-inositol synthesis flux and, therefore, the reported initial productions are unreachable by a real cell.

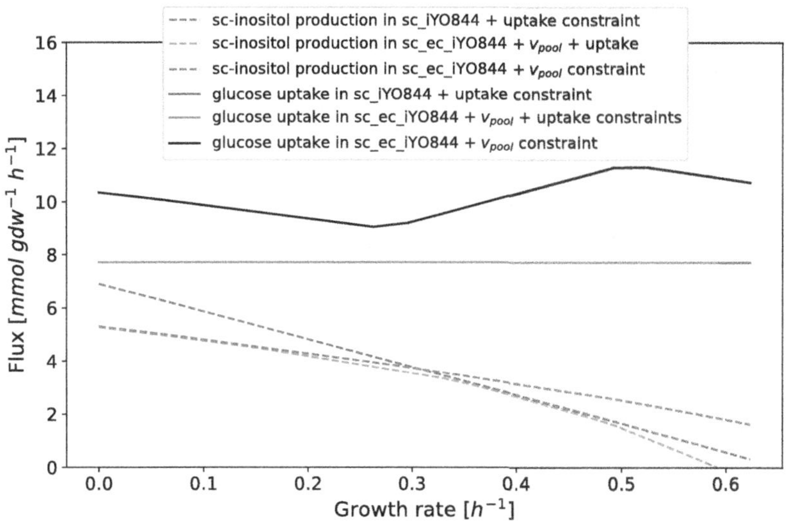

Fig. 8. Fluxes of *scyllo*-inositol synthesis (dashed lines) and glucose uptake (continuous lines) for growth rates in the interval $[0,\ 0.624]\ h^{-1}$ with the experimental glucose uptake in *B. subtilis* as an additional constraint.

As shown in Fig. 7, the glucose uptake rate varies in an interval between 9.06 and 11.30 $mmol\ g_{DW}^{-1}\ h^{-1}$ when implementing the whole set of enzymatic constraints in the model, so our enriched model predicts a similar but slightly higher uptake rate than the experimental one even if it is not explicitly constrained in the model. Therefore, this adjusted uptake rate solely limited by the enzyme availability leads to a more reduced flux for the *scyllo*-inositol production and a more realistic value in comparison to the one obtained with the original model sc_iYO844. To further support the predictive capacity of the enriched sc_ec_iYO844 + v_{pool}, we calculated the conversion efficiency from glucose to *scyllo*-inositol (the ratio between the *scyllo*-inositol synthesis flux and the glucose uptake rate) and compared it with the available bibliography. In

Tanaka et al. [2017], a conversion efficiency of 0.55 with *myo*-inositol as substrate was reported, and in Michon et al. [2020], the reported value using glucose as substrate was 0.10. Notice that these values are closer to the conversion efficiencies predicted using sc_ec_iYO844 + v_{pool}, see Table 2 (Annex).

In the end, we conducted a further analysis to uncover how the glucose uptake rate and the kinetic and proteomic constraints affect the *scyllo*-inositol production.

In Fig. 8, the green and black lines refer to the *scyllo*-inositol production and the glucose uptake rate, respectively, in the sc_ec_iYO844 + v_{pool} model; the blue and red lines correspond with the *scyllo*-inositol production and the glucose uptake rate, respectively, in the sc_ec_iYO844 + uptake constraint model; the orange and purple lines represent the *scyllo*-inositol production and glucose uptake rate when utilizing the sc_ec_iYO844 + v_{pool} + uptake constraint model. Here, the results indicate that when higher levels of *scyllo*-inositol are produced, which implies low growth rates, the enzymatic constraints are predominant over the glucose uptake constraint, being the enzyme availability the bottleneck for the *scyllo*-inositol synthesis. However, there is a point, when the growth rate is approximately 0.32 h^{-1}, where the glucose uptake rate constraint imposes over the enzymatic constraints. This happens exactly when the blue and green lines intersect. This can also be noted in the black line: when growth rate reaches 0.32 h^{-1}, the glucose uptake rate needs to increase because it becomes the limiting constraint instead of the kinetic and proteomic ones.

The Python script that performs the reported optimizations is available at: https://github.com/jlazaroibanezz/scyllo-inositol-bacillus.

8 Conclusions

Flexible Nets (FNs) can integrate naturally different enzymatic constraints in order to produce a realistic model of metabolism. In this work, two different types of enzymatic constraints were implemented in the same FNs model: the constraints that limit individually the availability of an enzyme (GECKO) and a global constraint that restricts the total amount of enzymes (sMOMENT). Among other properties such as having a simple and intuitive graphical representation, FNs offer the possibility of modeling the activation or inhibition of a certain reaction depending on the concentration of a metabolite, see Fig. 2, and can approximate non-linear dynamics (Júlvez et al. [2018]), making this approach more versatile in several aspects than its recently updated analogue GECKO 3.0 (Chen et al. [2024]). Moreover, the integration of the constraints seems more natural in FNs because they directly constrain the rates of real biological processes (biochemical reactions). In contrast, GECKO 3.0 does the same but less directly and consequently, in a harder to interpret way.

An FN has been developed to study the production of *scyllo*-inositol in a genome-scale metabolic model of *Bacillus subtilis*. It has been shown that more realistic levels of *scyllo*-inositol production are computed when the model is enriched with enzymatic constraints. In particular, the model predicts that the *scyllo*-inositol production is limited by the enzyme availability and kinetics when the metabolic flux is mainly directed to maximize *scyllo*-inositol production. Nevertheless, the limiting constraint changes to the glucose uptake rate as the growth rate increases and, consequently, the *scyllo*-inositol production decreases.

9 Annex

Table 1. This table gathers data from Massaiu et al. [2019], and includes the kinetic and proteomic information of three of the enzymes participating in the *scyllo*-inositol synthesis pathway (IMP, IolG and IolW).

Reaction name	EC	$k_{cat}[s^{-1}]$	$[E][mmol/g_{DW}]$	Organism of kcat data
PGI	5.3.1.9	126	$1.55 \cdot 10^{-5}$	*E. coli*
TPI	4.1.1.31	150	$1.28 \cdot 10^{-5}$	*E. coli*
GAPD_NAD	1.2.1.12	70	$5.77 \cdot 10^{-5}$	*B. subtilis*
PGK	2.7.2.3	329	$3.60 \cdot 10^{-5}$	*E. coli*
PGM	5.4.2.12	765.9	$8.85 \cdot 10^{-6}$	*B. subtilis*
ENO	4.2.1.11	130.4	$3.17 \cdot 10^{-5}$	*B. subtilis*
G6PDH	1.1.1.49	174	$8.05 \cdot 10^{-6}$	*E. coli*
CS	2.3.3.16	49	$2.51 \cdot 10^{-5}$	*B. subtilis*
ICDHy	1.1.1.42	82	$1.10 \cdot 10^{-4}$	*B. subtilis*
FUM	4.2.1.2	283.3	$7.29 \cdot 10^{-6}$	*E. coli*
MDH	1.1.1.37	177.1	$1.06 \cdot 10^{-4}$	*B. subtilis*
PTAr	2.3.1.8	651.6	$8.49 \cdot 10^{-6}$	*B. subtilis*
LDH_L	1.1.1.27	6416.6	$3.60 \cdot 10^{-6}$	*B. subtilis*
PGCDr	1.1.1.95	14.56	$1.90 \cdot 10^{-5}$	*B. subtilis*
OXADC	4.1.1.2	59	$6.21 \cdot 10^{-7}$	*B. subtilis*
MICITL	4.1.3.30	19	$6.80 \cdot 10^{-8}$	*E. coli*
OXGDC	4.1.1.71	0.2	$6.80 \cdot 10^{-8}$	*B. subtilis*
IMP	3.1.3.25	6.5	$6.80 \cdot 10^{-8}$	*E. coli*
IOLG	1.1.1.18	21.84	$6.80 \cdot 10^{-8}$	*B. subtilis*
IOLW	1.1.1.371	84.63	$6.80 \cdot 10^{-8}$	*B. subtilis*

Table 2. Conversion efficiency values in the sc_iYO844 and the sc_ec_iYO844 + *vpool* models depending on the fixed growth rate.

	Conversion efficiency	
Growth rate (h^{-1})	**sc_iYO844**	**sc_iYO844 + vpool**
0.0	0.931	0.513
0.032	0.928	0.505
0.065	0.924	0.496
0.098	0.921	0.487
0.131	0.918	0.477
0.164	0.915	0.468
0.197	0.912	0.457
0.229	0.909	0.447
0.262	0.906	0.436
0.295	0.903	0.409
0.328	0.899	0.374
0.361	0.896	0.341
0.394	0.893	0.309
0.426	0.890	0.280
0.459	0.886	0.253
0.492	0.883	0.227
0.525	0.880	0.207
0.558	0.876	0.189
0.591	0.873	0.170 v
0.624	0.870	0.150

References

Bekiaris, P.S., Klamt, S.: Automatic construction of metabolic models with enzyme constraints. BMC Bioinform. **21**, 1–13 (2020)

Chang, A., et al.: BRENDA, the ELIXIR core data resource in 2021: new developments and updates. Nucleic Acids Res. **49**(D1), D498–D508 (2020). ISSN 0305-1048. https://doi.org/10.1093/nar/gkaa1025

Chen, Y., et al.: Reconstruction, simulation and analysis of enzyme-constrained metabolic models using gecko toolbox 3.0. Nat. Protoc. **19**(3), 1–39 (2024)

Domenzain, I., et al.: Reconstruction of a catalogue of genome-scale metabolic models with enzymatic constraints using gecko 2.0. Nat. Commun. **13**(1), 3766 (2022)

Durot, M., Bourguignon, P.-Y., Schachter, V.: Genome-scale models of bacterial metabolism: reconstruction and applications. FEMS Microbiol. Rev. **33**(1), 164–190 (2008). ISSN 0168-6445. https://doi.org/10.1111/j.1574-6976.2008.00146.x

Ebrahim, A., Lerman, J.A., Palsson, B.O., Hyduke, D.R.: COBRApy: COnstraints-based reconstruction and analysis for python. BMC Syst. Biol. **7**(1), 74+, (2013). ISSN 1752-0509. https://doi.org/10.1186/1752-0509-7-74

Goelzer, A., et al.: Quantitative prediction of genome-wide resource allocation in bacteria. Metab. Eng. **32**, 232–243 (2015). ISSN 1096-7176. https://doi.org/10.1016/j.ymben.2015.10.003, https://www.sciencedirect.com/science/article/pii/S1096717615001317

Gotsmy, M., et al.: Sulfate limitation increases specific plasmid DNA yield and productivity in E. coli fed-batch processes. Microb. Cell Fact. **22**(1), 242 (2023)

Gu, C., Kim, G.B., Kim, W.J., Kim, H.U., Lee, S.Y.: Current status and applications of genome-scale metabolic models. Genome Biol. **20**, 1–18 (2019)

Huang, Q., Szklarczyk, D., Wang, M., Simonovic, M., von Mering, C.: PaxDb 5.0: curated protein quantification data suggests adaptive proteome changes in yeasts. Mol. Cell. Proteomics **22**(10), 100640 (2023)

Júlvez, J., Dikicioglu, D., Oliver, S.G.: Handling variability and incompleteness of biological data by flexible nets: a case study for Wilson disease. NPJ Syst. Biol. Appl. **4**(1), 7 (2018)

Júlvez, J., Oliver, S.G.: Steady state analysis of flexible nets. IEEE Trans. Autom. Control **65**(6), 2510–2525 (2020). https://doi.org/10.1109/TAC.2019.2931836

Kim, H.U., Kim, T.Y., Lee, S.Y.: Genome-scale metabolic network analysis and drug targeting of multi-drug resistant pathogen acinetobacter baumannii aye. Mol. BioSyst. **6**(2), 339–348 (2010)

Lázaro, J., et al.: Combination of genome-scale models and bioreactor dynamics to optimize the production of commodity chemicals. Front. Mol. Biosci. **9**, 855735 (2022)

Mao, L., Nicolae, A., Oliveira, M.A.P., He, F., Hachi, S., Fleming, R.M.T.: A constraint-based modelling approach to metabolic dysfunction in parkinson's disease. Comput. Struct. Biotechnol. J. **13**, 484–491 (2015). ISSN 2001-0370. https://doi.org/10.1016/j.csbj.2015.08.002. https://www.sciencedirect.com/science/article/pii/S2001037015000379

Massaiu, I., et al.: Integration of enzymatic data in bacillus subtilis genome-scale metabolic model improves phenotype predictions and enables in silico design of poly-γ-glutamic acid production strains. Microb. Cell Fact. **18**, 1–20 (2019)

Michon, C., Kang, C.-M., Karpenko, S., Tanaka, K., Ishikawa, S., Yoshida, K.: A bacterial cell factory converting glucose into scyllo-inositol, a therapeutic agent for alzheimer's disease. Commun. Biol. **3**(1), 93 (2020)

Milo, R.: What is the total number of protein molecules per cell volume? A call to rethink some published values. BioEssays **35**(12), 1050–1055 (2013)

Murata, T.: Petri nets: properties, analysis and applications. Procs. of the IEEE **77**(4), 541–580 (1989)

Norsigian, C.J., et al.: BiGG models 2020: multi-strain genome-scale models and expansion across the phylogenetic tree. Nucleic Acids Res. **48**(D1), D402–D406 (2019). ISSN 0305-1048. https://doi.org/10.1093/nar/gkz1054

O'brien, E.J., Lerman, J.A., Chang, R.L., Hyduke, D.R., Palsson, B.: Genome-scale models of metabolism and gene expression extend and refine growth phenotype prediction. Mol. Syst. Biol. **9**(1), 693 (2013)

Orth, J.D., Thiele, I., Palsson, B.: What is flux balance analysis? Nat. Biotechnol. **28**(3), 245–248 (2010)

Orth, J.D., et al.: A comprehensive genome-scale reconstruction of Escherichia coli metabolism-2011. Mol. Syst. Biol. **7**(1), 535 (2011)

Pedreira, T., Elfmann, C ., Stülke, J.: The current state of SubtiWiki, the database for the model organism Bacillus subtilis. Nucleic Acids Res. **50**(D1), D875–D882 (2021). ISSN 0305-1048. https://doi.org/10.1093/nar/gkab943

Senger, R.S.: Biofuel production improvement with genome-scale models: the role of cell composition. Biotechnol. J. **5**(7), 671–685 (2010)

Shaw, R., Cheung, C.Y.M.: Integration of crop growth model and constraint-based metabolic model predicts metabolic changes over rice plant development under water-limited stress.

in Silico Plants **3**(2), diab020 (2021). ISSN 2517-5025. https://doi.org/10.1093/insilicoplants/diab020

Silva, M.: Introducing Petri Nets. Practice of Petri Nets in Manufacturing, pp. 1–62, Chapman & Hall, London (1993)

Tanaka, K., Natsume, A., Ishikawa, S., Takenaka, S., Yoshida, K.: A new-generation of bacillus subtilis cell factory for further elevated scyllo-inositol production. Microb. Cell Fact. **16**, 1–8 (2017)

Wittig, U., Rey, M., Weidemann, A., Kania, R., Müller, W.: SABIO-RK: an updated resource for manually curated biochemical reaction kinetics. Nucleic Acids Res. **46**(D1), D656–D660 (2017). ISSN 0305-1048. https://doi.org/10.1093/nar/gkx1065

Zorrilla, F., Buric, F., Patil, K.R., Zelezniak, A.: metaGEM: reconstruction of genome scale metabolic models directly from metagenomes. Nucleic Acids Res. **49**(21), e126–e126 (2021)

Reverse Engineering of Renal Tubule Networks in the High-Dimensional Regime

Roberto Pagliarini[1,2](✉)

[1] Molecular Basis of Cystic Kidney Disorders Unit, Division of Genetics and Cell Biology, IRCCS, San Raffaele Scientific Institute, Milan, Italy
roberto.pagliarini@uniud.it
[2] Department of Mathematics, Computer Science, and Physics University of Udine Italy, Udine, Italy

Abstract. The high-dimensional regime is a condition that arises in many large-scale data applications, like omics, economy, and ecoinformatics. In this scenario, the number of n acquired samples is far fewer than the number of p observed variables, and then the classical statistical approaches can not be applied. In this work, with the aim to infer direct interactions for all the proteins in the kidney tubule segments, we point out our attention on the reverse-engineering of conditional networks in this regime. Particularly, the mammalian renal tubule is made up of at least 14 segments, containing at least 16 distinct epithelial cell types. Each cell type has its own characteristic set of cellular functions, which have been elucidated largely over the past 50 years since the development of single-tubule microdissection approaches and the expansion of the micropuncture technique to mammalian physiology. New developments in protein mass spectrometry have resulted in a marked increase in sensitivity of protein detection and quantification. In order to address this problem, we developed an algorithm rooted on Graphical Gaussian Models to infer a network, and on a simple method to establish the interactions different from zero. We found that the proposed approach outperforms others methods in terms of accuracy of diagnostic tests, particularly in the high-dimensional regime. Moreover, we suggest that it can be applied to several types of data.

Keywords: Graphical Gaussian Models · Kidney tubule segments · Network inference · High-dimensional regime

1 Introduction

High-throughput techniques have allowed to obtain huge amounts of data about biological processes, but, at the same time, the paradox of Systems Biology is that this huge amount of data often creates more confusion than understanding. Researchers proposed plausible approaches to reverse-engineering the mechanisms underlying the phenomena generating these data. Continuing studies

R. Gori et al. (Eds.): CMSB 2024, LNBI 14971, pp. 155–173, 2024.
https://doi.org/10.1007/978-3-031-71671-3_12

have dramatically expanded the types of inference algorithms and their applications to experimental data. Still, the development of inference methods that can be applied to biological phenomena remains a challenging and active area of research. Challenges arise from several reasons. First, the number of variables that come into play is very high and there is normally no sufficient biological knowledge to restrict the analysis to a subset of core elements for a given biological process. Second, there is no standard model of the regulatory mechanisms, except for a generic cause-effect relationship between transcription factors and corresponding binding sites. Third, data are typically noisy, high dimensional, and significantly under-sampled.

Specifically, in large-scale data applications like omics, we are in presence of a regime where the number of n acquired samples is far fewer than the number of p observed variables (genes, proteins, metabolites). In this work, we point our attention in the so called "high-dimensional regime", which represents the situation in which $n << p$ [1]. Our goal will be the inference of large scale networks modelling the strength of direct conditional dependence among variables.

The proposed approach is rooted on Concentration Graph Models (CGMs) [17], which define undirected graph structures through the inverse of the covariance matrix. This matrix is related to partial correlation (PC). PC differs for the Pearson product-moment correlation (PPMC), because it is a method used to describe the relationship between two variables whilst taking away the effects of another variable, or several other variables, on this relationship. Differently, a high PPMC can indicate both direct relation or indirect dependence through the influence of other variables or due to transitive nature of interactions. Particularly, the effect of indirected edges become more dramatic as the graph size grows, and this results in an inaccurate graph reconstruction. Mathematically, both these correlation indexes are computed starting from the covariance matrix, and a very elegant result shows how correlation and concentration graphs are related in the scenario $n > p$ [27]. In this situation, the simplest way to reconstruct a CGM is based on the sample covariance matrix which is easy to compute.

However, in the high-dimensional regime the sample covariance matrix is not a good estimator of the population covariance matrix [19]. We will address this problem by considering the Oracle Approximating Shrinkage estimator of high dimensional covariance matrices proposed in [4]. Different data sets, ranging from metabolic gene-regulatory to species interaction networks, suggest that these networks are sparse [11], that is, the percentage of the active interactions scales inversely proportional to the system size. One way to reconstruct a sparse graph is to threshold the covariance matrix. In this case, a selection of the threshold is hard to tackle analytically, and several methods have been developed to infer it from data [27]. However, these methods have been developed for the case $n > p$ and do not perform well in the high-dimensional regime. We will overcome this problem by employing a test procedure based on the quantile PC and PPMC to select important entries of the covariance matrix in high dimensionality.

We have seen that the proposed approach outperforms, in terms of accuracy of diagnostic tests, other methods in the high-dimensional regime scenario, and then we used it to reverse-engineer conditional networks for the 14 distinct segments of renal tubule. They consist of epithelial cells with different transports and metabolic functions. Recent studies have identified and quantified proteins expressed in each segment of the kidney [18]. Identifying the protein networks mediating each function is crucial to gaining an overall understanding of kidney physiology and pathophysiology. According to the National Institutes of Health, the overall prevalence of chronic kidney disease is approximately 14% [28]. Therefore, the inference of proteomics conditional networks for each type of these epithelial cells represents a crucial step to uncover new insights into the kidney regulative mechanisms.

2 Methods

2.1 Concentration Graph Models

Given experimental data $\mathbf{Y} = [y_1, y_2, \dots, y_n]^T \in \mathbb{R}^{n \times p}$ for n samples and p molecules, the profile of each sample is assumed to be independent and follows a Gaussian distribution $\mathcal{N}(\mu, \Sigma)$, where μ is the mean and $\Sigma \in \mathbb{R}^{p \times p}$ is the covariance matrix.

Reconstructed and true graphs, also indicated as networks (in this manuscript graph and network will be used as synonymous), are written in terms of weighted undirected graphs $G = (V, E, w)$, where $V = \{v_1, v_2, \dots, v_p\}$ is a set of p variables representing the nodes of G, $E = \{(i,j) | v_i \in V \wedge v_j \in V\}$ is a set of pairs of indexes, the edges of G, modelling links between elements of V that are conditional independent, and $w : E \rightarrow \mathbb{E}$ is a function such that $w_{i,j}$ denotes the weight of $(i,j) \in E$.

Our networks will be based on the concentration matrix, the inverse of the covariance matrix. Covariance indicates the relationship of two variables whenever one variable changes. If an increase in one variable results in an increase in the other one, both variables have a positive covariance. Decreases in one variable also cause a decrease in the other. Both variables move together in the same direction when they change. Decreases in one variable resulting in the opposite change in the other one are referred to as negative covariance. However, covariance models both direct and indirect relationships. Differently, the concentration matrix $\Omega = \Sigma^{-1} \in \mathbb{R}^{p \times p}$ is a positive and symmetric matrix, and a zero entry $\omega_{i,j}$ represents that nodes v_i and v_j are conditional independent given the other variables. Non-zero elements of the concentration matrix correspond to partial correlations through the relation:

$$\rho_{i,j} = \frac{-\omega_{i,j}}{\sqrt{\omega_{i,i}\omega_{j,j}}} \tag{1}$$

and this allows us to conclude that the concentration matrix coincides with the graph structure defining the network of the p variables. $\rho_{i,j}$ describes the correlation between v_i and v_j on all the remainder of the variables. The partial correlation coefficients allow for several interpretations [19]. In our approach, it can be shown that two variables are conditional independent given the remaining variables if and only if the corresponding partial correlation vanishes. This implies that learning the elements of the set E is equivalent to infer the entries of Ω statistically different to zero.

2.2 Network Inference

With the aim to infer a CGM from a given dataset of n samples and p variables in the high-dimensional regime, we employed an algorithm composed of several steps. The main problem is that in the high-dimensional regime the sample matrix Σ is not a good estimator of the population covariance matrix. For this reason, we first employ a shrinkage estimator of the covariance matrix, then the concentration matrix is computed, and finally we determine which entries of the concentration matrix are significantly different from zero. However, in this scenario, it is not possible to use the standard correlation statistical tests. Therefore, we use a quantile-correlations test to select important causative links and prune the concentration matrix. Finally, a weighted adjacency matrix is obtained to model the observed correlations among graph elements. We will call this algorithm the *"ORA-Quantile"*.

Step1-Covariance Matrix Computation. The Oracle Approximating Shrinkage estimator of covariance matrix Σ, that we will call ORA, is computed by applying the method developed in [4]. The authors proposed to estimate the covariance matrix by iteratively approximating the shrinkage with $\hat{\Sigma} = \lambda \hat{F} + (1 - \lambda)\hat{S}$ where $\lambda \in (0, 1)$ is a control parameter, $\hat{S}$ is the empirical covariance matrix, and $\hat{F} = tr(\hat{S}/p)I_{p\times p}$ is the target matrix, with $tr()$ representing the trace of a matrix, and $I_{p\times p}$ is the identity matrix of dimension p.

Step2-Precision Matrix Computation. The precision matrix Ω is obtained by computing the Moore-Penrose pseudo-inverse of $\hat{\Sigma}$, a generalization of the standard matrix inverse based on the singular value decomposition that can be also applied to a non-invertible matrix [21]. After that, since Ω is not invariant to scaling of the component variables of $\mathbf{Y}$, a scale invariant of this matrix is obtained by:

$$P = diag(\Omega)^{-1/2}\Omega diag(\Omega)^{-1/2} \in \mathbb{R}^{p\times p} \tag{2}$$

where, for a square matrix B, $diag(B)$ is a matrix formed from the diagonal elements of B and, for a diagonal matrix D with non-zero entries, $\{d_{i,i}\}_{i=1}^{p}$, $D^{-1/2}$ represents the matrix with entries $\{1/\sqrt{d_{i,i}}\}_{i=1}^{p}$. As corollary of the Cauchy-Schwarz inequality [30], the diagonal elements of P are equal to 1 and the off-diagonal ones are between -1 and 1. This scale invariance property is important when one is interested in the measure of dependency that is not a function of the units used to represent the different variables.

Step3-Network Pruning. A filter based on quantiles of correlation and partial-correlation is applied. The scale invariant version R of the correlation matrix is computed, that is:

$$R = diag(\Sigma)^{-1/2}\Sigma diag(\Sigma)^{-1/2} \in \mathbb{R}^{p\times p}. \tag{3}$$

Then, since R and P are symmetric matrixces, we consider only their components above the main diagonal, that are stored in vectors $\hat{R}$ and $\hat{P}$, and their quantiles for the cumulative probabilities 0.05 and 0.95 are computed. Let $q^R_{0.05}, q^R_{0.95}, q^P_{0.05}, q^P_{0.95}$ be these quantiles. They can be used to find the variables having a strong conditional dependences, by following the hypothesis that a conditional link between v_i and v_j exists only if they have a strong PC and PPMC. Then, the set of couples of variable indexes having strong conditional dependencies is obtained as follows:

$$E = \{(i,j)|(r_{i,j} \in \hat{R} \wedge (q^R_{0.05} > r_{i,j} > q^R_{0.95})) \wedge (p_{i,j} \in \hat{P} \wedge (q^P_{0.05} > p_{i,j} > q^P_{0.95})\} \tag{4}$$

and all the elements $p_{i,j} \in P$ for which $(i,j) \notin E$ are set to zero.

Step4-Weighted Adjacency Matrix Selection. Finally, in order to capture the observed direct correlative information among variables, a weighted square adjacency matrix A of dimension p is obtained from R, where the elements of the matrix indicate whether pairs of vertices are adjacent or not in the graph, while the magnitude of the entries models the strength of the edges. That is, an entry $a_{i,j}$ of A is obtained by the following rule:

$$a_{i,j} = \begin{cases} r_{i,j} \text{ if } (i,j) \in E \\ 0 \text{ if } (i,j) \notin E. \end{cases} \tag{5}$$

3 Evaluating the Approach Performance

3.1 Dataset

In order to evaluate the performance of an inference algorithm, it must be tested on expression data from "gold standard networks", for which all the edges are already known, and scored based on its ability to recover the true structure from the input data. We will report the results obtained by evaluating the performance of our procedure on the StatSeq compendium, which consists of 72 datasets originated from 9 different *in-silico* gene networks, each simulated under 8 different parameter settings such as population sizes, marker distances, and heritability [22]. For each of the 72 datasets there are two matrices: *i)* the gene expression matrix and *ii)* the genotype matrix which represents the mutated genes.

3.2 Evaluation Procedure

We computed the Area Under the Receiver operating characteristic curve (AUROC), a scoring metric commonly applied in machine learning community, to test the efficiency of a binary classifier [5]. Particularly, we compared our results with those obtained the following algorithms, that we implemented in Matlab:

- *Schafer-Quantile*: precision matrix has been estimated starting from a covariance matrix computed according to [26], and filtered by applying the quantile approach described in this paper;
- *Schafer-LocalFDR*: precision matrix has been obtained starting from a covariance matrix computed according to [26], and filtered by employing the statistical testing problem of non-zero partial correlations developed in [25];
- *Schafer-Zscore*: precision matrix has been obtained starting from a covariance matrix computed according to [26], and then filtered by selecting as different from zero the partial-correlations having a $|z - score| \geq 1.96$ [16];
- *Schafer-2sigma*: precision matrix has been computed starting from a covariance matrix computed according to [26], and selecting as different from zero the partial-correlation having an absolute value $\geq 2\sigma$, where σ represents the mean of the partial-correlation values;
- *ORA-LocalFDR*: precision matrix has been obtained starting from a covariance matrix computed according to [4], and filtered by employing the statistical testing problem of non-zero partial correlations developed in [25];
- *ORA-Zscore*: precision matrix has been obtained starting from a covariance matrix computed according to [4], and then filtered by selecting as different from zero the partial-correlations having a $|z - score| \geq 1.96$ [16];
- *ORA-2sigma*: precision matrix has been computed starting from a covariance matrix computed according to [4], and selecting as different from zero the partial-correlation having an absolute value $\geq 2\sigma$, where σ represents the mean of the partial-correlation values;
- *Regression*: a different way to compute the elements of the precision matrix Ω can be found in [19]. The least squares regression coefficient $\hat{\beta}^i_j$ for each of the p linear regression models $Y_i = \beta^i_0 + \sum_{j \neq i} \beta^i_j + \epsilon$ is inferred, where ϵ is the error model. Then, each entry of Ω is computed as:

$$\omega_{i,j} = \begin{cases} sign(\hat{\beta}^i_j)\sqrt{\hat{\beta}^i_j \hat{\beta}^j_i} & \text{if } sign(\hat{\beta}^i_j) = sign(\hat{\beta}^j_i) \\ 0 & \text{if } sign(\hat{\beta}^i_j) \neq sign(\hat{\beta}^j_i). \end{cases} \tag{6}$$

- *Lasso*: the method proposed in [29] has been employed to compute the precision matrix, where the least squares estimates are replaced by a regularised variant that uses the Lasso procedure[1] [8]. Then, each entry of Ω is computed as:

$$\omega_{i,j} = \begin{cases} sign(\hat{\beta}^i_j)\min(1, \sqrt{\hat{\beta}^i_j \hat{\beta}^j_i}) & \text{if } sign(\hat{\beta}^i_j) = sign(\hat{\beta}^j_i) \\ 0 & \text{if } sign(\hat{\beta}^i_j) \neq sign(\hat{\beta}^j_i). \end{cases} \tag{7}$$

- *Stepwise regression*: the least square estimates, used to compute the entries of the precision matrix, have been computed by using the stepwise-regression, a method of fitting regression models in which the choice of predictive variables is carried out by an automatic procedure [3,12].

[1] (http://www.cs.ubc.ca/~schmidtm/Software/L1precision.html).

3.3 Results

In order to assess our method, we only used the expression matrix of the StatSeq dataset for gene networks of size 1000. We first evaluated the algorithms on 100 samples for each gene (Fig. 1). Then, with the aim to simulate different situations of high-dimensional regime, we considered only 10 (Fig. 2), 5 (Fig. 3), and 3 (Fig. 4) samples randomly selected. The four pictures show the box-plots representing the AUROC for each inference method. Firstly, from Fig. 1, it is important to highlight the good performances of the Schafer-Quantile method, that improves the results obtained by using the Schafer-LocalFDR. Entering into the details, this is the case in which the shrinkage approach developed in [26] is associated with our network-pruning step. This result points out the efficiency of the network-pruning approach based on quantile filtering that we proposed in this manuscript, which can be associated to several shrinkage methods for covariance estimators. Secondly, as shown in Fig. 1, the Schafer-Zscore and the ORA-Zscore resulted in poor performances for the case of 100 samples. Therefore, they were not evaluated for the experiments with a lower number of samples (Fig. 2, Fig. 3, and Fig. 4). It is also possible to observe that ORA-Quantile outperforms all the others methods, in terms of accuracy of diagnostic test, when we move towards the very high-dimensional regime. While the results of the ROC analysis support our approach, interpreting these results requires more nuance. In fact, there is a cost in accuracy due to the challenges introduced by the inference in the high-dimensional regime [13] and the pruning approaches [24]. This is emphasized in our problem, where the modelling of direct conditional dependence among variables may be affected by the prevalence of redundant connections. Then, the ORA-Quantile can indeed be useful in our setting. Another important aspect has to be commented. Figure 4 shows that the performances of ORA-Quantile and of ORA-LocalFDR are very close. This is due to the fact the local fdr approach, developed in [25,26], takes account of the dependencies among estimated partial correlation coefficients studied in [6]. These dependencies are also considered in our approach. As a first consequence, the filter based on quantiles of correlation and partial-correlation and the local fdr approach are interchangeable. This point will require further theoretical studies.

4 Application to Kidney Tubules Expression Atlas

4.1 Dataset

Data have been extracted from [18], where authors identified and quantified proteins expressed in each of the 14 segments of rat kidney tubules, and used these data to create an online information resource, the Kidney Tubule Expression Atlas[2]. The renal tubule's 14 distinct segments consist of epithelial cells with different transports and metabolic functions. Over the 44 samples analyzed, the average number of quantified proteins per segment was 4234, while the number

[2] (https://esbl.nhlbi.nih.gov/KTEA/).

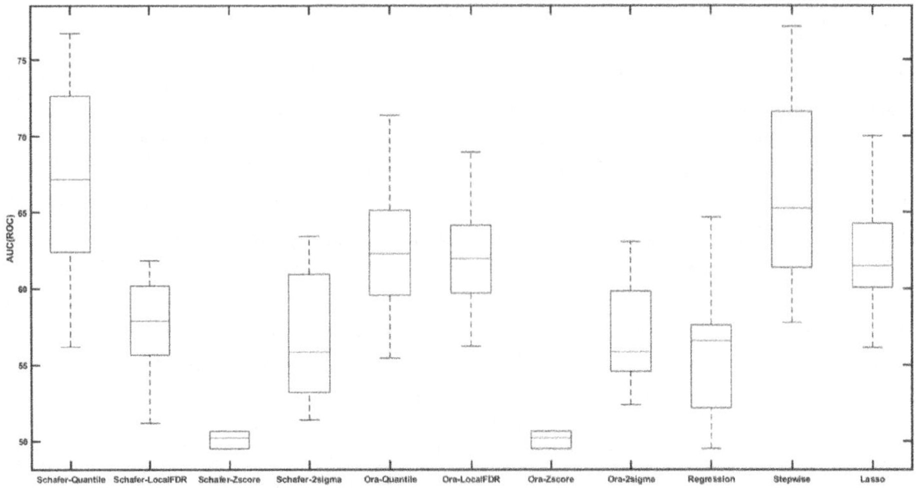

Fig. 1. Boxplots representing AUROC values for each dataset and each inference methods using 100 samples.

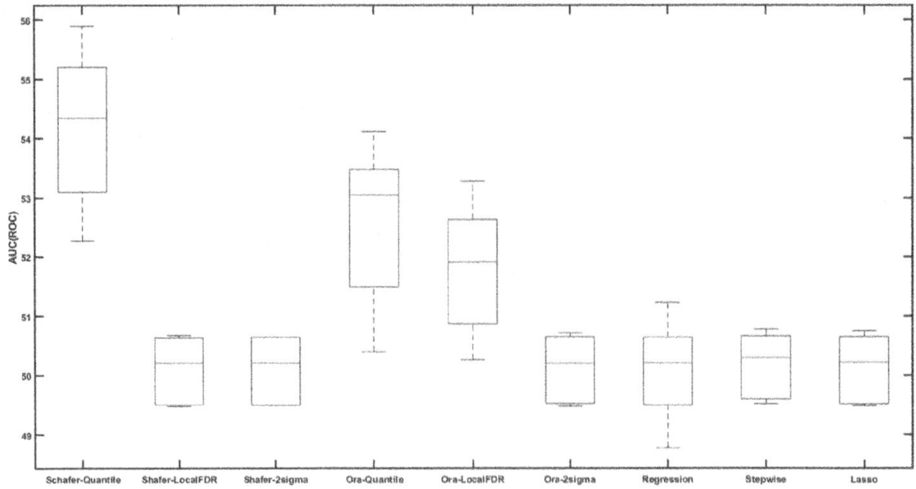

Fig. 2. Boxplots representing AUROC values for each dataset and each inference methods using 10 samples.

of replicates ranges from 3 to 4 (Table 1). The analysis reported in [18] indicates that the obtained proteome depth is likely to account for at least the 99% of the total protein mass in each cell type. With the aim to apply our algorithm, we removed, for each segment, all the proteins whose expression levels are zero in at least one sample.

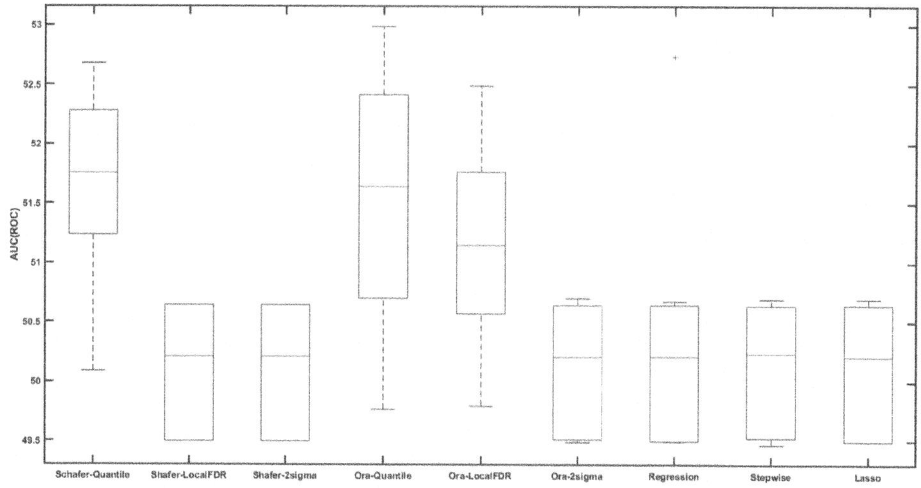

Fig. 3. Boxplots representing AUROC values for each dataset and each inference methods using 5 samples.

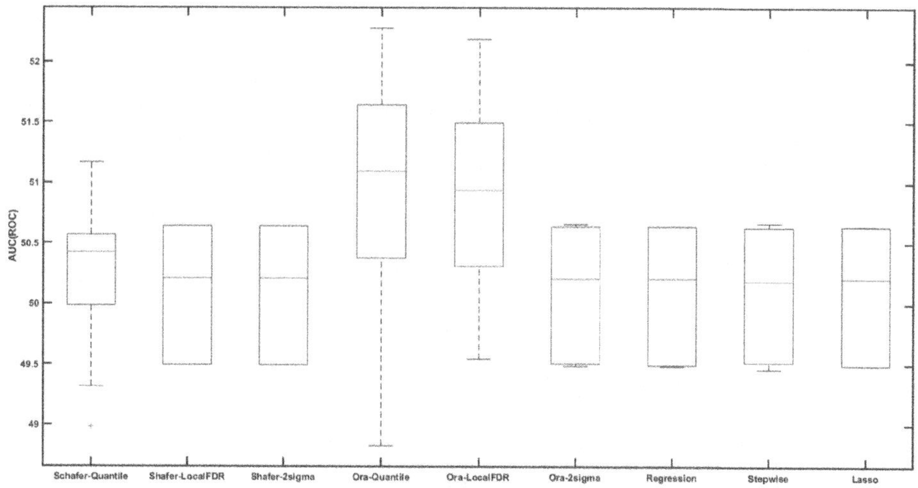

Fig. 4. Boxplots representing AUROC values for each dataset and each inference methods using 3 samples.

4.2 Segment-Specific Networks Comparison

We considered all the 14 networks, associated with the renal segments of Table 1, inferred by employing the ORA-Quantile algorithm. With the aim to quantify the similarity between the topology of all the networks, the adjacency matrix of every graph has been extended in order to consider the set of $p = 7429$ proteins

Table 1. Some details of the renal segments studied. The number of proteins quantified is indicated as mean ± standard deviation.

Full Name	Short Name	Origin	Samples	Proteins Quantified
S1	First segment of proximal tubule	Cortical labyrinth	3	2734 ± 103
S2	Second segment of proximal tubule	Cortical medullary ray	3	4750 ± 415
S3	Third segment of proximal tubule	Outer strip of outer medulla	3	3508 ± 287
DTL1	Descending thin limb type 1	Inner strip of outer medulla	3	4228 ± 19
DTL2	Descending thin limb type 2	Inner strip of outer medulla	3	3362 ± 312
DTL3	Descending thin limb type 3	Inner medulla	3	4426 ± 51
ATL	Ascending thin limb	Inner medulla	3	3285 ± 62
mATL	Medullary thick ascending limb	Inner stripe of outer medulla	4	3494 ± 413
cATL	Cortical thick ascending limb	Cortical medullary ray	3	4762 ± 1228
DCT	Distal convoluted tubule	Cortical labyrinth	3	4854 ± 808
CNT	Connecting tubule	Cortical labyrinth	3	5120 ± 286
CCD	Cortical collecting duct	Cortical medullary ray	4	5265 ± 744
OMCD	Outer medullary collecting duct	Inner stripe of outer medulla	3	3977 ± 216
IMCD	Inner medullary collecting duct	Inner medulla	3	5421 ± 216

obtained by the union of all the molecules quantified in all the segments. Each network has been treated as weighted, and the difference between two adjacency matrices A and B has been computed by means of Frobenius norm as follow:

$$\|A - B\|_F = \sqrt{\sum_{i=1}^{p}\sum_{j=1}^{p}(a_{i,j} - b_{i,j})^2}. \tag{8}$$

The matrix norm (8) is equal to zero if the topology of two networks, also considering the edge weights, is equivalent, and tend to infinity as the similarity of the graphs decreases. Figure 5 shows the distance matrix related to the segments in Table 1. It can be seen that the topology of the 14 networks can be very different, and there are not similar ones.

Topological Differences Between Conditional Networks of IMCD and S1. The distance matrix represented in Fig. 5 can be used as a starting point for pairwise comparison between networks. For example, it is possible to observe

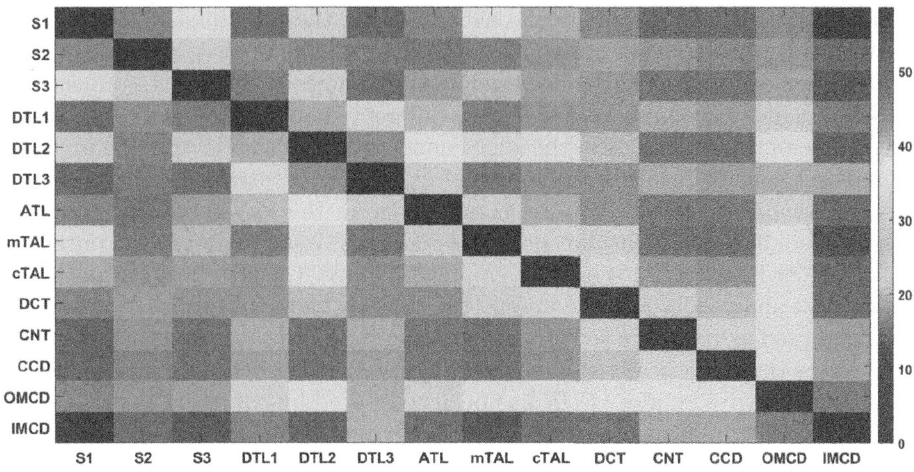

Fig. 5. Distance matrix for the 14 renal tubule segments in rat.

that the biggest difference is between the segments S1 and IMCD. This is mainly due to the functional differentiation between this two segments. The function of the proximal tubule is essentially reabsorption of filtrate in accordance with the needs of homeostasis, whereas the distal part of the nephron and collecting duct are mainly concerned with the detailed regulation of water, electrolyte, and hydrogen-ion balance [20]. Here, as an example, we will analyze more carefully this difference. First, edge density index has been used to check the sparsity of the networks. For undirected graphs, this index is defined as:

$$D = \frac{|E|}{|V|(|V|-1)}. \tag{9}$$

The maximum number of edges is $|V|(|V|-1)$, so the maximal density is 1 for complete graphs, and the minimal density is 0 for empty networks. In our case, $D_{IMCD} = 0.0086$ and $D_{S1} = 0.0645$, reflecting the sparsity of these networks.

After that, we used DyNet [10] to analyze and visualize differences between the two networks. It is a Cytoscape application that provides a range of functionalities for the visualization, real-time synchronization and analysis of large multi-state dynamic molecular interaction networks enabling users to quickly identify and infer the most rewired nodes across many network states. The use of this tool allows to understand how molecular interaction networks change in their composition and connectivity in response to temporal, spatial, cellular and environmental signals. Figure 6 shows a reference network obtained by the union of the graphs modelling S1 and IMCD. This reference network fulfills a role as a canvas for the visualization of comparative node/edge statistics.

In order to study the main players regulating the functional differentiation between this two segments, we applied a variance analysis based on node degrees to identify most rewired vertexes based on the changes in their edge connections. In Fig. 6, nodes have been coloured with different shades of red according to the magnitude of their variance in the edge connections. This node rewiring analysis, rooted on a specific rewiring metric called D_n-score [10], makes it easy to identify proteins that have widely varying connectivity in the two networks. With the aim to gain mechanistic insight into the nodes most rewired and most conserved, ranked according the D_n-score, we employed ShinyGO [9] to perform pathways enrichment analysis, where the union of KEGG [14] and Reactome [7] databases has been used as reference set of rattus norvegicus pathways. In Fig. 7, it is possible to see the statistically significant ($FDR \leq 0.05$) enriched pathways for the top 50 most rewired nodes (FDR has been computed by adjusting the $p-values$ of ShinyGO for multiple-comparison by applying the Benjamini and Hochberg procedure [15]). This result suggests that the most rewired and reprogrammed pathways are related to metabolism and nutrient acquisition in order to meet a shift in bioenergetic, biosynthetic, and redox demands.

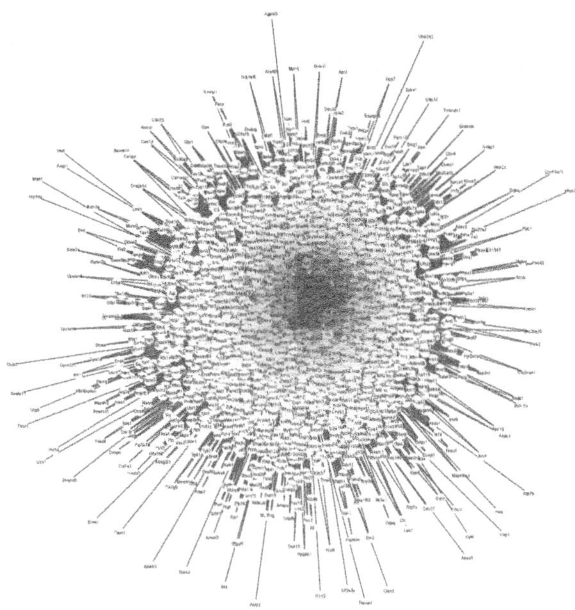

Fig. 6. Reference network obtained by the union of the graphs of S1 and IMCD. It highlights nodes which are most rewired (more red = higher variation) via the D_n-score. (Color figure online)

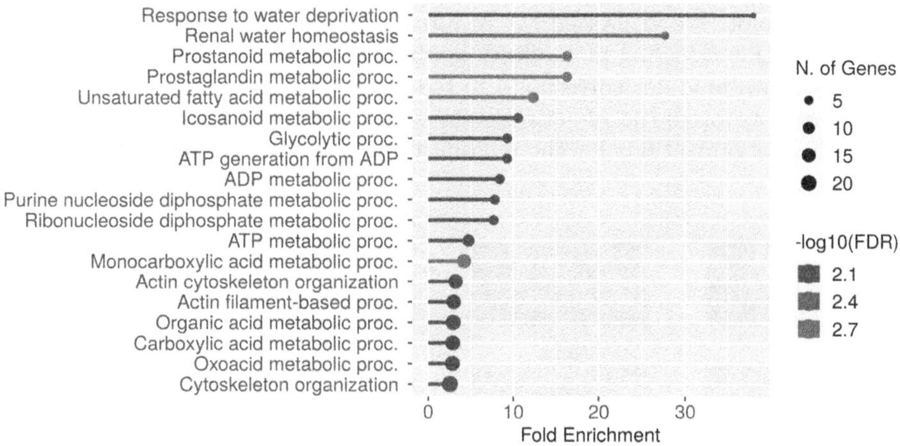

Fig. 7. Statistically significant enriched pathways for the top 50 most rewired nodes in reference network of Fig. 6.

4.3 Hierarchical Organization of Modularity in Segment-Specific Networks

Several studies have shown that proteins having a higher topological overlap are more likely to belong to the same functional class or may be part of the same biological pathway than proteins having a lower topological overlap [2,23,31]. Thus, an important task in biological network analysis is to identify modules of densely interconnected nodes. In networks involving few nodes, these groups can easily be identified, but for graphs involving hundreds of nodes, it is useful to provide a different point of view. In order to analyze the hierarchical organization of modularity in the 14 segment-specific networks that have been inferred, the Matlab implementation[3] of the first order generalized topological overlap measure (GTOM) has been employed [32]. Each undirected weighted adjacency matrix has been converted in an unweighted adjacency matrix A, where $a_{i,j} = 1$ if vertexes i and j are connected, and $a_{i,j} = 0$ otherwise. Then, for each adjacency matrix GTOM has been used to compute the topological overlap matrix T [32], and a dissimilarity matrix D has been obtained in a way such that each entry is computed as $d_{i,j} = 1 - t_{i,j}$. After that, matrix D has been used as the input of the average linkage hierarchical clustering procedure, and the modules has been defined as the discrete branches of the clustering tree with cut-off equals to 0.4. Thus the modules have been found by inspection; a height cut-off equals to 0.4 has been chosen in the dendrogram such that some of the resulting branches correspond to the discrete diagonal blocks (modules) in

[3] (https://it.mathworks.com/matlabcentral/fileexchange/17668-gtom-generalized-topological-overlaping-measure).

the GTOM plot. We visualized the results of the clustering procedure by using multi-dimensional scaling plots. Since modules are sets of nodes with high topological overlap, modules correspond to red squares along the diagonal. Figure 8 shows the plots for all the 14 networks, by which it is possible to observe different hierarchical organizations for the modularity of the segment networks.

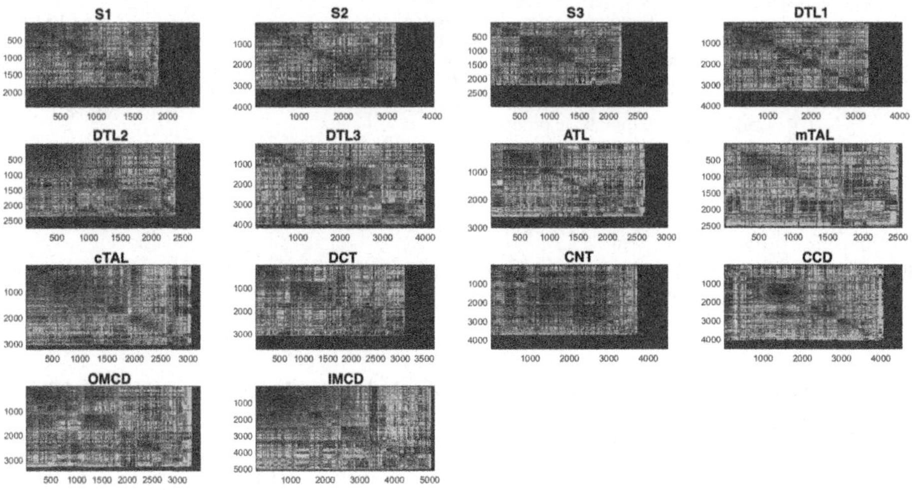

Fig. 8. Topological overlap matrix plots for the 14 renal tubule segments networks. Red diagonal blocks represent protein modules. (Color figure online)

As an example, we zoom in on the modularity of IMCD network. It is possible to observe that this graph has 5 modules (Fig. 9), that are enriched for set of pathways that are disjoint among them (Fig. 10 module A, Fig. 11 module B, Fig. 12 module D, and Fig. 13 module E. *FDR* computed by adjusting the $p-values$ of ShinyGO for multiple-comparison by applying the Benjamini and Hochberg procedure [15].). For proteins that belong to module C no pathways result enriched in a statistically significant way. This lack of enrichment can be as informative as positive results. In fact module of proteins with no enrichment might play fundamental roles in cellular processes that remain constant regardless of external conditions.

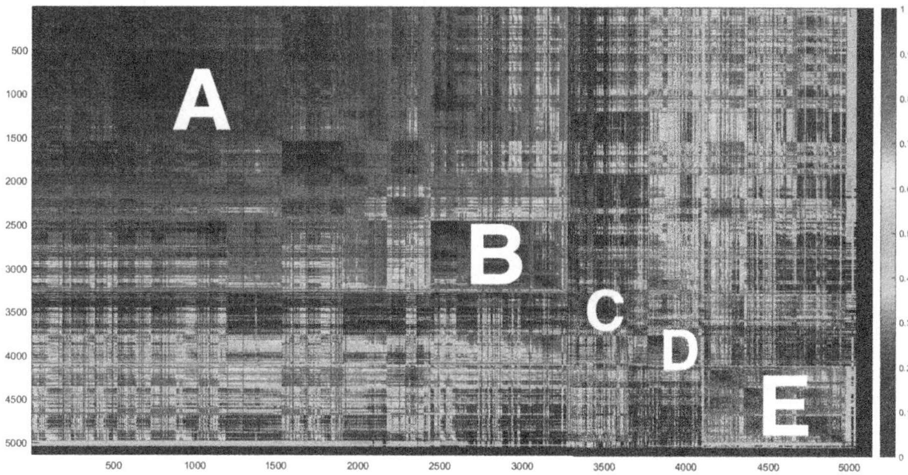

Fig. 9. Topological overlap matrix associated with IMCD, where five diagonal blocks are reported. These blocks represent protein modules.

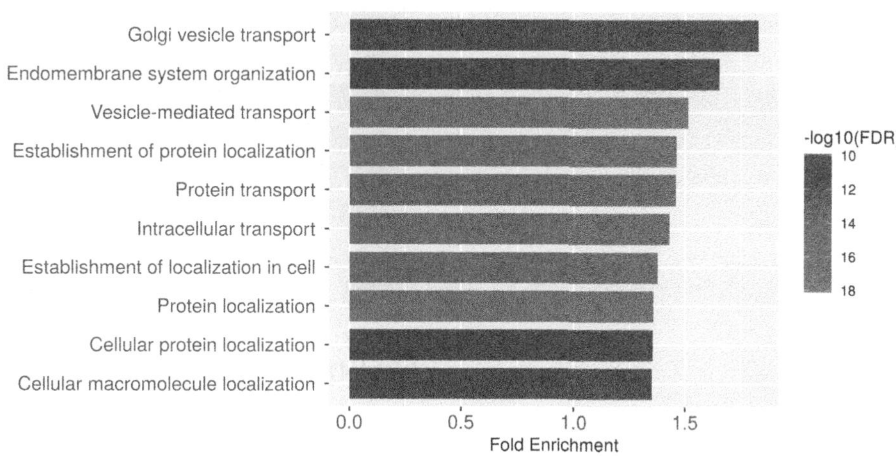

Fig. 10. Statistically significant enriched pathways for module A of Fig. 9.

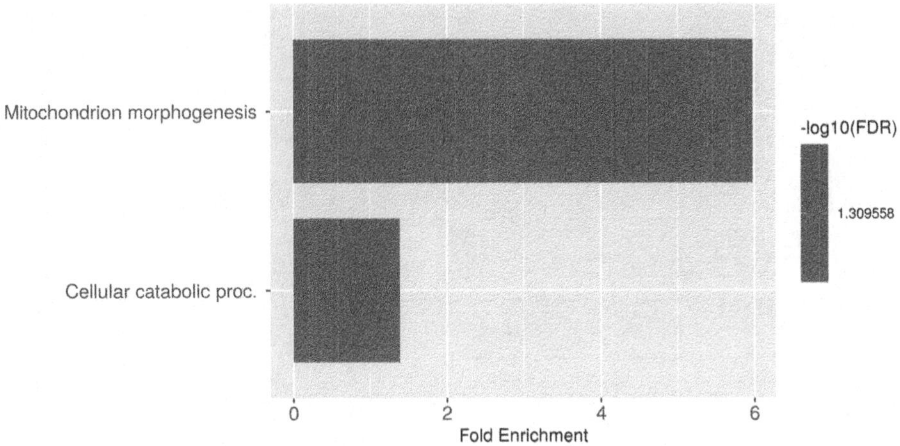

Fig. 11. Statistically significant enriched pathways for module B of Fig. 9.

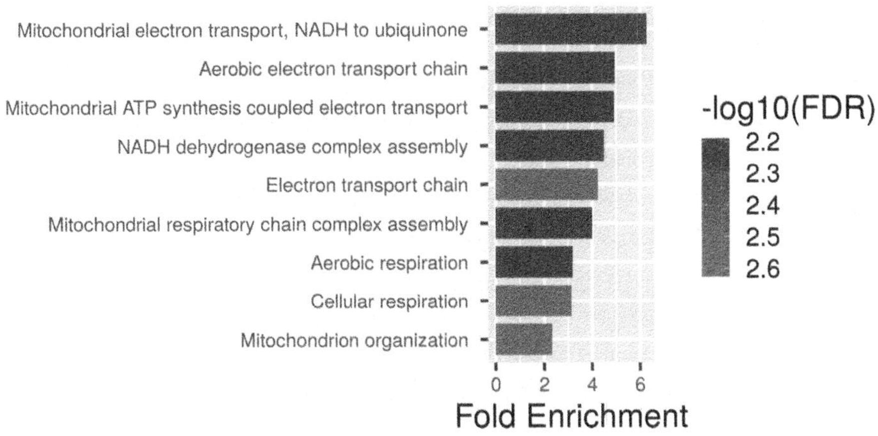

Fig. 12. Statistically significant enriched pathways for module D of Fig. 9.

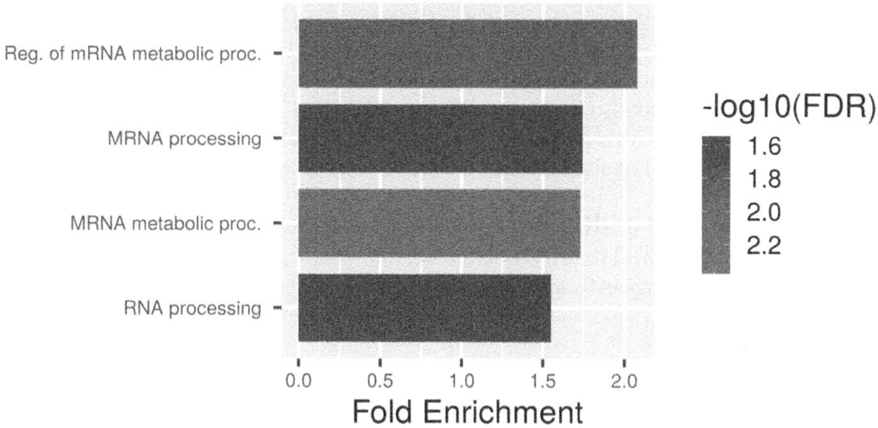

Fig. 13. Statistically significant enriched pathways for module E of Fig. 9.

5 Conclusions

Biological networks are extremely complicated processes composed of multiple molecular entities that interact among them in a conditional way and directly or indirectly regulate each other's expression levels. In this work, we focused our attention on the high-dimensional regime, a typical situation in biological dataset, in which the number of samples is much smaller than the number of variables. We proposed an algorithm to infer large scale networks modelling the strength of direct conditional dependence among variables. We showed that our approach outperforms all the others methods when the scenario moves towards the very high-dimensional regime. Then, we used it to inferred proteomics conditional networks for all the renal tubule segments, by using a dataset of proteins expressed in the segments of rat kidney tubules. Our results pave the way for the use of these graphs to uncover new insights into how interaction networks in mammalian cells are physically rewired in the different segments and/or stimuli. Moreover, due to the nature of the applied methods, the proposed approach can also be applied to others datasets and several types of data.

Our analyses also suggest that the inference of the degree of association between variables, with the effect of a set of controlling variables removed, may have a negative effect on overall accuracy that has not previously studied. Further work will include an investigation into *why* this problem leads to a drop in accuracy, in order to develop a strategy that achieves the goal of false positive minimisation without sacrificing the accuracy.

Acknowledgments. I am grateful to Alessandra Boletta leader of the Molecular Basis of Cystic Kidney Disorders Unit, Division of Genetics and Cell Biology, IRCCS, San Raffaele Scientific Institute, Milan (Italy), and to all the other members of her lab for helpful discussions. This work was funded by the Italian Ministry of Health (GR-2016-02364851 to R.P.), and by the National Recovery and Resilience Plan (NRRP),

Mission 4 Component 2 Investment 1.4 - Call for tender No. 3138 of 16 December 2021, rectified by Decree n.3175 of 18 December 2021 of Italian Ministry of University and Research funded by the European Union - NextGenerationEU. Award Number: project code $CN\_00000033$, Concession Decree No. 1034 of 17 June 2022 adopted by the Italian Ministry of University and Research, CUP G23C22001110007, Project title "National Biodiversity Future Center - NBFC".

Data and code availability. All original code and the inferred networks have been deposited and publicly available at the following link. https://github.com/RobertoPagliarini/Reverse-Engineering-of-Renal-Tubule-Networks-In-the-High-dimensional-Regime

References

1. Bühlmann, P., van de Geer, S.: Statistics for high-dimensional data methods, theory and applications. Springer Series in Statistics, Springer, Heidelberg (2011). https://doi.org/10.1007/978-3-642-20192-9
2. Carlson, M., Zhang, B., Fang, Z., Mischel, P.S., Horvath, S., Nelson, S.F.: Gene connectivity, function, and sequence conservation: predictions from modular yeast co-expression networks. BMC Genomics **7**, 40–40 (2006). https://api.semanticscholar.org/CorpusID:17089546
3. Castellini, A., Franco, G., Pagliarini, R.: Data analysis pipeline from laboratory to MP models. Nat. Comput. **10**(1), 55–76 (2011). http://dblp.uni-trier.de/db/journals/nc/nc10.html#CastelliniFP11
4. Chen, Y., Wiesel, A., III, A.O.H.: Shrinkage estimation of high dimensional covariance matrices. In: ICASSP, pp. 2937–2940. IEEE
5. Davis, J., Goadrich, M.: The relationship between precision-recall and ROC curves. In: ICML '06: Proceedings of the 23rd International Conference on Machine Learning, pp. 233–240. ACM, New York, NY, USA (2006). https://doi.org/10.1145/1143844.1143874, http://portal.acm.org/citation.cfm?id=1143874
6. Efron, B.: Correlation and large-scale simultaneous significance testing. J. Am. Stat. Assoc. **102**, 103–93 (2007). https://api.semanticscholar.org/CorpusID:805125
7. Fabregat, A., et al.: The reactome pathway knowledgebase. Nucleic Acids Res. **46**(D1), D649–D655 (2018). https://doi.org/10.1093/nar/gkx1132
8. Fu, W.J.: Penalized regressions: the bridge versus the lasso. J. Comput. Graph. Stat. **7**, 397–416 (1998). https://api.semanticscholar.org/CorpusID:123095463
9. Ge, S.X., Jung, D., Yao, R.: ShinyGO: a graphical gene-set enrichment tool for animals and plants. Bioinformatics **36**(8), 2628–2629 (2019). https://doi.org/10.1093/bioinformatics/btz931, https://doi.org/10.1093/bioinformatics/btz931
10. Goenawan, I.H., Bryan, K., Lynn, D.J.: DyNet: visualization and analysis of dynamic molecular interaction networks. Bioinformatics **32**(17), 2713–2715 (2016). https://doi.org/10.1093/bioinformatics/btw187
11. Grigorov, M.G.: Global properties of biological networks. Drug Discov. Today **10**(5), 365–372 (2005). https://doi.org/10.1016/S1359-6446(05)03369-6
12. Hocking, R.R.: A biometrics invited paper. the analysis and selection of variables in linear regression. Biometrics **32**(1), 1–49 (1976). http://www.jstor.org/stable/2529336
13. III, A.O.H., Rajaratnam, B.: Foundational principles for large-scale inference: illustrations through correlation mining. Proc. IEEE **104**(1), 93–110 (2016). http://dblp.uni-trier.de/db/journals/pieee/pieee104.html#HeroR16

14. Kanehisa, M., Sato, Y., Kawashima, M., Furumichi, M., Tanabe, M.: KEGG as a reference resource for gene and protein annotation. Nucleic Acids Res. **44**(D1), D457-62 (2016). https://doi.org/10.1093/nar/gkv1070
15. Korthauer, K., et al.: A practical guide to methods controlling false discoveries in computational biology. Genome Biol. **20**(1), 118 (2019). https://doi.org/10.1186/s13059-019-1716-1
16. Kyrgidis, A., Triaridis, S.: Methods and biostatistics: a concise guide for peer reviewers. Hippokratia **14**(Suppl 1), 13–22 (2010)
17. Lauritzen, S.L.: Graphical Models. Oxford University Press, Oxford (1996)
18. Limbutara, K., Chou, C.L., Knepper, M.A.: Quantitative proteomics of all 14 renal tubule segments in rat. J. Am. Soc. Nephrol. **31**(6), 1255–1266 (2020). https://doi.org/10.1681/ASN.2020010071
19. Madigan, D.: Graphical models in applied multivariate statistics, by j. whittaker, john wiley & sons, new york, 1990, 448 pp. Networks **24**(2), 125 (1994). http://dblp.uni-trier.de/db/journals/networks/networks24.html#Madigan94
20. Mitchell, G., Robson, J.S., Vay, D.L.: Renal System. Encyclopedia Britannica (2023). https://www.britannica.com/science/human-renal-system
21. Penrose, R.: A generalized inverse for matrices. In: Proceedings of the Cambridge Philosophy Society. vol. 51, pp. 406–413 (1955)
22. Pinna, A., Soranzo, N., Fuente, A., Hoeschele, I.: Simulation of the benchmark datasets. In: de la Fuente, A. (eds.) Gene Network Inference. Springer, Berlin, Heidelberg (2013). https://doi.org/10.1007/978-3-642-45161-4_1
23. Ravasz, E., Somera, A., Mongru, D., Oltvai, Z., Barabási, A.: Hierarchical organization of modularity in metabolic networks. Science **297**(5586), 1551 (2002). https://doi.org/10.1126/science.1073374
24. Saint-Antoine, M.M., Singh, A.: Evaluating pruning methods in gene network inference. In: Baruzzo, G., Daberdaku, S., Camillo, B.D., Furini, S., Giordano, E.D., Nicosia, G. (eds.) CIBCB. pp. 1–7. IEEE (2019). http://dblp.uni-trier.de/db/conf/cibcb/cibcb2019.html#Saint-AntoineS19
25. Schäfer, J., Strimmer, K.: An empirical bayes approach to inferring large-scale gene association networks. Bioinformatics **21**(6), 754–764 (2005)
26. Schäfer, J., Strimmer, K.: A shrinkage approach to large-scale covariance matrix estimation and implications for functional genomics. Stat. Appl. Genet. Mol. Biol. **4**(1), 32 (2005). https://doi.org/10.2202/1544-6115.1175, http://www.uni-leipzig.de/~strimmer/lab/publications/journals/shrinkcov2005.pdf
27. Sulaimanov, N., Koeppl, H.: Graph reconstruction using covariance-based methods. EURASIP J. Bioinform. Syst. Biol. **2016**(1), 19 (2016)
28. Sundström, J., et al.: Prevalence, outcomes, and cost of chronic kidney disease in a contemporary population of 2·4 million patients from 11 countries: the CaReMe CKD study. Lancet Reg. Health Eur. **20**, 100438 (2022). https://doi.org/10.1016/j.lanepe.2022.100438
29. Tibshirani, R.: Regression shrinkage and selection via the lasso. J. R. Stat. Soc. Ser. B-Methodol. **58**, 267–288 (1996). https://api.semanticscholar.org/CorpusID:16162039
30. Xiang, J.X.: A note on the Cauchy-Schwarz inequality. Am. Math. Monthly **120**(5), 456–459 (2013)
31. Ye, Y., Godzik, A.: Comparative analysis of protein domain organization. Genome Res. **14**(3), 343–53 (2004). https://api.semanticscholar.org/CorpusID:399313
32. Yip, A.M., Horvath, S.: Gene network interconnectedness and the generalized topological overlap measure. BMC Bioinform. **8**, 22–22 (2007). https://api.semanticscholar.org/CorpusID:9460567

Causal Model Discovery in Cancer Guided by Cellular Pathways

Rodrigo Henrique Ramos[1,2,3(✉)], Adenilso Simao[1], and Mohammad Reza Mousavi[2]

[1] University of São Paulo, Av. Trab. São Carlense, 400, São Carlos, São Paulo, Brazil
[2] King's College London, Strand, London WC2R 2LS, England
[3] Federal Institute of São Paulo, Estrada Municipal Paulo Eduardo de Almeida, São Carlos - São Paulo, Brazil
ramos@ifsp.edu.br

Abstract. Cancer is a genetic disease caused by mutations accumulated over time that corrupt cell pathways. Factors such as smoking, sunlight exposure, and age can influence the emergence of the disease and contribute to its heterogeneity. Identifying sub-groups within cancer populations is crucial for tailored treatments, given varied genetic responses to therapies. The increasing availability of (public) cancer data enables the development of many statistical studies exploring the relationships between mutations, pathways, and risk factors. Causal discovery can enhance these relationships into causal associations. Our paper presents an approach for investigating causal models in cancer, considering patient data, tumor stage, and pathways while addressing mutational diversity through patient sub-groups. Utilizing data from lung cancer, we use cancer driver genes and pathways to group patients associated with 12 cellular functions. We design and apply a causal discovery method to extract causal associations from the patient level to the genomic level, offering an overview of the entire set of patients. We also construct causal models for the 12 sub-groups of patients, exploring the different causal associations found within the same cancer type. Our results considering all patients show expected causal associations between patients and tumor data and new insights for causality among cellular pathways, especially Programmed Cell Death and Metabolism of Proteins. Also, sub-groups of patients within the same cancer type have distinct causal associations. These findings highlight the importance of considering mutational heterogeneity and personalized characteristics when exploring the complex landscape of causal discovery for cancer.

Keywords: Causality · Causal Discovery · Cancer Genomics · Cancer Patient Data · Cellular Pathways

This research was supported by King's College London, University of São Paulo, São Paulo Research Foundation (FAPESP), and the Brazilian Federal Foundation for Support and Evaluation of Graduate Education (CAPES). M.R. Mousavi was partially supported by UKRI and EPSRC (grants EP/V026801/2 and EP/Y005244/1.

R. Gori et al. (Eds.): CMSB 2024, LNBI 14971, pp. 174–195, 2024.
https://doi.org/10.1007/978-3-031-71671-3_13

1 Introduction

The causes for the emergence and spread of cancer are a central question in cancer research [31]. Causality [26], as a formal approach to identifying relations beyond simple correlation, has evolved over the last few decades. This has led to the creation of methods and algorithms to establish causal links between random variables in order to create causal models (CMs) [14]. Causal discovery from observational data goes beyond statistical correlation alone, e.g., by excluding confounding factors that influence both the purported causes and effects. In the worst case, this may require combinatorial calculations that are exponential in the number of variables [21,33]. The computational cost of causal discovery often limits the number of random variables analysed [14]. This limitation is a problem in the context of cancer research, where it is necessary to consider information from large patient-, tumor-, pathway- and gene datasets.

The temporal cellular processes of genetic mutations, driver mutations, pathway corruption, and cancer development form a causal network [16,31]. However, there are also indirect factors for cancer development, such as phenotype, environment, lifestyle, and diet [3,8,35]. When analysing the direct and indirect factors for cancer development, it is also important to consider sub-groups of patients to address the mutational heterogeneity of subcohorts in the same cancer type [4,20]. Examining the multifaceted aspects of causality in cancer can improve our understanding of the disease's initiation, progression, and spread [31]. To the best of our knowledge, no compositional causal analysis technique has been employed to explore those direct and indirect links. By "compositional", we refer to a (potentially mechanizable) technique that can divide the dataset into sub-groups, discover causal relations in an efficient manner, and then generalise the results up to the global data set as much as warranted by a causal analysis, while annotating causal relations that are specific to sub-groups.

In this study, we develop an approach for discovering causal models in cancer. We consider patient data, tumor stage, and pathways while addressing mutational heterogeneity through patient sub-groups. For each patient, we only use mutations in known cancer driver genes. We group the patients considering biological pathways that intersect with mutated drivers, enabling meaningful association between patients and cellular processes. This reduction from mutations in all genes to only driver genes and from drivers to pathways significantly decreases the number of variables, enabling the execution of the causal model discovery algorithm while creating sub-groups of the whole population. We create causal models for the sub-groups of patients in order to explore the nuances and specificities of causal associations found within the same cancer type. Since the same patients are associated with more than one group, we also create a causal model considering all patients and their associations with the pathways.

Analysis of the entire patient cohort revealed expected causal connections between patients and tumor data, while also uncovering novel insights into causal relationships within cellular pathways, particularly with Programmed Cell Death and Metabolism of Proteins. Since we made the causal discovery from observational data without prior knowledge, the expected associations found increased the method's credibility. Additionally, causal models constructed for patient sub-groups unveiled distinct and recurrent causal associations. A consensus causal model was then developed to consolidate findings from the sub-groups, emphasizing recurring patterns. These findings underscore the significance of accounting for mutational diversity and personalized characteristics while exploring the intricate landscape of cancer causality. By creating causal models that incorporate patient, tumor, and genomic data from both the entire cohort and sub-groups, this study offers complementary insights into varying levels of causal associations within cancer.

To summarise, the contributions of this paper are twofold:

1. Designing a scalable and compositional causal discovery pipeline for a composition of hetergenous dataset with three major components: variable reduction, decomposition, and aggregation; and
2. Applying the methodology to a composition of three types of cancer datasets and obtaining novel causal models and derive novel insights from them.

The remainder of this paper is organized as follows: Sect. 2 provides the background information for the rest of the paper. Section 3 present our methodology for a scalable and compositional causal discovery pipeline. Section 4 presents the results of applying our methodology and presents the gained insights. Section 5 concludes the paper.

2 Background

2.1 Cancer Data

Cancer is a complex disease that emerges from the intricate interplay of mutations that disrupt the normal regulation of genes and cellular pathways [16]. Advances in DNA sequencing methods have led to the availability of a large volume of cancer data [24,25]. Databases and portals such as TCGA and cBioPortal provide public data from multiple cancer studies. The specific information may change from study to study and also depends on the type of cancer (i.e. smoking in lung cancer and menopause in breast cancer) [27]. At a global level, all studies tend to present at least three types of datasets:

1. datasets with de-identified and anonymized *patient information*,
2. datasets with *tumor information* for each patient, and
3. datasets containing aggregated *mutation information*, including detailed genomic analysis of mutations.

The National Institutes of Health (NIH) offers detailed information about format of the last type of datasets, called the Mutation Annotation Format (MAF) [18]. A MAF file lists all mutations found in each gene for each patient [23]. Most of these mutations are unrelated to cancer development, thus called passenger mutations [22]. On the other hand, driver mutations are responsible for cancer development and progression [16,22]. The search for drivers led to the creation of curated lists of known driver genes, such as NCG [10], and the development of many computational methods that generally use the MAF file combined with protein networks and pathways [6,7,9,17,30].

2.2 Pathways

Cellular pathways are complex networks of molecular reactions that govern essential biological processes within cells. These pathways encompass a wide array of functions, such as signal transduction, DNA replication, metabolism, protein synthesis, and others. Researchers rely on databases such as Reactome, KEGG, and WikiPathways to study these pathways. Understanding these pathways is crucial for deciphering cellular behaviour's complexities and elucidating the mechanisms underlying diseases [19,36].

Reactome is a curated knowledgebase of biological pathways that focuses on human biology and offers a hierarchical organization of cellular processes, enabling researchers to explore interconnected molecular events at different levels of complexity [13]. Reactome uses the concept of Super Pathways to organize its pathways hierarchically. There are 27 Super Pathways (SPs) representing the main cellular process and 2,654 sub-pathways related to the main processes. SPs have different sets of genes and display complex and distinct topology when modelled as networks [28]. Figure 1 shows the Reactome pathway hierarchy for the SP Programmed Cell Death. Programmed Cell Death is a main cellular process and has 215 genes. This process is divided into two sub-pathways, Apoptosis with 181 genes and Regulated Necrosis with 61, meaning these two pathways share 27 genes. The SP hierarchical tree spreads in depth and width. The Programmed Cell Death has 41 sub-pathways, with smaller sets of genes and more specific cellular functions at the bottom.

2.3 Causal Discovery

Causal discovery is a crucial pursuit in a variety of fields. Its goal is to reveal the causal relationships that underlie the observed data [26]. This involves using methods and algorithms to infer causal structures from observational or experimental data, often relying on probabilistic reasoning and statistical dependencies [14]. By identifying causal relationships, researchers can uncover the mechanisms that drive phenomena of interest, leading to informed decision-making and predictive modelling [26]. Ultimately, causal discovery enriches our comprehension of complex systems and has numerous practical applications.

Identifying causal relationships from observational data is complex and computationally intensive, especially for large-scale or high-dimensional data [14]. One of the most prominent approaches in this field is the Peter-Clark (PC) algorithm, which identifies directed acyclic graphs (DAGs) by testing conditional independence assumptions iteratively [21]. The algorithm initiates with a complete undirected graph, representing the maximum potential set of edges without prior knowledge of the underlying causal relationships. The algorithm then refines the graph by increasing conditional independence tests with edge removal based on the results. This process ultimately leads to a minimal graph, called skeleton, that reflects the observed conditional independence relationships in the data. After obtaining the skeleton, the PC algorithm orients edges by applying rules based on the skeleton topology to determine causality between

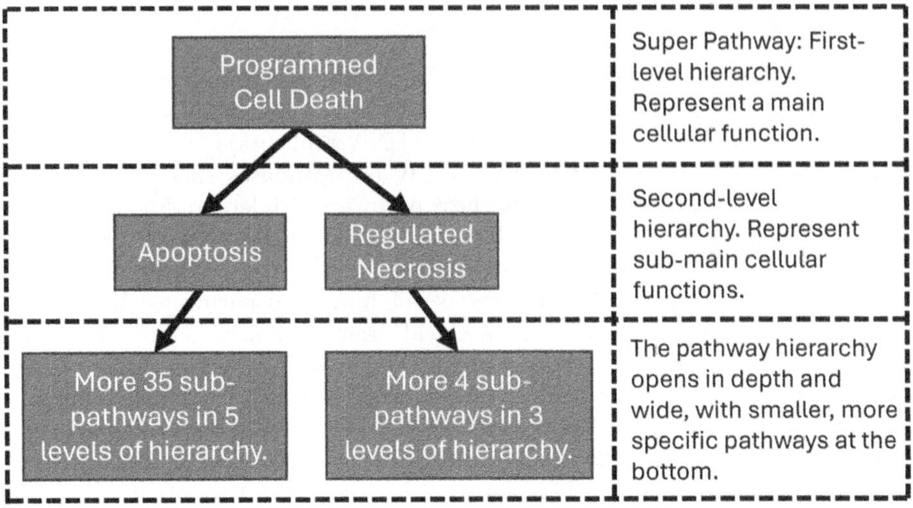

Fig. 1. Reactome Super Pathways Hierarchy. Reactome organizes its pathways in a tree-like structure. With the Super Pathway representing a main cellular function as root and more generic and smaller pathways as descendants.

variables. A detailed explanation of the rules is presented in the work of Le Thuc Duy [21]. The algorithm transforms the skeleton into a directed acyclic graph (DAG) representing causal relationships among variables. When an edge cannot be oriented, it remains undirected, indicating correlations between variables that may be affected by common causes or confounding factors but not a direct causal relationship. When edges remain undirected, the resulting graph may not necessarily be acyclic. As the number of variables increases, the complexity of conditional independence tests increases exponentially. Scalability has been traditionally improved by parallelization and approximation strategies, but it still faces challenges related to its time complexity in high-dimensional data [14,21].

3 Causal Discovery Pipeline

Our pipeline starts by exploring and preprocessing the patient and tumor data. We removed outliers and discretised continuous variables as usual. Due to space limitation, we included the details of these steps in the Appendix. After the preprocessing, we divide our data preparation into three steps: variable reduction, decomposition, and aggregation. Finally we apply the causal discovery algorithm on the reduced, decomposed, and aggregated outcomes and compose them back again.

3.1 Variable Reduction and Decomposition

These two steps, respectively, address the following two requirements: 1) We must reduce the number of variables to run the causal discovery algorithm. 2) We must address the mutation heterogeneity by creating sub-groups of patients.

Considering variables of interest in cancer study from lung cancer [5], in the patient dataset, there are five variables: Age, Sex, Smoking History, Smoking Packing Years (SPY), and Status (alive or deceased). The tumor dataset has the Stage variable. Given the time complexity of causal discovery methods, these six variables are manageable if we want to analyse just patient and tumor data. The problem arises when we consider the hundreds of cellular pathways and the approximately 20,000 protein-coding genes [32]. The MAF file from the study mentioned above has 17,959 protein-coding genes. Since most of those mutations are passengers, the first step is to consider only known drivers. The NCG database presents a list of 3,347 drivers. This procedure removes all false positive genes for cancer (passengers) at the expense of removing possible unknown drivers. Figure 2-(A) exemplifies the passenger removal from the MAF file, a step that reduces the number of genes from 17,959 to 3,347. This reduction is significant and also removes false positives, but it is still unmanageable for combinatorial causal discovery methods. We assume a causal relation in cancer between drivers genes and SPs [16], then we associate each patient with the Reactome's SPs by intersecting their genes with patients' mutated drivers.

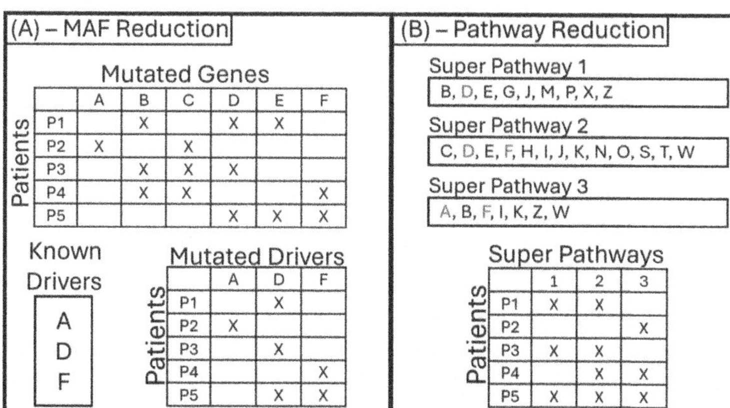

Fig. 2. Reducing the number of variables. In (A), we use a curated list of known cancer driver genes from NCG to remove passenger genes while keeping all the patients. In (B), we use the Mutated Drivers table from (A) to decompose the data and associate each patient to pathways harbouring one or more altered drivers.

3.2 Aggregation

Figure 2-(B) shows how we intersect the Mutated Drivers table from Fig. 2-(A) with Reactome's SPs. P1 in Fig. 2-(A) has mutations on the genes B, D, and E. The pathways in Fig. 2-(B) intersect with these three genes, but since only D is a known driver, we associated P1 with the two first pathways. We need to test the statistical significance of these intersections between pathway genes and patients drivers. For each patient and for each SP, we test if the intersection between genes is statistically significant (p-value ≤ 0.05). If it is, we add the patient in the SP group. If the SP group contains less than 221 statistically significant patients (20% of all patients), we search in the SP hierarchy for a sub-pathway that statistically significantly comprises more than 220 patients. We use the Fisher Exact Test [1] to establish the statistical significance between gene sets, in a similar way used to find significant associations among two sets of genes in mutual exclusivity or co-occurrence [2,12]. We calculate $(Neither * Both)/(AnotB * BnotA)$, where A is the patient genes, and B is the SP or sub-pathway genes. To calculate *Neither*, we use all genes in the MAF file. Figure 3-(A) illustrates the intersection and Fig. 3-(B) the contingency table used in the statistical test.

(A) - Genes Sets

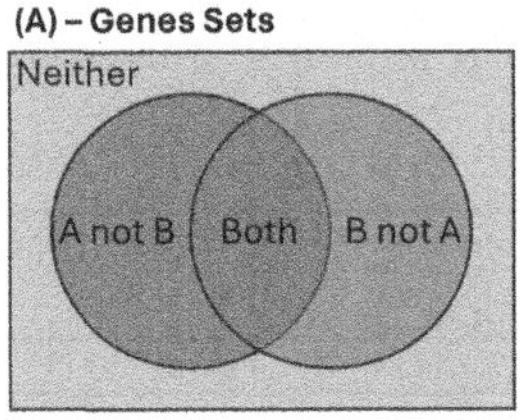

(B) - Contingency table

		Patient Gene Set	
		Yes	No
Pathway Gene Set	Yes	Both	A not B
	No	B not A	Neither

Fig. 3. Significant intersection. A is the patient set of genes, and B is the SP or sub-pathway set genes. *Neither* is the set of all other genes found in the MAF.

After this first step, we considerably reduce the number of variables in the genomic level from 17,959 to, at most, 27. This reduction makes the application of combinatorial causal discovery algorithms possible. The association of patients and SPs or sub-pathways using a statistical test enables the meaningful creation of sub-groups of patients. These groups address the mutational heterogeneity found in cancer and have biological significance.

3.3 Datasets for Causal Model Discovery

In this third step, we prepare the datasets to discover causal models. We expand the final table from Fig. 2 regarding genomic level information to include patient and tumor level data. This dataset stores information about all patients and SPs or sub-pathways. Since the patients may be grouped in multiple pathways, we can extract causal links between them together with patient and tumor data, offering an overview of the cancer type. It is possible to divide the dataset considering the pathways, creating sub-groups of patients sharing driver genes in the same biological process. These smaller datasets keep the six patient and tumor level data and one pathway at a time. Figure 4 exemplify the datasets.

(A) – All Patients Dataset

	Patient and Tumor level data						Genomic Level Data			
Patient	Sex	Age	Smoking	SPY	Status	Stage	SP 1	SP 2	SP 3	SP 4
p1	Male	38-43	Reformed >=15	15-29	Living	IA	2	0	0	0
p2	Male	48-54	Reformed <=15	0-14	Deceased	IIIA	0	5	0	1
p3	Female	64-69	Reformed	0-15	Living	IIIB	2	0	7	0
p4	Female	80-85	Current	29-44	Deceased	V	1	0	0	1
p5	Male	64-69	Reformed >=15	29-44	Living	IIA	3	2	1	0

(B) – Patients' Sub-groups Datasets

Patient	Patient and Tumor level data	SP 1
p1	Patient and Tumor level data	2
p3	Patient and Tumor level data	2
p4	Patient and Tumor level data	1
p5	Patient and Tumor level data	3

Patient	Patient and Tumor level data	SP 2
p2	Patient and Tumor level data	5
p5	Patient and Tumor level data	2

Patient	Patient and Tumor level data	SP 3
p3	Patient and Tumor level data	7
p5	Patient and Tumor level data	1

Fig. 4. Datasets for causal model discovery. In (B), we suppress the variable's names and the sub-group for SP 4 to save space in the plot, but all datasets have six variables: Sex, Age, Smoking, SPY, Status, and Stage.

Figure 4-(A) shows the dataset for all patients and SPs or sub-pathways that significantly group patients. The variables can be divided into two types: Patient and Tumor Level Data, and Genomic Level Data. The latter shows how many drivers each patient has on each SP or sub-pathway. For the SP or sub-pathway be included as column, it must significantly group 221 patients (20% of all patients) or more. For this reason, it is not guaranteed that all 27 SPs or sub-pathways will be in the dataset.

Figure 4-(B) displays datasets for subgroups of patients with driver mutations occurring in the same biological process. These sub-groups provide a means of categorizing the MAF file based on mutational diversity. It is possible to investigate the causal associations of each group and develop a consensus causal model, which shows all the associations identified in each group, placing more weight on recurring edges. Figure 5 illustrates these models.

In Fig. 5 (B), and (C), we exemplify how different types of patient and tumor data may interact. Dashed edges indicate a strong correlation that persists through independent sets but not causation. Direct edges indicate causation. In Fig. 5 (C) the edge's weight helps identify the strength of correlation or causation in different sub-groups of patients.

4 Results

Using data from lung cancer, we found 5 SP and 7 sub-pathways that significantly grouped more than 220 patients (20% of the 1,107 patients after preprocessing).

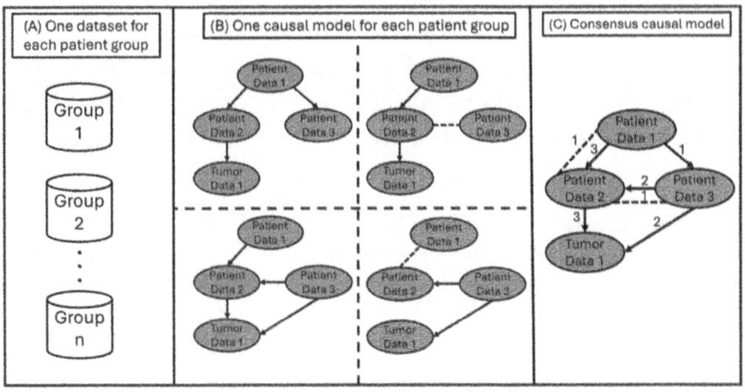

Fig. 5. Sub-groups and consensus causal model. In (A), we show the patients datasets from Fig. 3. In (B) we create one causal model for each patient group that may vary on the edges. In (C), we combine all causal models, adding weights for repeating edges.

Table 1 shows the pathway names and number of patients in each group. Sub-pathways are indicated by "- -", with the SP name before the "- -" and the sub-pathway name after it. Each group in Table 1 harbours, on average, 35% of all patients. (For details, we have included Fig. 14 in the Appendix, which shows the intersection of patients in one group (rows) with all other groups (columns).) Cell-Cell Communication has 339 patients, of which 29% are also on Cellular responses to stimuli - - Oncogene Induced Senescence (with 277 patients), and 64% are in Development Biology (with 589 patients).

Table 1. Patient groups. All SP or sub-pathways, indicated by "- -", which significantly grouped more than 220 patients.

Super Pathway groups	Patients
Cell-Cell communication	339
Cellular responses to stimuli - - Oncogene Induced Senescence	277
Chromatin organization - - PKMTs methylate histone lysines	237
Developmental Biology	589
Extracellular matrix organization	518
Gene expression (Transcription) - - Transcriptional regulation by RUNX3	246
Immune System - - Dectin-2 family	279
Metabolism of proteins - - SUMOylation	330
Neuronal System	428
Programmed Cell Death - - Activation of NOXA and translocation	656
Signal Transduction	518
Vesicle-mediated transport - - Scavenging by Class H Receptors	235

4.1 Sub-groups and Consensus Causal Models

With 12 groups of patients associated with cellular functions, we executed the PC algorithm from the TETRAD library [29] and plotted the networks using the NetworkX library [15]. Section *"Causal Diagrams"* on the Appendix presents the 12 causal models. Figure 6 shows 4 causal models with distinct topology. Each causal model is from a sub-group of patients and has a dataset equal to Fig. 4-(B), with 6 variables from patients and tumor, and 1 variable representing the SP or sub-pathway.

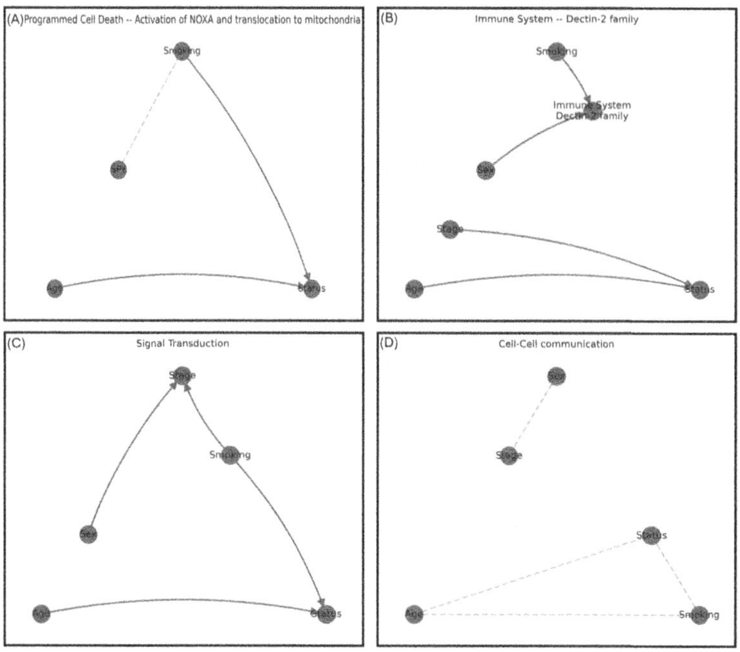

Fig. 6. Four causal models. Grey dashed edges represent significant correlation, and solid green edges causation. The sub-plots (C) and (D) shows the causal models for two SP, Signal Transduction and Cell-Cell Communication. The sub-plots (A) and (B) shows causal models for sub-pathways. (Color figure online)

It can be observed that not all 7 variables are included in the causal models, and the nodes have few edges. This is because the PC algorithm searches for conditional independent sets, so the absence of nodes and edges is significant. Age and Smoking are linked only in the causal model for Cell-Cell Communication, but in the other three, this connection was eliminated because some combination of the other 5 variables made them independent. Age and Status are associated in all models, meaning this association persist independently of any combination of the other variables.

In Fig. 6-(A), the group of patients with driver mutations on a sub-pathway of Programmed Cell Death have a causal association between Smoking and Status, and Age and Status. In these patients, age and the habit of smoking cause death or survival. In Fig. 6-(D), these three variables are also associated, but not in a causal way. In Fig. 6-(B), the corresponding node to the sub-pathway Dectin-2 Family within the hierarchy of the SP Immune System is caused by Smoking and Sex. The V-Structure $Smoking \rightarrow Pathway \leftarrow Sex$ indicates that the corruption of the pathways by mutated drivers depends at the same time of smoking and sex. This finding brings insights into the complex dependencies in cancer development and the interplay of direct and indirect factors in cancer. The rest of this section further explores these associations.

All 12 causal models are unique, indicating that the process of grouping patients to reduce complexity and address mutational heterogeneity successfully extracted distinct sub-groups of patients. Albeit unique, the causal models have re-occurring edges. Figure 7 shows a consensus causal model with all associations found in the 12 models, with edge weight indicating the occurrences numbers.

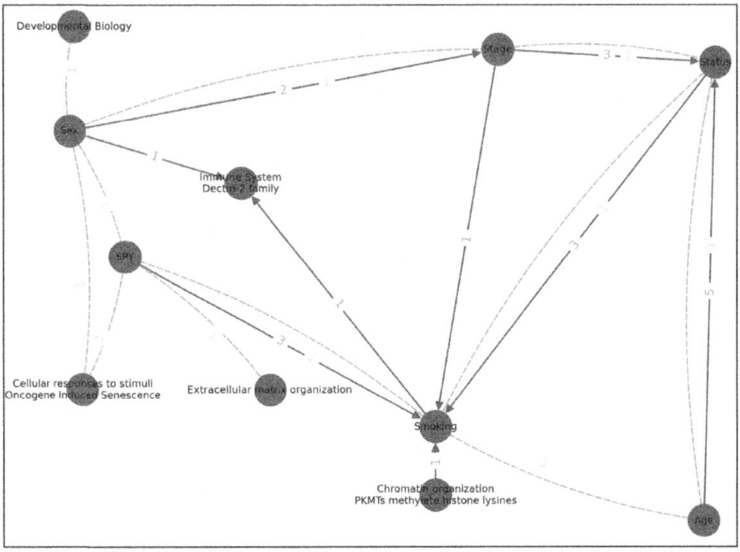

Fig. 7. Consensus Causal Model. Grey dashed edges represent significant correlation, and solid green edges are causation. Edge weight shows the edge occurrences in the 12 sub-groups causal models. (Color figure online)

In Fig. 7, Age and Status appear 8 times, 3 as correlation and 5 as Age causing Status. We understand that patient age impacts the chance of survival, and this association was found in 8 causal models. We also see another expected association when Stage cause Status in 3 cases and is correlated in 2. We can also question some causal edge orientations. Status is correlated with Smoking in 1 case, and in 3 cases, Status causes Smoking, also Stage causes Smoking in 1 case. In both cases, we believe the orientation should be inverted. The PC algorithm extracts causal associations from observational data without previous knowledge and expects the distributions to be Gaussian. These few inaccurate edge orientations can be manually fixed by cancer specialists to improve the model's accuracy. The consensus causal model in Fig. 7 offers an aggregated summary for causal associations found in sub-groups of patients, and place Smoking and Status as key variables among the others due the number of connections.

4.2 All Patient's Causal Model

The lung cancer study used in this paper has 1,107 patients after our preprocessing. Each patient appears on average in 4.5 groups, with a median of 4 and a standard deviation of 2. Since each group represents a cellular process, we can explore the causal association among pathways. Figure 8 presents two causal models created using all patients, considering the dataset in Fig. 4-(A). Figure 8-(A) was constructed using only data from the patient and tumor level. Figure 8-(B) adds data from the genomic level. Comparing the patient and tumor variables in the two models, the only difference is the absence of the correlation between Sex and Stage in Fig. 8-(B). Some combination of pathways made them conditional independent. On the other hand, any combination of pathways made the other patients and tumor variables independent. It is important to point out the relevance of Smoking in connecting the genomic level to the patient and tumor level.

Age and Smoking are associated in 3 causal models for sub-groups of patients and are not associated with SPY in any of them. In the causal model for all patients, Age and Smoking are intermediated by SPY, meaning that once you have information about the SPY, one does not need the Age to know about the time a patient smokes (i.e. the Smoking variable). Status is conditionally independent from all other nodes once we have information about the patient's Age and tumor Stage. Overall, variables from the patient and tumor level show the expected causal associations with each other, which increases the method's credibility when analysing the association at the genomic level.

Figure 8-(B) shows us the corruption of the pathway within the SP Programmed Cell Death by mutated drivers genes is caused by the time the patient smokes and also by the corruption of 3 other cellular processes. Corruption in a sub-pathway within the Metabolism of Protein is also caused by 3 pathways but is not directly or indirectly caused by Cellular responses to stimuli – Oncogene Induced Senescence in Fig. 8-(B). The interplay of associations and disassociations with pathways may be related to the known cancer phenomena of co-occurrence and mutual exclusivity among driver genes. Both phenomena have applications in cancer therapy [11]. Figure 8 presents an intricate network of causal associations from the data at different levels, giving us insights into the complex landscape of cancer causality.

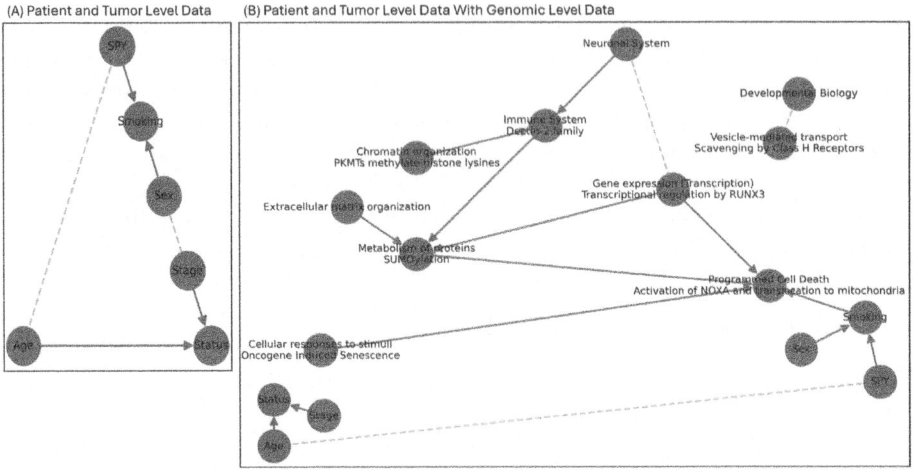

Fig. 8. All patient causal models. (A) shows a causal model with only patients and tumor level data. We see expected causal association among variables. (B) shows a causal model considering also the genomic-level data. We observe causal associations among pathways, which brings insights into the complex interactions of pathways in cancer.

4.3 Discussion

The methodology developed in this study enabled the creation of a causal discovery pipeline in cancer. The variable reduction, decomposition, and aggregation steps offer a scalable approach to deal with causal discovery on multidimensional datasets from complementary levels (patients, tumor, and genomic). The use of cancer drivers genes and pathways reduced the number of variables from 17,959 to 12, removed false positives and increased the biological significance of sub-groups.

Causal model analysis reinforced known results while unveiling new insights. Expected causal associations between patient and tumor data were observed, thereby validating the methodology's efficacy. Notably, variables such as smoking and status emerged as pivotal nodes to validate causal associations. Novel insights include distinct causal associations within sub-groups of patients, as in the group Immune System – Dectin-2 family, where the sex variable causes the pathway's corruption. However, this association is not found in any other sub-group. Patterns like this highlight the importance of considering mutational heterogeneity for personalized treatments. Also, the causal model for all patients and pathways brings insights into pathways' complex association and disassociation. Programmed Cell Death – Activation of NOXA and translocation to mitochondria is caused by four other variables but is not causing another variable. This pathway's corruption is the final result of a series of causes.

5 Conclusion

Our work aimed to create causal models in cancer, considering data from patient, tumor, and genomic levels while addressing mutational heterogeneity with subgroups of patients. Using data from more than 1,000 lung cancer patients, we integrate biological knowledge, namely cancer driver genes and pathways, to guide the grouping of patients into biologically relevant cohorts. This procedure reduced the number of variables from the genomic level from 17,959 to 12, making possible the execution of a combinatorial causal discovery method. We created 12 causal models from distinct sub-sets of patients. The models have unique and recurring causal associations. We also created a composed causal model considering all 1,107 patients, with 6 variables from patient and tumor level added with 12 cellular pathways.

Causal models for sub-groups of patients highlight the presence of heterogeneity in causal associations within patients from the same cancer type. The construction of a combined causal model summarized the sub-groups associations, emphasizing recurring edges. Notably, variables such as smoking and survival emerge as critical nodes, exhibiting a higher number of causal associations with other variables in the model.

The causal model with all patients and variables provides an overview of the cancer type. Variables at the patient and tumor levels showed expected causal associations, increasing the method's credibility. Associations and disassociation among pathways might be related to the phenomenon of mutual exclusivity and co-occurrence of cancer driver genes.

In addition to an approach to join cancer data from different levels, the paper bring insights into the complex dependencies in cancer development and the interplay of direct and indirect factors in cancer. In the future, we plan to apply this approach to other types of cancer, comparing causal models from distinct and similar types of cancer. Providing a generic methodology that allows for reusing data across neighbouring cohorts and similar disease types remains an interesting avenue for research. Furthermore, we intend to use the causal models we have developed to perform further causal inference analysis, deepening our understanding of the underlying mechanisms of cancer.

Acknowledgments. We acknowledge the informative and constructive comments of anonymous reviewers. M.R. Mousavi has been partially supported by by the UKRI Trustworthy Autonomous Systems Node in Verifiability, Grant Award Reference EP/V026801/2, the EPSRC project on Verified Simulation for Large Quantum Systems (VSL-Q), grant reference EP/Y005244/1 and the EPSRC projecton Robust and Reliable Quantum Computing (RoaRQ), Investigation 009, grant reference EP/W032635/1. Also King's Quantum grants provided by King's College London are gratefully acknowledged.

Disclosure of Interests. The authors have no competing interests to declare that are relevant to the content of this article.

Appendix

Preprocessing of Patient and Tumor Data

This section presents the preprocessing performed on patient and tumor variables. We show distribution plots for each variable before and after preprocessing and a textual consideration.

Figures for continuous or discrete variables shows acronyms in the title: Q1, Q1, and Q3 are the quartiles, Avg is the average value, and Max and Min are the maximum and minimum values found on the distribution.

AGE - In Fig. 9, the first figure shows the age distribution of patients. Since most other variables are categorical, we discretize the age into ten groups, as shown in the second figure (Fig. 10).

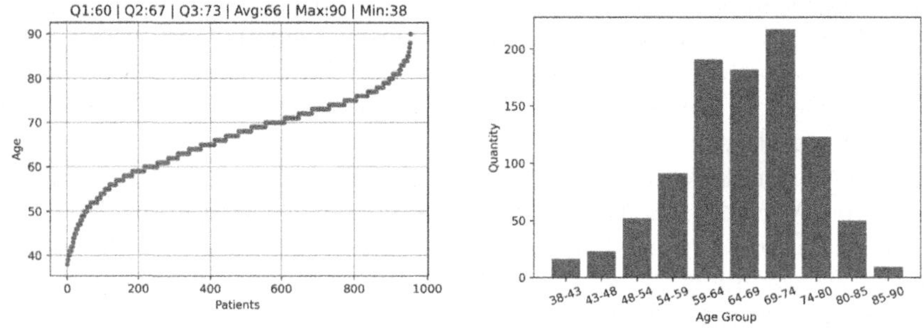

Fig. 9. Age original distribution and discretized

SEX - No preprocessing was made on the sex variable

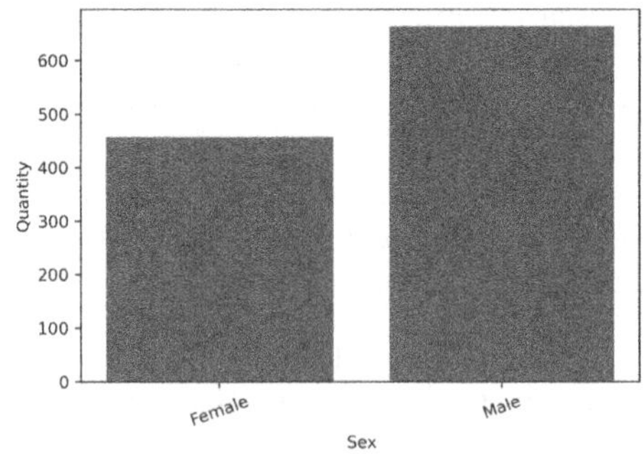

Fig. 10. Patient's sex distribution

SMOKING PACK YEARS (SPY) - Each number in this variable represents the smoking of one pack of cigarettes a day for one year. For instance, a

SPY of 20 indicates a person smoked one pack per day for 20 years, two packs per day for 10 years, or half a pack per day for 40 years. The two figures on the left show the discrete and categorical distribution before the removal of outliers, and the two figures on the right show the distribution after the removal of outliers. To identify outliers, we remove patients whose values are greater than a threshold, calculated by $Q3 + 4.5 \times IQR$, where Q3 and IQR are the third quartile and the interquartile range of the distribution, a common strategy to remove outliers in cancer datasets [27,34]. This procedure removed only 6 patients, while keeping the quartiles the same and reducing the maximum value from 240 to 147

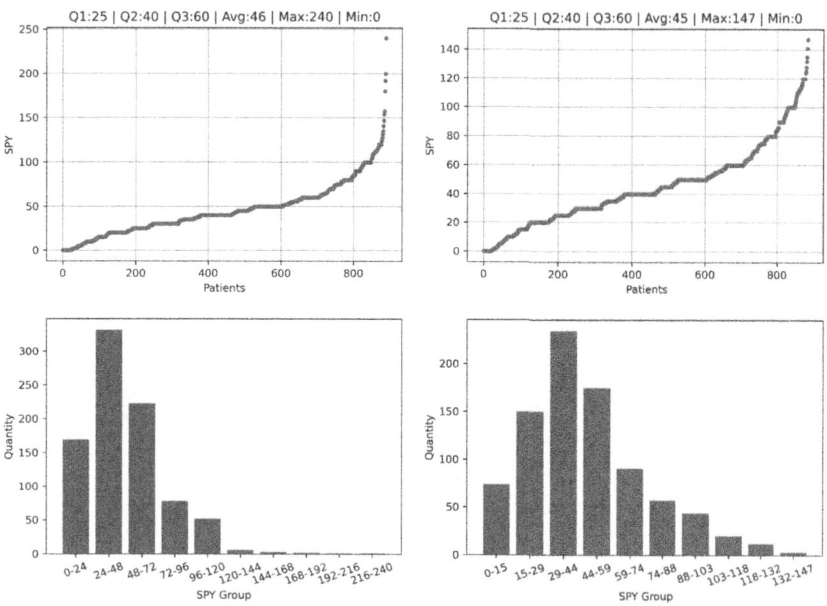

Fig. 11. Patient's smoking pack years

STATUS - Shows if the patients died or survive during the study that collected the data. No preprocessing made.

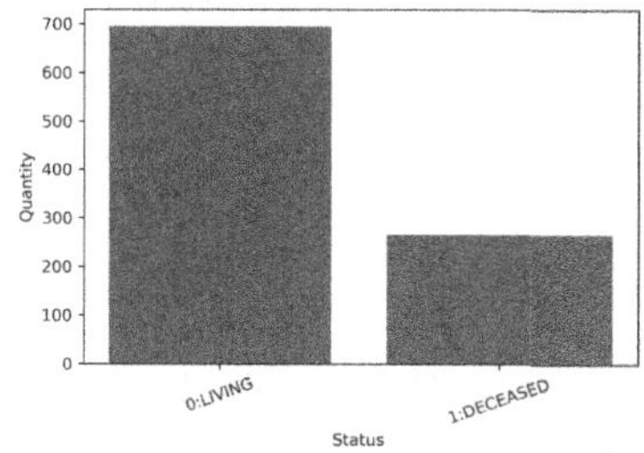

Fig. 12. Patient Status

STAGE - Represents the cancer stage. The left figure shows the original distribution. Pre-processing consisted of removing stages I, II and III, as they correspond to very small groups (Figs. 11, 12 and 13).

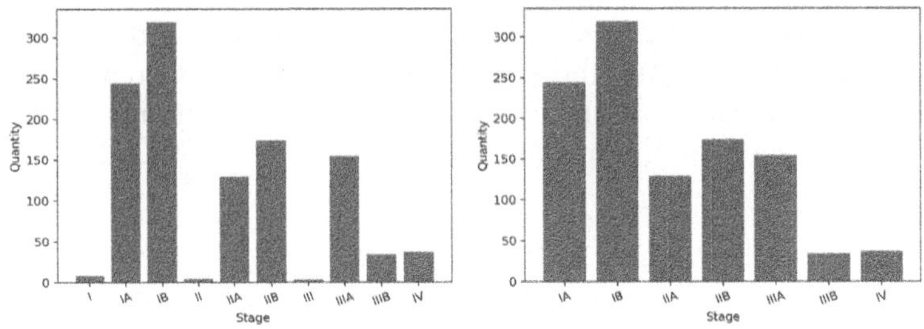

Fig. 13. Cancer Stage

Patient groups' intersections

This section presents the Fig. 14 showing how the sub-groups of patients intersect.

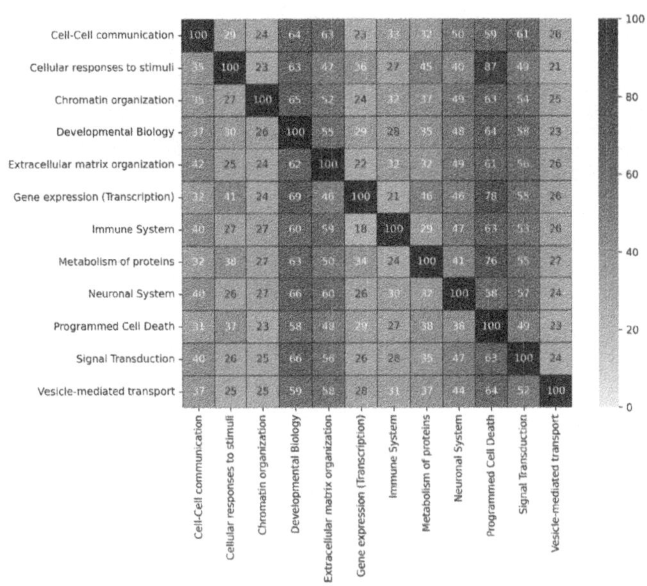

Fig. 14. Patient groups' intersections. Shows in percentuals how many patients from one group (rows) are also present in other groups. To save space on the plot, we present only the SP name.

Causal Diagrams

This section presents the causal diagrams for the 12 groups of patients (Figs. 15, 16 and 17).

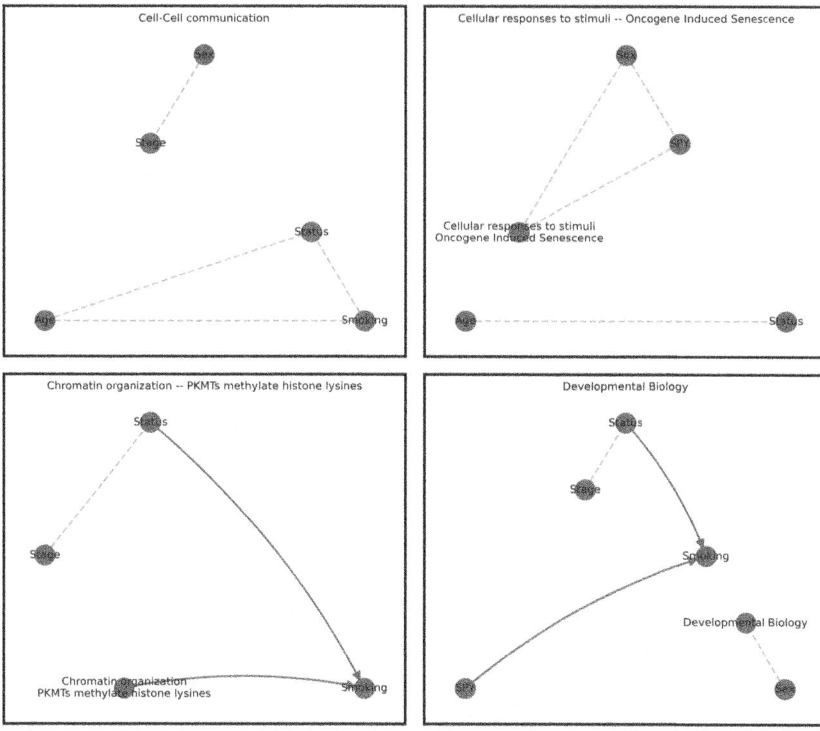

Fig. 15. Causal Models for Cell-Cell communication, Cellular responses to stimuli – Oncogene Induced Senescence, Chromatin organization – PKMTs methylate histone lysines, and Developmental Biology.

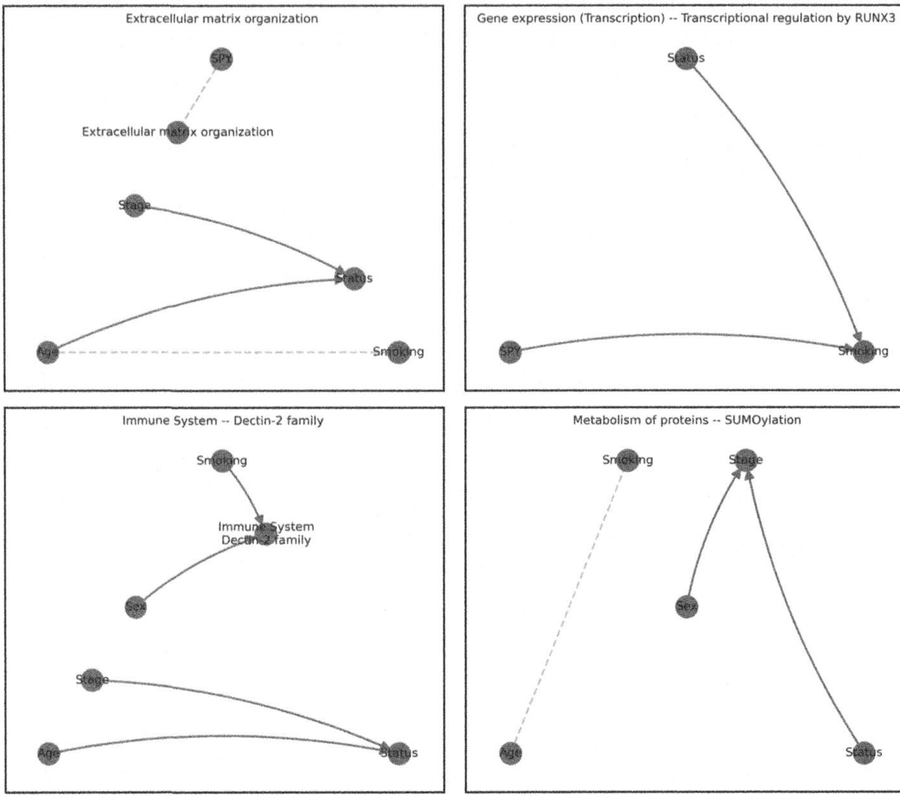

Fig. 16. Causal Models for Extracellular matrix organization, Gene expression (Transcription) – Transcriptional regulation by RUNX3, Immune System – Dectin-2 family, and Metabolism of proteins – SUMOylation.

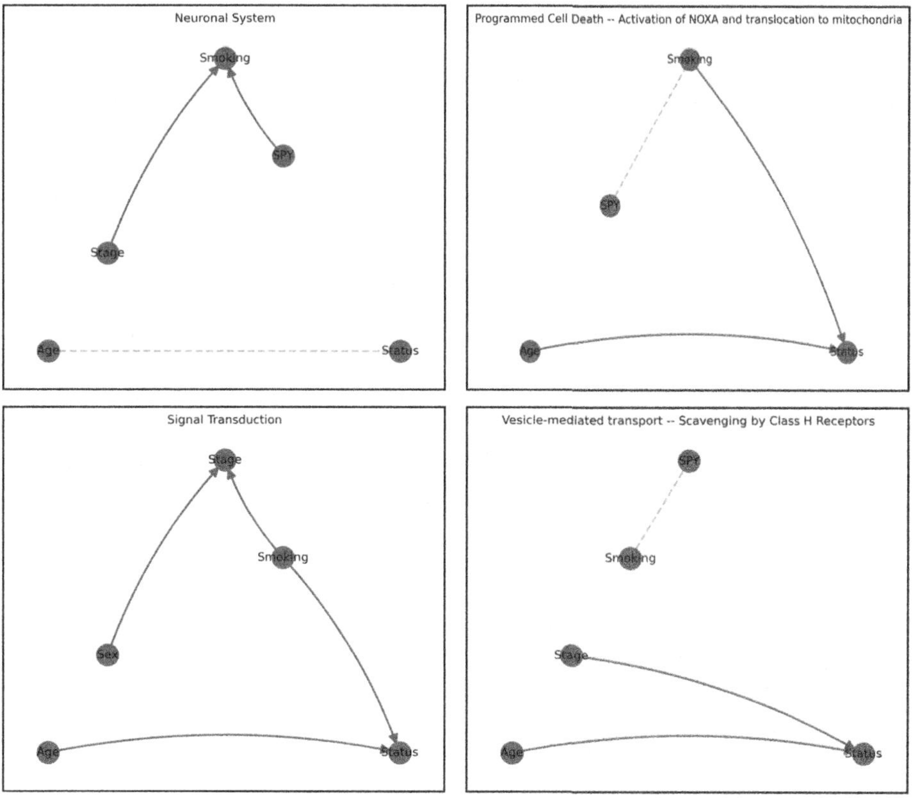

Fig. 17. Causal Models for Neuronal System, Programmed Cell Death – Activation of NOXA and translocation to mitochondria, Signal Transduction, and Vesicle-mediated transport – Scavenging by Class H Receptors.

References

1. Agresti, A.: Categorical Data Analysis, vol. 792. Wiley, Hoboken (2012)
2. Ahmed, R., Erten, C., Houdjedj, A., Kazan, H., Yalcin, C.: A network-centric framework for the evaluation of mutual exclusivity tests on cancer drivers. Front. Genet. **12**, 746495 (2021)
3. Alberts, B., Johnson, A., Lewis, J., Raff, M., Roberts, K., Walter, P.: The preventable causes of cancer. In: Molecular Biology of the Cell. 4th edition. Garland Science (2002)
4. Amatya, A.K., et al.: Subgroup analyses in oncology trials: regulatory considerations and case examples. Clin. Cancer Res. **27**(21), 5753–5756 (2021)
5. Campbell, J.D., et al.: Distinct patterns of somatic genome alterations in lung adenocarcinomas and squamous cell carcinomas. Nat. Genet. **48**(6), 607–616 (2016)
6. Cutigi, J.F., Evangelista, A.F., Reis, R.M., Simao, A.: A computational approach for the discovery of significant cancer genes by weighted mutation and asymmetric spreading strength in networks. Sci. Rep. **11**(1), 1–10 (2021)
7. Cutigi, J.F., Evangelista, A.F., Simao, A.: Approaches for the identification of driver mutations in cancer: a tutorial from a computational perspective. J. Bioinform. Comput. Biol. **18**(03), 2050016 (2020)
8. Danaei, G., Vander Hoorn, S., Lopez, A.D., Murray, C.J., Ezzati, M.: Causes of cancer in the world: comparative risk assessment of nine behavioural and environmental risk factors. Lancet **366**(9499), 1784–1793 (2005)
9. Dimitrakopoulos, C.M., Beerenwinkel, N.: Computational approaches for the identification of cancer genes and pathways. Wiley Interdisc. Rev. Syst. Biol. Med. **9**(1), e1364 (2017)
10. Dressler, L., et al.: Comparative assessment of genes driving cancer and somatic evolution in non-cancer tissues: an update of the network of cancer genes (NCG) resource. Genome Biol. **23**(1), 35 (2022)
11. El Tekle, G., et al.: Co-occurrence and mutual exclusivity: what cross-cancer mutation patterns can tell us. Trends Cancer **7**(9), 823–836 (2021)
12. Gao, J., et al.: Integrative analysis of complex cancer genomics and clinical profiles using the cbioportal. Sci. Signal. **6**(269), pl1–pl1 (2013)
13. Gillespie, M., et al.: The reactome pathway knowledgebase 2022. Nucleic Acids Res. **50**(D1), D687–D692 (2022)
14. Glymour, C., Zhang, K., Spirtes, P.: Review of causal discovery methods based on graphical models. Front. Genet. **10**, 524 (2019)
15. Hagberg, A., Swart, P., S Chult, D.: Exploring network structure, dynamics, and function using networkx. Tech. rep., Los Alamos National Lab.(LANL), Los Alamos, NM (United States) (2008)
16. Hanahan, D.: Hallmarks of cancer: new dimensions. Cancer Discov. **12**(1), 31–46 (2022)
17. Hristov, B.H., Chazelle, B., Singh, M.: uKIN combines new and prior information with guided network propagation to accurately identify disease genes. Cell Syst. **10**(6), 470–479 (2020)
18. Institute, N.C.: GDC MAF format v.1.0.0. https://docs.gdc.cancer.gov/Data/File_Formats/MAF_Format/
19. Khatri, P., Sirota, M., Butte, A.J.: Ten years of pathway analysis: current approaches and outstanding challenges. PLoS Comput. Biol. **8**(2), e1002375 (2012)
20. Kuipers, J., et al.: Mutational interactions define novel cancer subgroups. Nat. Commun. **9**(1), 4353 (2018)

21. Le, T.D., Hoang, T., Li, J., Liu, L., Liu, H., Hu, S.: A fast PC algorithm for high dimensional causal discovery with multi-core PCs. IEEE/ACM Trans. Comput. Biol. Bioinf. **16**(5), 1483–1495 (2016)
22. Martínez-Jiménez, F., et al.: A compendium of mutational cancer driver genes. Nat. Rev. Cancer **20**(10), 555–572 (2020)
23. Mayakonda, A., Lin, D.C., Assenov, Y., Plass, C., Koeffler, H.P.: Maftools: efficient and comprehensive analysis of somatic variants in cancer. Genome Res. **28**(11), 1747–1756 (2018)
24. Meldrum, C., Doyle, M.A., Tothill, R.W.: Next-generation sequencing for cancer diagnostics: a practical perspective. Clin. Biochem. Rev. **32**(4), 177 (2011)
25. Nair, S.V., Madhulaxmi, T.G., Ankathil, R.: Next-generation sequencing in cancer. J. Maxillofac. Oral Surg. **20**, 340–344 (2021)
26. Pearl, J., et al.: Causality: Models, reasoning and inference. Cambridge, UK: Cambridge University Press **19**(2), 3 (2000)
27. Ramos, R., Cutigi, J., Ferreira, C., Evangelista, A., Simao, A.: Analyzing different cancer mutation data sets from breast invasive carcinoma (BRCA), lung adenocarcinoma (LUAD), and prostate adenocarcinoma (PRAD). In: Anais do XX Simpósio Brasileiro de Computação Aplicada à Saúde, pp. 37–48. SBC (2020)
28. Ramos, R.H., Cutigi, J.F., de Oliveira Lage Ferreira, C., Simao, A.: Topological characterization of cancer driver genes using reactome super pathways networks. In: Stadler, P.F., Walter, M.E.M.T., Hernandez-Rosales, M., Brigido, M.M. (eds.) BSB 2021. LNCS, vol. 13063, pp. 26–37. Springer, Cham (2021). https://doi.org/10.1007/978-3-030-91814-9_3
29. Ramsey, J., Andrews, B.: Py-tetrad and RPy-tetrad: a new python interface with r support for tetrad causal search. In: Causal Analysis Workshop Series, pp. 40–51. PMLR (2023)
30. Reyna, M.A., Leiserson, M.D., Raphael, B.J.: Hierarchical HotNet: identifying hierarchies of altered subnetworks. Bioinformatics **34**(17), i972–i980 (2018)
31. Rondeau, E., Larmonier, N., Pradeu, T., Bikfalvi, A.: Characterizing causality in cancer. Elife **8**, e53755 (2019)
32. Salzberg, S.L.: Open questions: how many genes do we have? BMC Biol. **16**(1), 94 (2018)
33. Squires, C., Uhler, C.: Causal structure learning: a combinatorial perspective. Found. Comput. Math. **23**(5), 1781–1815 (2023)
34. Tamborero, D., et al.: Comprehensive identification of mutational cancer driver genes across 12 tumor types. Sci. Rep. **3**(1), 2650 (2013)
35. Wu, S., Zhu, W., Thompson, P., Hannun, Y.A.: Evaluating intrinsic and non-intrinsic cancer risk factors. Nat. Commun. **9**(1), 3490 (2018)
36. Zhao, K., Rhee, S.Y.: Interpreting omics data with pathway enrichment analysis. Trends Genet. **39**(4), 308–319 (2023)

Deciphering Structural Selective Constraints: A Comparative Evolutionary Analysis of RNA Hairpin Structures

Alexey N. Spiridonov and Svetlana A. Shabalina(✉)

National Center for Biotechnology Information, National Library of Medicine, National Institutes of Health, Bethesda, USA
shabalin@ncbi.nlm.nih.gov

Abstract. RNA molecules often rely on specific, stable secondary structures to fulfill their functions. Examples include microRNA (miRNA) precursors in eukaryotes and terminator signals in prokaryotes, which require stable single-stem hairpin substrates to interact with proteins or complexes. By analyzing the sequence similarity and folding patterns of highly conserved hairpin structures, we identify critical signals and features indicative of selective constraints on RNA structure formation. We classify fixed mutations according to various modes of RNA selection pressure. In this study, we introduce a comparative evolutionary approach to quantify selective pressure on RNA sequences pairing in hairpins based on their structural requirements. This approach compares substitution patterns between natural and control sequences designed to nullify some structural selection while minimizing sequence changes. We evaluate evidence of structural selective constraint by quantifying the destabilization of hairpin structures in its absence. Our analysis reveals significantly higher rates of double compensatory base-pair substitutions in functional hairpins than expected. Additionally, we observe significant differences in the frequency of structure-preserving and structure-destroying single substitutions in pre-miRNAs compared to randomized sequences or a simple theoretical model. This approach also enables the quantification of structural selective constraints in stable elements of long noncoding RNAs and mRNAs in different organisms, where such constraints are prominent.

Keywords: RNA Folding · Structural Selective Constraint · RNA Hairpin Evolution · Compensatory Mutations

1 Introduction

RNA folding on a precise structure plays a crucial role for different molecules of RNA, many of which have a variety of functions in biological systems. The fraction of fixations driven by compensatory evolution and positive selection of substantially advantageous mutations remains uncertain for non-coding RNAs (ncRNAs) with conserved and stable secondary structures despite the decades of debates about the importance of these forces. It was shown earlier that mutations in stems of rRNAs and tRNAs are generally

R. Gori et al. (Eds.): CMSB 2024, LNBI 14971, pp. 196–208, 2024.
https://doi.org/10.1007/978-3-031-71671-3_14

accompanied by compensating mutations that maintain their base pairing [1–3]. The rates of substitution at interacting sites are decreased relative to those at independent sites in RNA secondary structures [4, 5]. However, a few exceptions to Kimura's compensatory neutral model were described [6]. In this model, deleterious and compensatory mutations can arise and occasionally drift to fixation as a pair within a single genome [7]. Compensatory mutations can be fixed at an appreciable rate even if their individual deleterious effects are large, although genomes containing single deleterious mutations are not likely to become fixed in this process. Application of Kimura's model of compensatory evolution at interacting sites to ten local stable RNA secondary structures showed that the ratio of transition to transversion substitutions [8] at interacting sites should be equal to the square of the ratio at independent sites. In fact, the difference in substitution rates at interacting and independent sites has been used to predict RNA secondary structures [9, 10].

miRNAs are one abundant class of small ncRNAs that regulate translation in eukaryotes. miRNAs must fold to single-stem hairpins, and compensatory evolution might play a role in the evolution of miRNA stem regions. Thus, miRNA hairpins are especially simple models for studying the evolutionary effects of such structural constraints. The metazoan miRNA pathway derived from a basal RNA-silencing pathway named as RNA interference (RNAi), which appears to have been present in the last common ancestor of eukaryotes and continues to defend against viruses and transposons [11]. The function and processing of miRNAs depends on complex intra- and inter-molecular RNA-RNA interactions. In both animals and plants, the mechanism through which miRNAs are generated is dictated by the highly conserved secondary structures of their transcripts that form stable imperfectly paired hairpins. These hairpins are processed by RNAi machinery into miRNA precursor duplexes [12, 13]. However, the miRNA hairpins in animals have a higher level of structural conservation between species than plant pre-miRNA hairpin. Thus, structurally conserved hairpins with allowed juxtapositions likely provide flexibility for different mechanisms of miRNA compensatory evolution in vertebrates [11]. In this paper, we concentrate on the structural selective pressure which maintains functionally optimal and stable RNA secondary structures in vertebrates. We present a new classification of distinct types of mutations allowing us to discuss the signatures of different modes of selection in non-coding structurally conserved RNAs. We also quantify the amount, and type of selective pressure resulting from the need to maintain a functional RNA secondary structure. Our results are in good agreement with recently published data on signatures of purifying and local positive selection in human miRNAs as well as with estimation of evolutionary rates [14, 15]. The presented mutation/substitution statistics from diverse species lead us to conclude that selective pressure on RNA folding formation in miRNAs significantly increases the observed levels of RNA selective constraint compared to random controls of miRNA-like sequences. Our approach to estimating selective constraint for miRNAs genes considers stable hairpins and RNA secondary structure requirements. Consequently, this technique might be helpful not only for estimation of structural selective constraint (SSC) for small RNAs, but also for any RNA molecules with stable structural elements.

2 Comparing Both RNA Sequences and Structures: Proposed Approach

2.1 Computational Methods: Aligning and Folding

To estimate the selective pressure on RNAs arising from structural constraints of stable hairpins or stem-loop structures, both sequence and structure conservation must be considered. This involves simultaneously aligning both the sequences and the hairpin structures of interest in the approach proposed in this study. For example, in the case of pre- miRNA hairpins, we analyzed several organisms all together (see below: Data for application) but focused on triple alignments of the "reference-other-outgroup" sequences of the analyzed genomes to assess structural selective constraints (Fig. 1A). Human and zebrafish were used as reference and outgroup genomes, respectively, with zebrafish being the natural outgroup when applied to pre-miRNA hairpins, since all "other" members are comprised of the tetrapod's clade. Thus, such triple alignments allow us to measure the selective constraint from humans to every other organism in the set.

We created triple alignments and used them for mutations statistics, but computed consensus folding only after omitting the outgroup:

i) Triple alignments were created for "reference" vs "other" vs "outgroup" sequences for all genomes analyzed (see Fig. 1A). For application to pre-miRNA hairpins, all orthologous sequences were retrieved from the UCSC Genome Browser using LiftOver [16] and validated by the Infernal program [17] against the Rfam database [18]. We used the Muscle program [19] to make the triple alignments "reference-other-outgroup" for all "other" organisms in the analyzed sets.
ii) The outgroup sequence was dropped and the secondary structures of two-fold alignments as well as three-fold alignments together with "outgroup" as a control were computed. We used the RNAalifold program from the Vienna RNA Package [20] to simultaneously fold pairs of orthologous sequences. The parameter "-p" was used to obtain a matrix of pairing probability. Positions that were paired with a probability of at least 50% were considered to be paired (numbers of miRNAs with known paired positions were used as a control set for the parameters selection and validation). If two such pairings conflicted each other, we took the one with the higher probability. For overlapping pairings or duplexes (for example, pseudoknots), we discarded the one that was not part of the main stem. To determine the main stem, we looked for coordinates $a < b$, such that $b < pair(b) < pair(a)$, which first maximizes the number of pairings from (a, b) to (pair(b)-pair(a)), and then minimize b-a.
iii) The pattern of substitutions in the folding and alignment was analyzed using custom software.

The reasons for performing these calculations were as follows:

i) Applying RNAalifold to multiple alignments produces rather low-quality structures, so we do only two-fold alignments.
ii) A single outgroup is useful, whereas a complete multiple alignment is too easily disrupted by low-quality sequences and too difficult to analyze.

iii) We fixed a "reference" and an "outgroup", particularly, human and zebrafish, for all alignments in pre-miRNA application, because their sequences were "well annotated and well known" and sufficiently divergent.

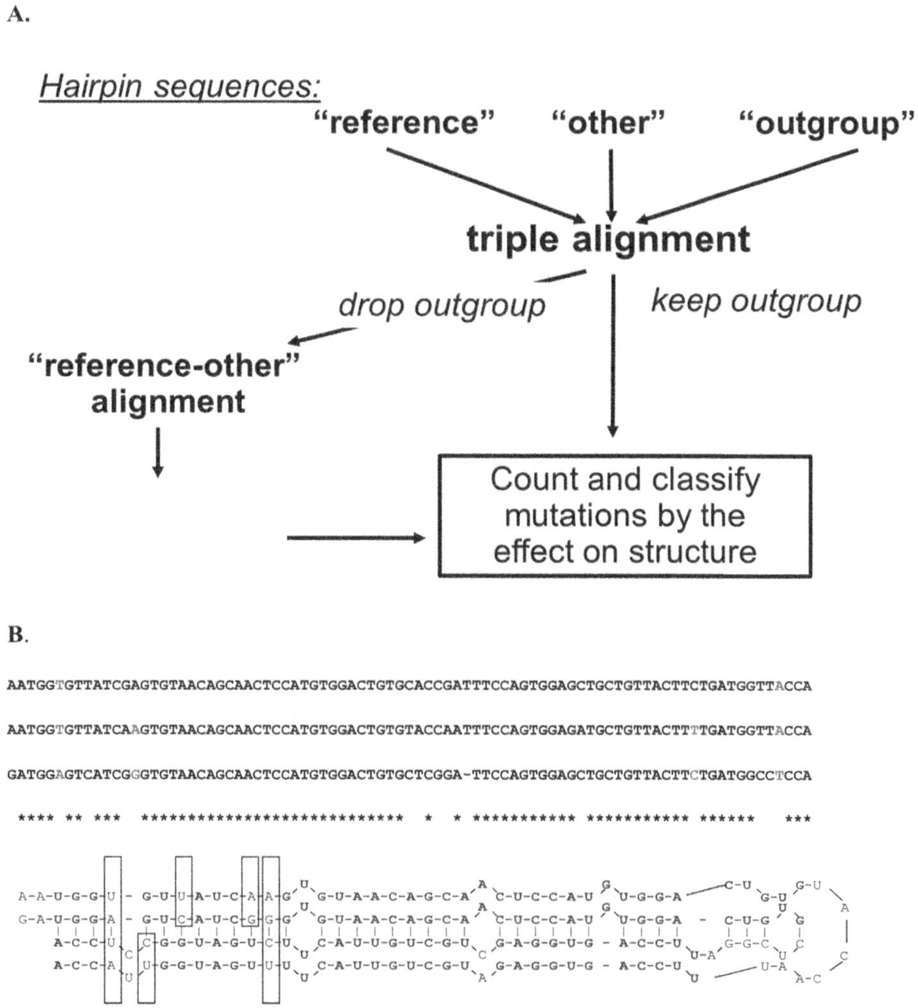

Fig. 1. Comparing both RNA hairpin sequences and structures: a schematic presentation of our triple alignment and consensus folding procedure. **A.** We excluded the zebrafish outgroup from the consensus structure computation, since our trial of 3-organism structure inference gave less biologically sensible folding, without offering a robust improvement in how we understood the evolutionary history of the structure. **B.** Comparative analysis of sequences (upper panel) and structures (low panel) for estimating selective constraint in hairpins. Compensatory mutations are shown in red.

2.2 Data for Application

The miRNA database miRBase [21] was used to create a manually validated set of alignments for triple orthologs—human-"others"-zebrafish. We analyzed pre-miRNA sequences from 12 organisms: human, mouse, zebrafish, cow, horse, cat, dog, chicken, platypus, lizard, opossum, and frog; in total 1119 alignments of the three orthologs were created, using between 59 and 213 sequences per organism. For example, 213 human-mouse-zebrafish orthologs were considered for mouse and curated using synteny. We carefully selected the correct ortholog from a set of recent duplicates and omitted sequences for which there was no clearly correct choice. We used human miRNA coordinates from miRbase using UCSC LiftOver to obtain coordinates for 10 other organisms. We then extracted the corresponding miRNA sequences using the UCSC Genome Browser [16]. The sets of orthologs were also validated against the Rfam database [18].

2.3 Spontaneous Substitutions

There are two important kinds of mutation rates: i) spontaneous mutation rates – DNA replication and repair errors; ii) post-selection mutation rates – selection acts as a filter, making some of the spontaneous events more or less likely to persist until the present day. If selection were absent, the current mutation pattern would reflect only the rates of spontaneous mutation. A simplistic model of spontaneous mutations was: i) the probability of a mutation at any site is independent of the current sequence, the nucleotide position, or other mutations; ii) the probability of a substitution is low, and all three possible replacements were equally likely. We analyzed only substitutions, because deletions and insertions are rare in the data to which the proposed approach was applied.

2.4 Control Models

Figure 2 shows how a control was made from the alignment of the three-organisms. We did not change the sequence of the "outgroup" (zebrafish) and made a heuristic ancestral alignment by taking a random half of the divergent nucleotides from the "reference" (human), and the other half from the "other" sequences. We computed the distribution of mutations going from the ancestor to the "reference"; in this case, it is: C- > T:1.

We introduced the exact same mutations at random positions and perform the same operation for the "other" to get a control alignment. Insertions and deletions were similarly shuffled, but this barely changed the results, so we omit the details. We split the RNA hairpins into stem and non-stem parts using the computed secondary structure and applied the control separately to these two parts. We also tried an alternative control preserving the divergent nucleotides, content, and the substitution distribution exactly, using an algorithm lacking the symmetry between the "reference" and the "other".

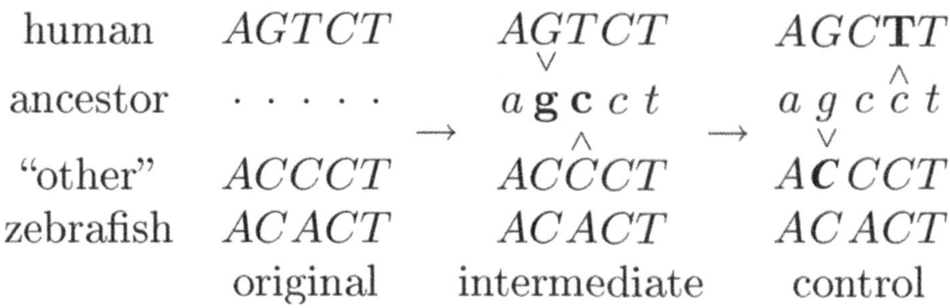

Fig. 2. Creating randomized control sequences from a three-organism alignment. First, we make up a fake ancestor sequence, by randomly choosing divergent nucleotides from either the human or the "other" organism. The zebrafish outgroup is left unchanged. Then, to make the controls, we start with the fake ancestor, and apply exactly the same substitutions (human: c → T, "other": g → C). To destroy position-dependent effects, we mutate random fake-ancestral positions containing the same source nucleotides.

2.5 Substitutions that Preserve Secondary Structure

A structure-preserving mutation is one or possibly two nucleotide substitutions, which allow the secondary structure to remain paired, as it was before the mutation.

How did we detect and measure these mutations? We simultaneously folded a 2-organism (later, a 3-organism) alignment. In a tight stem, even unpaired positions typically have an unambiguous "partner" on the complementary strand. Figure 1B shows what can happen because of substitutions at two nucleotides that are thus juxtaposed by the secondary structure we were interested in the two cases where the position is paired in both organisms, but with different nucleotides. This can lead to an estimate of "structural selective constraint" (SSC) for miRNAs.

Firstly, consider the case when both juxtaposed positions differ between the organisms (presumably due to substitutions), but are bound by secondary structure in both. This could be interpreted as a compensatory mutation: the exact nucleotides are not important to an organism's fitness, but their structural bond (pairing) is. A substitution that breaks the pairing is thus a deleterious mutation. So, it creates selective pressure to restore the pairing, either via the reverse mutation, or via a compensatory mutation in the juxtaposed nucleotide.

Our only "witnesses" of such compensatory activity are the "0 match, 2 Paired" substitutions (see Fig. 1B and Fig. 3). However, we must demonstrate that there is pressure to compensate: it is possible that all these double - substitutions are accounted for by chance. To argue the contrary, we can estimate the number of "0 match, 2 paired" substitutions without selective pressure. We can do this in two ways: via a Monte-Carlo simulation of mutations with all parameters taken from real folding of alignments, and with a slightly naive closed-form theoretical analysis. We were also interested in single-nucleotide substitutions that preserve pairing. Again, we have a single "witness" of this type of mutation - the "1 match, 2 paired" case from Fig. 3. As for the compensatory case, we can estimate the number of such mutations in the absence of selective pressure. A higher-than-chance rate of these "witnesses" also demonstrates SSC.

Case		Organism 1	Paired?	Organism 2
		No mutations		
2	match	A	↔	T
2	paired	A	↔	T
2	match	A		G
0	paired	A		G
		Potentially compensatory mutations		
0	match	A	↔	C
2	paired	T	↔	G
		Structure-preserving mutation		
1	match	T	↔	G
2	paired	C	↔	G
		Non-compensatory mutation(s)		
< 2	match	C		C
0	paired	A		G
any	match	A	↔	T
1	paired	C		T

Fig. 3. All the possible cases for two juxtaposed nucleotides in two aligned organisms. Every two rows represent one case. Each case considers two positions in two organisms (rows 1 and two of the case). We count the number of matches between organisms (0, 1, or 2). Then we count the number of paired positions in the two organisms (again, 0, 1, or 2).

3 Results: Application to Pre-miRNA Hairpins

3.1 Spontaneous vs Observed Substitutions: Selective Pressure in RNA Hairpins

We compared the divergence rates in miRNAs and in pseudogenes and showed that miRNAs are under selective constraint (see Table 1). However, we cannot say how much of it is structural, because a reduction in mutation rates could have other causes: *i*) target selection – a miRNA regulated numerous targets, which considerably constrains mutations in the seed positions 1–8 and likely elsewhere [11, 15]; or *ii*) protein interactions during the miRNA biogenesis (Dicer, Microprocessor and cofactors) can also constrain the pre-miRNA sequences [13, 14]. Considering that the portion of the substitutions located in seed regions of mature miRNAs is extremely low in our subset of data (less than 0.01), we can disregard target selection from our consideration in the analysis. We also did not include in the analysis overhang nucleotides that can be involved in miRNA biogenesis.

We computed the number of visible substitutions/nucleotide/10^9 years. We ignored indels and complex mutations because those are rare: 10×–25× less frequent [22]. We used years rather than generations because average generation length is difficult to establish (e.g., even mouse/human) and generation length is probably less important

Table 1. A comparison of the divergence rates in miRNAs with the spontaneous divergence rate measured pseudogenes.

	subs/pos x 10^9	*u/v*
spontaneous	4 [24]	3.5–5.2 [22, 24]
miRNA stem	0.1 ± 0.002	5 ± 0.47
miRNA non-stem	0.5 ± 0.006	2 ± 0.15
miRNA whole	0.2 ± 0.003	3 ± 0.23

than the frequency of division of germline cells, which varies far less in our dataset than generation length [22]. There are several estimates of largely spontaneous mutations; these estimates vary mainly between 0.9 [23] and ~ 4 according to Li [24, p.184; reviewed in 22]. In order to get our estimate and to compare against the rate of spontaneous mutations, we analyzed the distinct structural elements of 1119 triplets of pre-miRNA sequences (917 CpG - stems, 434 CpG - loops) separately, and our results are following: i) stems are 0.19 ± 0.01, loops are 0.72 ± 0.02, ii) non-CpG: 0.15 ± 0.01 (stems) and 0.63 ± 0.02 (loops), iii) CpG: 0.85 ± 0.04 (stem) and 2.14 ± 0.1(loops). A universal trend was found for all categories—on average, the rate was over three times higher for stems than for loops.

3.2 Control Versus Real MiRNAs

Positional dependencies between mutations, average ΔG and pairing were estimated). Our model's parameter set is the distribution of divergent nucleotides: 16 counts of the form $C_{\text{human -" other"}}$. As a result, our control preserves the following quantities quite well: *i)* the number of divergent nucleotides (approximately); *ii)* nucleotide content of all aligned sequences (exactly); *iii)* di-, tri-, *k*-nucleotide content is little changed, because miRNAs are highly conserved; the distribution $C_{\text{human-"other"}}$ (approximately) – in particular, transition/transversion ratio (*u/v*). The model is almost as detailed as it can be without any positional information. That is exactly what we need to detect SSC (see Table 2).

We also tried a control, that mixed mutations between stem and non-stem, and it also supported our findings. The structural effects are considerably stronger than our more conservative estimate; mutation rates, *u/v*, and other mutation statistics are practically averaged over the two areas. Correspondingly, it also mixes together the effects of non-structural and structural constraint and is not helpful for isolating the latter. However, one result was suggestive: the fraction of compensatory mutations in this control is ¼ of the original, rather than 1/3 in the proper one – a drop of 25%. At the same time, *u/v* drops by 25% because of stem-non-stem mixing. This disparity corresponds to the prediction of a simple theoretical model estimating compensatory mutations in terms of *u/v*. So, the elevation of *u/v* in stem compared to non-stem may be due to structural selection as well.

Table 2. Positional dependencies between mutations (structural selective constraint) affect ΔG and pairing, even though our random control sequences exhibit the same substitution rate and transition/transversion ratio.

Type	subs/pos x 10^9	*u/v*	average	% paired
miRNA stem	0.1 ± 0.002	5 ± 0.47		81
miRNA non-stem	0.5 ± 0.006	2 ± 0.15		
miRNA whole	0.2 ± 0.003	3 ± 0.23	−37.2	
control stem	0.1 ± 0.002	5 ± 0.67		78
control non-stem	0.5 ± 0.005	2 ± 0.17		
control whole	0.2 ± 0.002	3 ± 0.21	−34.1	

3.3 Structural Selective Constraint

Our comparison of control versus real miRNAs showed that no mutation dependences beyond CpG and neighbor effects have been found in spontaneous mutations (there are no interactions at distances of more than 20 nucleotides). We locally rearranged mutations, while preserving as much of their statistics as possible – the intent is to destroy only the positional dependences. We also showed that with the dependences gone, structure degrades. This is strong evidence that the mutations in the original positions were not eliminated because they either preserved or enhanced the structure.

How much they preserved the structure is called the "position-dependent portion of the SSC". There are two numbers: $\Delta\Delta G$/nucleotide and Δ(% paired) that should be comparable between different alignments (danio – others) between the same organisms (compared to human; see Table 3) and could be normalized for comparison between different organism pairs (compared to human). For each of the 10 species, we computed several measures of SSC relative to human. We found that, depending on the species: *i*) up to 5% of the nucleotides in pairings are under SSC, *ii*) per stem nucleotide, SSC affects the structure's minimum free energy (MFE) by −0.02 to–0.08, *iii*) per local constrained pairing from *i*), SSC affects the MFE by −1.3 to–2.2 kcal/mol.

3.4 Classifying Substitutions by Their Structural Effects

In this study, we suggested a new classification of mutations, based on the systematic analysis of the mechanisms of intramolecular compensation in the evolution of vertebrate pre-miRNAs, and showed how it is related to different modes of selections. Looking at a human-other-outgroup triple alignment together with the human-other consensus model of a miRNA hairpin stem, we studied the positions (used as the total in our percentages) where: *i*) the human nucleotide differed from "other", *ii*) at least one of human or "other" had a pairing to the opposite side of the hairpin. We classified such mutation events by their structural effects: *compensatory, neutral, outgroup diverged, structure-creating, structure-destroying* (see definition in Table 4 legends). A careful reading of the above cases will show that our categories are disjoint, and that together they cover all the

Table 3. A comparison of consensus folding statistics between natural and control sequences for different "other" organisms*.

	hor	*cow*	*dog*	*cat*	*mus*	*opo*	*pla*	*chi*	*liz*	*fro*
$-100 * \Delta\Delta G$	1	1	2	2	3	3	4	4	5	7
Δ(% paired)	1	1	1	2	2	2	4	4	4	4

* *As the organisms become more evolutionarily distant, we see that the per-nucleotide free energy increase from natural to control sequences gets higher, corresponding to the fact that the structure of the sequence was preserved despite accumulated point mutations. Similarly, applying the randomized control causes a larger loss of paired positions in more divergent organisms. If one were to normalize these statistics by the evolutionarily distance to the human, they could be construed as metrics of structural selective constraint.*

positions we set out to classify. In other words, the counts of the 5 categories always add up to our "total" defined above.

The resulting mutation counts, compared to randomized controls and a theoretical model, showed very clear evidence of the SSC. Our most conservative randomized control preserved the original alignment almost exactly - including nucleotide content, dinucleotide content, number of mismatches, and the kinds of mismatches (e.g., human A - > others T). Only the positions of the mismatches were shuffled, with the constraint that miRNA stem mismatches stay in the stem, and the rest stay out of the stem. We also compared our results with a simple theoretical control, parametrized by the mutation rate, the transition/transversion ratio, and the nucleotide contents. We used the outgroup to classify stem mismatches by effect on structure (see Table 4). The two major differences are in *compensatory* mutations and in *structure-destroying* mutations. This suggests that the 3x reduction in *compensatory* mutations, and the corresponding 3x increase in *structure-destroying* mutations reflect many positions that were held paired due to structural selective constraint. The latter is evidence of negative selection: deleterious mutations are eliminated. However, if that was all that was going on, the remaining categories would increase by roughly the same percentage. That is not the case: *compensatory* mutations pick up most of the weight, which means that these mutations are being fixed at a faster rate. The results for the alternative control were extremely close (data are not shown).

Our comparative analysis showed that double compensatory base-pair substitution rates are approximately three times higher in pre-miRNA hairpins than expected. We found that the frequency of *structure-preserving* single substitutions and the ratio between *structure-destroying* and *structure-preserving* substitutions in pre-miRNAs are significantly higher than for randomized sequences and theoretically estimated. We demonstrate that secondary structure is preserved through a combination of purifying selection and compensatory evolution; in other words, it is maintained by both positive and negative selection. A particularly striking example of epistatic interactions and positive selection is the ~ 3-fold excess of substitutions that affected both strands of a stem pairing but preserving the pairing. Thus, substitution events in RNA sequences are not independent, and in fact exhibit middle-range (30–60nt +) dependencies caused by SSC.

Based on these results, we updated our previously published models for estimating selective constraints based only on sequence similarity alone [25–28], and our estimates showed that the fractions of selectively constrained nucleotides may be higher than previously estimated by 5–10% in non-coding regions of genomes. Our method is readily extended to propose new models to replace the previous theoretical models that do not quantitatively fit the data (in progress). This approach can be applied to other RNA sequences and should permit the calculation of structural selective constraint in any RNA molecule with cross-species orthologs. The signals of positive and negative selection could also potentially be used for detecting novel regions of evolutionarily stable structures.

Table 4. Classification and statistics of stem substitutions by different effect on hairpin structure*. Unlike natural sequences, the control's distribution of substitutions is largely blind to their position within the structure.

	Neutral	Compensatory	Creating	Destroying	Outgroup diverged
miRNA	47%	20%	8%	13%	12%
control	44%	6%	8%	35%	7%

**Classification of stem mismatches/substitutions by different effect on hairpin structure:*
neutral: an alignment position with a mutation, where both human and "other" are paired to the same location containing the same nucleotide on the opposite side of the hairpin— without regard to the outgroup nucleotides,
compensatory: an alignment position with a mutation, where both human and "other" are paired to the same location on the opposite side of the hairpin, which also has a mutation — without regard to the outgroup nucleotides,
structure-creating: an alignment position where the outgroup agrees with either human or "other", and where the organism in agreement with the outgroup has no pairing, while the mutated organism is paired,
structure-destroying: an alignment position where the outgroup agrees with either human or "other", and where the organism in agreement with the outgroup has a pairing, while the mutated organism is unpaired,
outgroup diverged: an alignment position such that exactly one of human/"other" is paired, and where the outgroup nucleotide agrees neither with human, nor "other", meaning that we have no good guess of the direction of the mutation.

4 Discussion

We presented new classification of mutations/substitutions in pre-miRNA hairpins which allows distinguishing various modes of selection pressure on RNA secondary structures. Here we also quantified the selective pressure on the pre-miRNA sequences, which arises from their structural requirements. Why is it important to define quantitatively the selective pressure related to structure requirements and preservation? Although the determination of what is and what is not a miRNA may conceptually seem a relatively straightforward task, this has been proven not to be the case [13, 15]. With the size of the transcribed non-protein-coding portion of the human genome and the large amount of new canonical and non-canonical miRNA studies, it has become increasingly difficult to determine the actual repertoire of miRNAs [13, 15]. Thus, our new classification of substitutions and comparison of substitution patterns between natural sequences and control sequences designed to nullify some of the structural selection, while changing the sequence as little as possible, could be extremely helpful for determining the structural selection on miRNA-like hairpins. In this study, we also found evidence of SSC, quantifying the decrease in stability of canonical miRNA hairpin structures in its absence.

We showed that loops evolve much faster than stems and mature miRNAs evolve significantly slower than a complementary star strands. Evolutionary analysis of structural elements - paired versus unpaired - showed that transition/transversion bias might be helpful for genome-scale prediction of hybrid zones and stable local RNA secondary structures. Our results suggest that the elevation of transition - transversion ratio in stems compared to non-stem is due to structural selective pressure. Significant difference in transition - transversion bias in stem and loop structures is commonly related to the frequent appearance of G-U and U-G pairs (intermediates) in the paired regions that are under structural pressure. Thus, their density in paired structures is higher than in unpaired. Tillier and Collins [1] claimed that some compensatory (rRNA) mutations happened "simultaneously", rather than sequentially. We tested the hypothesis in our study and found no pronounced effects of simultaneous substitutions in pre-miRNA hairpins.

Different modes of selection in miRNA molecules appeared in mature miRNAs and their complementary strands. The level of substitutions is dramatically higher in the complementary strand compared to the mature miRNAs with enrichment by *structure-preserving* and *structure-destroying* ones. Thus, we can conclude that patterns of position-specific and strand-specific substitutions reflect structural selective pressure. We showed that pairing distance and timing/evolutionary distance between species are tightly related and comparison of pairing in closely related, and relatively distinct species could be used for the SSC estimation and evolutionary scenario reconstruction.

5 Disclosure of Interests.

The authors have no competing interests to declare that are relevant to the content of this article.

Acknowledgments. This work was supported by the Intramural Research programs of the National Library of Medicine, National Institutes of Health [to S.A.S.].

References

1. Tillier, E., Collins, R.: High apparent rate of simultaneous compensatory base-pair substitutions in ribosomal RNA. Genetics **148**, 1993–2002 (1998)
2. Wilke, C., et al.: Compensatory mutations cause excess of antagonistic epistasis in RNA secondary structure folding. BMC Evol. Biol. **3**, 3 (2003)

3. Kern, A., Kondrashov, F.: Mechanisms and convergence of compensatory evolution in mammalian mitochondrial tRNAs. Nat. Genet. **36**, 1207–1212 (2004)
4. Stephan, W.: The rate of compensatory evolution. Genetics **144**, 419–426 (1996)
5. Innan, H., Stephan, W.: Selection intensity against deleterious mutations in RNA secondary structures and rate of compensatory nucleotide substitutions. Genetics **159**, 389–399 (2001)
6. Wang, H., Hickey, D.: Evidence for strong selective constraint acting on the nucleotide composition of 16S ribosomal RNA genes. Nucleic Acids Res. **30**, 2501–2507 (2002)
7. Kimura, M.: The neutral theory of molecular evolution. New Scientis **1464**, 41–46 (1985)
8. Knies, J., et al.: Compensatory evolution in RNA secondary structures increases substitution ratevariation among sites. Mol. Biol. Evol. **25**, 1778–1787 (2008)
9. Muse, S.: Evolutionary analyses of DNA sequences subject to constraints of secondary structure. Genetics **139**, 1429–1439 (1995)
10. Pedersen, J., et al.: An evolutionary model for protein-coding regions with conserved RNA structure. Mol. Biol. Evol. **21**, 1913–1922 (2004)
11. Shabalina, S., Koonin, E.: Origins and evolution of eukaryotic RNA interference. Trends Ecol. Evol. **23**, 578–587 (2008)
12. Du, T., Zamore, P.: MicroPrimer: the biogenesis and function of microRNA. Development **132**, 4645–4652 (2005)
13. Bartel, D.: Metazoan microRNAs. Cell **173**, 20–51 (2018)
14. Quach, H., et al.: Signatures of purifying and local positive selection in human miRNAs. Am. J. Hum. Genet. **84**, 316–327 (2009)
15. Fromm, B., et al.: A uniform system for the annotation of vertebrate microRNA genes and the evolution of the human microRNAome. Annu. Rev. Genet. **49**, 213–242 (2015)
16. USCS genome browser homepage. https://genome.ucsc.edu/
17. Nawrocki, E.: Annotating functional RNAs in genomes using Infernal. Methods Mol. Biol. **1097**, 163–197 (2014)
18. Rfam Homepage. https://rfam.org
19. Edgar, R.: MUSCLE: multiple sequence alignment with high accuracy and high throughput. NucleicAcids Res. **32**, 1792–1797 (2004)
20. Bernhart, S.: RNAalifold: improved consensus structure prediction for RNA alignments. BMC Bioinform. **9**, 474 (2008)
21. Kozomara, A., Griffiths-Jones, S.: MiRBase: annotating high confidence microRNAs using deep sequencing data. Nucleic Acids Res. **42**, D68-73 (2014)
22. Kondrashov, A.: Crumbling genome: the impact of deleterious mutations on humans. Wiley Blackwell (2017)
23. Kondrashov, A.: Direct estimates of human per nucleotide mutation rates at 20 loci Causing Mendelian diseases. Hum. Mutat. **21**, 12–27 (2003)
24. Li, W.-H.: Molecular Evolution. Sinauer Associates Inc, Publishers (1997)
25. Shabalina, S., et al.: Selective constraint in intergenic regions of human and mouse genomes. Trends Genet. **17**(7), 373–376 (2001)
26. Kondrashov, A., Shabalina, S.: Classification of common conserved sequences in mammalian intergenic regions. Hum. Mol. Genet. **11**, 669–674 (2002)
27. Matveeva, O., Shabalina, S.: Intermolecular mRNA-rRNA hybridization and the distribution of potential interaction regions in murine 18S rRNA. Nucleic Acids Res. **21**, 1007–1011 (1993)
28. Shabalina, S., Kondrashov, A.: Pattern of selective constraint in C. elegans and C. briggsae genomes. Genet Res. **74**, 23–30 (1999)

How Much Do DNA and Protein Deep Embeddings Preserve Biological Information?

Matteo Tolloso[1(✉)], Silvia Giulia Galfrè[1], Arianna Pavone[2], Marco Podda[1], Alina Sîrbu[1(✉)], and Corrado Priami[1]

[1] Department of Computer Science, University of Pisa, Pisa, Italy
m.tolloso@studenti.unipi.it,
{silvia.galfre,marco.podda,alina.sirbu,corrado.priami}@unipi.it
[2] Department of Mathematics and Computer Science, University of Palermo, Palermo, Italy
ariannamaria.pavone@unipa.it

Abstract. Models that map DNA and protein sequences into deep embeddings have been recently developed. While their ability to improve prediction in downstream tasks has been demonstrated, clear advantages and disadvantages of embedding types, and different means of applying them, are not yet available. In this paper we compare five different models (one for DNA, four for proteins) and different embedding aggregation methods with respect to their ability to preserve evolutionary and functional information, using a hierarchical tree approach. Specifically, we introduce a novel procedure that builds hierarchical clustering trees to assess the relative position of sequences in the embedding latent space, compared to the phylogenetic and functional similarities between sequences. The methods are benchmarked on five different datasets from various organisms. The ESM protein language model and DNABert emerge as best performers in different settings.

Keywords: DNA embeddings · protein embeddings · language models · AlphaFold · phylogeny · functional annotations

1 Introduction

Machine learning and deep learning have revolutionized the natural language processing field (NLP) [21], with extraordinary results in solving problems like named entity recognition, translation, and question answering. Some fundamental ideas developed in NLP have been successfully applied to the biological language, composed of DNA, RNA and amino acid sequences. One of these fundamental ideas is the concept of word embedding [28], which represents words as dense and compact vectors whose position in space reflects their semantic relationships. Typically, embeddings are obtained through self-supervised training of deep learning architectures.

R. Gori et al. (Eds.): CMSB 2024, LNBI 14971, pp. 209–225, 2024.
https://doi.org/10.1007/978-3-031-71671-3_15

In the biological domain, several methods were introduced to embed DNA and protein sequences, including DNA or protein language models (LM). For DNA, a first LM is DNABert [18], which adapted the idea of Bidirectional Encoder Representations from Transformers (BERT) [8], initially applied to natural language, to the genomic DNA setting. This was trained on human DNA sequences randomly extracted from the human reference genome. Several models have been introduced by different authors for proteins. Prose [5] is a multi-layer bidirectional Long Short Term Memory (bi-LSTM) network, trained on more than 76 million proteins from UniRef and the Structural Classification of Proteins (SCOP) database. SeqVec [15] is an adaptation of the standard ELMo configuration [31] to work with protein sequences, trained on 33 million proteins of UniRef50 [37]. Evolutionary Scale Modeling (ESM2) [25] is based on an underlying BERT-style encoder transformer, trained on UniRef50 and UniRef90 for a total of over 60M proteins. Biological sequence embeddings extracted from these models have been applied to different downstream tasks, such as identification of transcription factor binding sites, splicing sites, protein structural similarity prediction, subcellular localisation of proteins, protein structure prediction and others, and have been demonstrated to be superior to existing algorithms [5,24]. For three-dimensional structure prediction, an important step ahead was made with the AlphaFold model [20], which combines physical properties of proteins with information from multiple sequence alignments to obtain better structures. The AlphaFold model can also be used to generate protein embeddings.

While the contribution of these models has been crucial for advancements in various biological problems, it is not clear what are the advantages and disadvantages of each approach, as each model has been applied independently on different downstream tasks, concentrating on the task itself rather than on a comparison with other possible embeddings. Furthermore, evaluation of the amount of biological information captured by embeddings is generally limited. It is well known that sequence similarity does not always correspond to functional similarity [22], and since embeddings are extracted from sequence only (except AlphaFold), a more complete evaluation could be useful, to understand on one hand how much evolutionary information (based on sequence), and on the other hand how much functional information is preserved by embeddings. Furthermore, application of the models on new sequences is not always straightforward. Typically, the atomic units of information processed by these models are tokens, such as k-mers or single aminoacids/nucleotides for biological sequences. The models output an embedding (a vector) for each token of the input sequence. So, for each new DNA or protein sequence, we end up with a matrix representation, where each row represents the embedding of one token. It is important to note that the vectorial representation of a token of the sequence is contextual rather than fixed, i.e. it depends on all other tokens in the sequence. This is intuitively beneficial when operating on natural language, because a word can assume different meanings in different contexts [33]. Similarly, in biological sequences an amino acid/nucleotide assumes a certain role depending on its surrounding context [32] (e.g. they interact with each other to form secondary or tertiary structures). Furthermore, since the size of the context that a single token can attend to is limited, sequences longer than the context size need to be split into sub-sequences to prevent the loss of long range dependencies. Therefore, to represent the entire sequence, a three-dimensional tensor is needed in order to account for the different sub-sequences. Since sequences are of different length, the tensor has a different size for each sequence. In order to feed these tensor representations to downstream models, we need to aggregate them into fixed-size representations. In general, the aggregation can be performed through averaging, summation, maximum, or dimensionality reduction (e.g. PCA). While averaging is by far the most used method [5], there is no strong rationale as to why it should be preferred. A similar issue exists in NLP, when obtaining document embeddings from word embeddings [23].

The main objective of this work is to provide a comprehensive comparison of DNA and protein embeddings. This includes a comparison of different models, i.e. DNABert, Prose, SeqVec, ESM and AlphaFold, and a comparison of different aggregation methods, i.e. *sum*, *average*, *maximum*, *PCA*. The evaluation is based on a novel method to evaluate the biological information preserved by these models, using five different datasets as benchmark. Specifically, we compare the hierarchical clustering tree derived from the embeddings with two baselines: one obtained using standard bioinformatics tools for phylogenetic analysis based on multiple sequence alignments and another one based on Gene Ontology annotations. This enables the comparison of relative positions of the sequences in the embedding space versus their phylogenetic and functional similarities. We introduce two indicators based on the Adjusted Rand Index [17]. The first will evaluate the amount of relevant evolutionary sequence information preserved by the embeddings, while the second will evaluate the functional information preserved.

2 Methods

The overall methodology of our study is described in Fig. 1. We start from a set of genetic sequences, and we generate their embeddings with different models and aggregation methods. Subsequently, to evaluate the performance of the embeddings, for each model, we calculate a matrix of distances between sequences in the embedding space. We perform agglomerative clustering using this distance matrix to obtain an embedding-based hierarchical tree of the original sequences. We compare the resulting tree with two baselines: (1) a phylogenetic tree obtained from the sequences using the standard bioinformatics approach with sequence-based distances, and (2) a hierarchical tree obtained from a distance matrix based on Gene Ontology annotations of the sequences. The comparison yields an indicator based on the adjusted rand index (ARI) that shows how similar the hierarchical trees are. The analysis was performed in Python and all scripts are available on GitHub at https://github.com/matteotolloso/BioEmbedding.

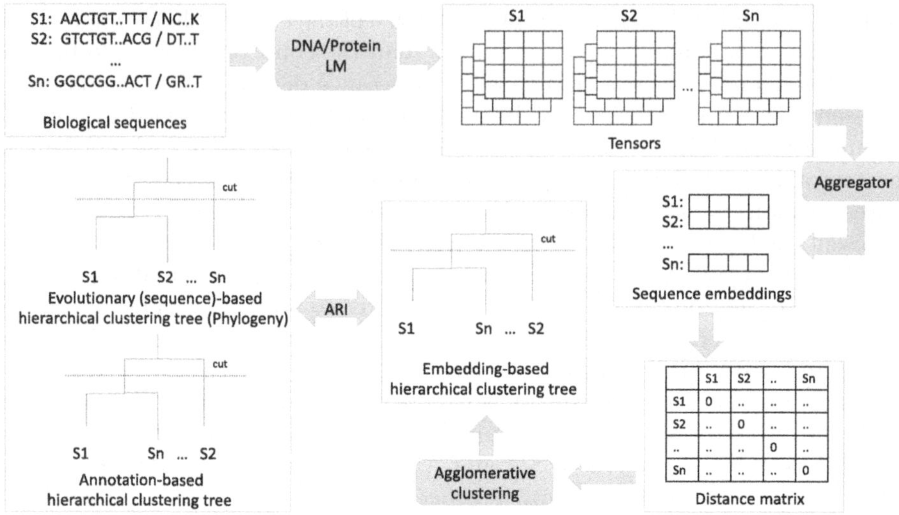

Fig. 1. Analysis pipeline.

2.1 Datasets

We selected five datasets for the analysis: *Hemoglobin*, *SATB*, *Mouse*, *N. meningitidis*, *Covid-19 WG*. These account for three different scenarios:

- The analysis of orthologous proteins over different species. In this case we opted for two datasets. *Hemoglobin* includes a set of genes from the Hemoglobin family and *SATB* includes the SATB1 and SATB2 which are two paralogue human DNA binding genes. Both datasets cover the genes from different species. In order to be able to compare DNA and protein embeddings, for each protein sequence we also extract the corresponding transcript.
- The analysis of the genome/proteome of one single species. We selected a mammal and a bacterium as case studies: *Mouse* and *N. meningitidis*, and downloaded a set of genes and corresponding proteins of these organisms.
- The analysis of multiple whole genomes/proteomes from the same species. In this case we opted for a virus: *Covid19 WG* (Whole Genome). DNA embeddings were obtained by analysing the whole genome sequence with our pipeline, while for protein embeddings we aggregated the embeddings for all individual proteins in the genome.

Data was exported by hand from the NCBI website and is available on GitHub together with the analysis code. Table 1 shows the number of sequences and the average sequence length (nucleotides) for each case study.

2.2 From Genetic Sequences to Embeddings

The DNA/protein sequences for each dataset were fed to the five pretrained models. Sequences larger than the model input were split into subsequences of

Table 1. Datasets used in the experiments.

Name	Description	Number of sequences	Average sequence length (DNA)
Hemoglobin	Hemoglobin for various organisms	74	689
SATB	SATB1 and SATB2 proteins/genes for various organisms	927	100146
Mouse	Genes/proteins from mouse	919	3045
N. Meningitidis	Genes/proteins from N. Meningitidis	903	788
Covid19 WG	Covid19 complete genomes	77	29831

Table 2. Models used in the experiments, their maximum input length and the dimension of the embedding produced.

Name	input length (subsequence)	embedding dimension
SeqVec [15]	1024	1024
DNABert [18]	512	768
Prose [5]	512	100
AlphaFold [20]	1024	384
ESM2 [25]	1024	1280

correct length. DNABert generated one embedding for each subsequence. For protein embeddings, the models yield one embedding per amino acid, multiplied by the number of subsequences. We tested four methods to join together the amino acid embeddings to produce a fixed-size embedding for the subsequence: *average*, *maximum*, *sum*, and principal component analysis (PCA). The same operator used to combine the amino acid embeddings was also used to combine the embeddings of the subsequences, obtaining thus a fixed-size representation for the entire sequence. The final size of the whole sequence representation is equal to the size of the embedding of one single token for each model, shown in Table 2.

In the rest of this subsection we briefly describe the main characteristics of the five models used in our experiments.

DNABert [18]**.** DNABert employs a Transformer, an attention-based architecture that achieves state-of-the-art performance in most natural language processing tasks. It uses the k-mer representation of the sequence with k=$\{3, 4, 5, 6\}$ as tokens for the model. The pretraining was performed with the token-masking technique on the human genome. The fine-tuning and the subsequent evaluation were performed on the tasks of: (i) predicting the gene promoter, (ii) identifying transcription factor binding sites, (iii) predicting splice sites, (iv) identifying functional genetic variants. In this paper we used the pretrained model with k=6 since it achieved the best performance.

SeqVec [15]. This model adapted the standard ELMo configuration [31] to work with protein sequences, and was tested on four different prediction tasks: secondary structure, intrinsic disorder, subcellular localization, and membrane protein classification. With respect to ELMo, the number of tokens and the unroll steps were modified. The network is composed of 1 CharCNN and 2 LSTM-Layers. Given a protein sequence of arbitrary length, it returns 3072 features for each residue derived by concatenating the outputs of the three layers of ELMo, each describing a token with a vector of length 1024. In order to obtain a smaller representation for each amino acid we computed the mean of the three layers (as suggested in the official repository). Given the architecture of the ELMo, these representations are context-dependent.

Prose [5]. The model structure is a multi-layer bidirectional Long Short Term Memory (bi-LSTM) network. Following the intuition that some aspects of protein structure and semantics cannot be discovered by statistical sequence models alone, the authors came up with the idea of multi-task learning with structural supervision. The learning tasks are: (i) classical masked language modeling task, (ii) residue-residue contact prediction, (iii) structural similarity prediction. This model showed its validity outperforming existing approaches in both transmembrane position labeling and phenotype prediction of sequence variants.

Alphafold [20]. This is the only approach that does not rely on a language model but is a combination of bioinformatics and physical approaches. It is based on large datasets of protein sequences that are similar enough to be aligned with high confidence but contain enough divergence to confidently infer statistical couplings between positions. It consists of two modules, the Evoformer module and the structure module. The Evoformer builds separate multiple sequence alignment (MSA) and residue-pairwise embedding spaces. The model was validated on the protein structure prediction task where it outperformed the previous approaches by a large margin. As it is structured, this system is not able to learn patterns across large-scale databases of possibly unrelated proteins [5]. In our experiments we used the "single" representation of the sequence, without the MSA and the templates due to limited computational resources.

ESM2 [25]. The underlying architecture is a BERT style encoder transformer [8] with modifications in the number of layers, number of attention heads, hidden size, and feed-forward hidden size. A notable further improvement, among the others, is the use of Rotary Position Embedding [36] that allows it to generalize beyond the context window it is trained on. The model is evaluated on several benchmarks for protein structure prediction. The main advantages compared to AlphaFold are the removal of the need for MSA and an increasing up to two orders of magnitude of the speed of the prediction pipeline. We used the version with 33 layers and 650B parameters due to limited computational resources, even if bigger (and presumably better) models exist.

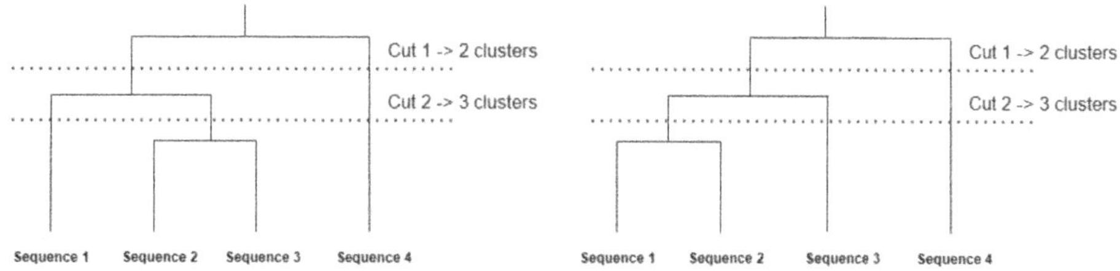

Fig. 2. Comparison of hierarchical clustering trees: we cut the trees at all possible levels and obtain an ARI value for each.

2.3 Evaluation of Embeddings

We study whether the placement of sequence embeddings in the embedding space corresponds with properties of the sequences.

Preservation of Evolutionary Information. Firstly, we are interested in how much evolutionary information is preserved. Therefore, we build a hierarchical clustering tree from the distance matrix in the embedding space, and compare it with the phylogenetic tree obtained using standard bioinformatics tools. The embedding-based distance matrix was obtained by computing the euclidean distance between embeddings, after an initial PCA phase to reduce dimensionality. We tested 10 and 50 principal components, as well as the option of no PCA analysis. We also tested the cosine distance between embeddings, however results were similar so they are not shown. The hierarchical tree was built using agglomerative clustering with average linkage.

The phylogenetic tree was obtained directly from the sequence information, using multiple sequence alignment followed by phylogenetic tree construction. Multiple sequence alignment was performed with Clustal Omega [34], which provides an evolutionary-related distance between sequences [35]. The phylogenetic tree, instead, was built using the UPGMA algorithm, a bottom up hierarchical clustering algorithm specific to phylogenetic analysis [9], and very similar to agglomerative clustering with average linkage.

We adopt an iterative approach to compare the hierarchical trees (see Fig. 2). We cut the tree at all possible levels, and we compute, at each level, the similarity between the clusters obtained, using the adjusted rand index (ARI) [17], a measure of similarity previously shown to be superior to others [29]. We then study the series of ARI values, and compare them across embedding types.

The comparison could have been done directly on the distance matrices, using for example, the DISTATIS [1] or the Mantel test. However, the method we use allows us to compare the structure of the two distance spaces at different resolutions, making also considerations related to the actual biological structure of the data.

Biological Annotations. Secondly, we want to understand how higher level annotations such as Gene Ontology (GO) [2,39] are distributed in the embedding space. The Gene Ontology is a collection of standardised annotations of genes and gene products (proteins). In this vocabulary, each gene/protein is associated a set of terms that describes our knowledge of the sequence, concerning molecular function, cellular component and biological process.

For uniformity with the evolutionary analysis, we decided to produce another distance matrix based on annotations, and a corresponding annotation-based tree, and compare this with the embedding-based tree using the ARI, as before. Given the sets A and B of annotations of two sequences, we derived a distance starting from a similarity metric: the overlapping coefficient [42]. In details:

$$d = 1 - oc(A, B) = 1 - \frac{|A \cap B|}{\min\{|A|, |B|\}}$$

This distance varies between 0 and 1, and reaches 1 when one set is a subset of the other.

We build an annotation-based tree using the same agglomerative clustering algorithm on the new distance matrix, and compare it with the embedding-based tree using the same procedure we employed for the evolutionary analysis. The ARI series obtained shows how well annotations are distributed in the embedding space at different resolutions.

3 Results

The five models for embedding DNA/protein sequences and the four methods to aggregate embeddings of tokens/subsequences were combined and tested on the five datasets, using PCA before computing distances for the agglomerative clustering. The resulting trees for all datasets were compared with phylogenetic trees extracted from sequences to quantify the evolutionary information preserved by embeddings. For two datasets, Mouse and N. Meningitidis, the trees were also compared with trees obtained from GO annotations. The annotation analysis was performed only for the Mouse and N. Meningitidis datasets because these include a set of different genes/proteins from the same organism, which have different functions/locations identifiable through GO annotations. The other 3 datasets refer either to only one gene/protein family or to entire genomes, so GO annotations do not apply.

3.1 Evolutionary Information

Our evaluation results in a series of ARI values for each method, which tells us how similar clusters are when the cutting the two hierarchical clustering trees at different levels. Tables 3, 4 and 5 show the 90th percentile of the ARI values obtained for each method and dataset. This gives an indication on the amount of evolutionary information preserved by each method. For each dataset, Fig. 3 shows the actual ARI series obtained for one model for each embedding type. The

Table 3. SATB & Hemoglobin datasets: ARI 90th percentile when comparing the evolutionary and embedding-based trees.

aggre-gator	dimen-sional PCA	SATB					Hemoglobin				
		dna-bert	seqvec	prose	alpha-fold	esm2	dna-bert	seqvec	prose	alpha-fold	esm2
pca	10	0.381	0.319	0.154	0.275	0.278	0.673	0.849	0.848	**0.898**	0.889
	50	0.421	0.265	0.145	0.204	0.226	0.759	0.855	0.860	0.890	0.943
	all	0.424	0.216	0.145	0.201	0.214	0.759	0.854	0.844	0.890	0.924
avg	10	0.342	0.297	0.154	0.270	0.261	0.782	0.806	0.847	0.826	0.889
	50	0.368	0.242	0.155	0.205	0.216	0.807	0.852	0.853	0.872	0.927
	all	0.367	0.200	0.155	0.209	0.202	0.807	0.852	0.847	0.871	0.935
sum	10	0.329	**0.352**	0.228	0.360	0.323	0.633	0.834	0.858	0.850	0.845
	50	0.337	0.288	0.218	0.348	0.288	0.665	0.859	**0.860**	0.896	0.929
	all	0.337	0.252	0.218	0.345	0.270	0.677	0.859	0.855	0.896	0.929
max	10	0.371	0.327	**0.246**	**0.397**	**0.340**	0.726	0.746	0.791	0.735	0.834
	50	0.485	0.300	0.219	0.340	0.303	**0.839**	**0.923**	0.831	0.851	0.965
	all	**0.520**	0.233	0.213	0.243	0.233	0.837	0.880	0.826	0.853	**0.967**

model shown is the one with best results when changing aggregation method and PCA parameters (in bold in the corresponding tables). The baseline phylogenetic trees and the hierarchical tree for the best performing method for each dataset (underlined in the corresponding tables) are available on GitHub.

We observe that results vary greatly between datasets. On the Hemoglobin dataset, all models have very good performance, with a large agreement between the embedding-based and the sequence-based trees. The best model appears to be ESM2, and the best aggregation method the *max* function, with no PCA. The ARI series in Fig. 3 indicates that a good agreement is maintained at all resolutions, i.e. the embeddings preserve both information on wide organism clusters but also on closely related homologues. For SATB, instead, agreement is lower. Here, the number of sequences and their length is much larger, requiring thus more aggregation and resulting in a longer ARI series (since we have taller trees). The dataset contains higher variability compared to the *Hemoglobin dataset*, as SATB1 and SATB2 proteins do not have the same molecular function but are just paralogues. While Hemoglobins are all binding and transporting oxygen, SATB proteins bind to DNA in specific moments of the differentiation, therefore they might be specialised for target mechanisms. This intrinsic variability of the dataset could explain partly the lower agreement between the embedding-based and the sequence-based tree. The best performing model appears to be DNABert, with *max* aggregation and no PCA. The figure shows that for protein embeddings the models appear to preserve information at high level, i.e. when the number of clusters is small, hence it is able to group well large families of proteins. DNABert instead appears to perform best at medium resolution.

For the Mouse dataset, ranges are comparable to those from SATB, with best results obtained by the ESM2 model with *average* aggregation and no PCA. Best agreement is obtained with medium and large number of clusters, i.e. the models are able to capture better the close relations between genes/proteins. For

N. Meningitidis, agreement is very low for all models. This could be due to the fact that these are bacterial DNA/proteins. The best performing model is ESM2 with *average* aggregation and no PCA, which appears to show some agreement with a high number of clusters, i.e. preserve close relations between proteins. For high number of clusters, DNABert shows slightly better performance than other models, however ARI values barely exceed 0.2.

Table 4. Mouse & N. Meningitidis datasets: ARI 90th percentile when comparing the evolutionary and embedding-based trees.

aggre-gator	dimen-sional PCA	Mouse					N.Meningitidis				
		dna-bert	seqvec	prose	alpha-fold	esm2	dna-bert	seqvec	prose	alpha-fold	esm2
pca	10	0.197	0.272	0.250	0.206	0.256	0.066	0.043	0.086	0.044	0.059
	50	0.241	0.280	0.260	0.330	0.415	0.093	0.065	0.084	0.045	0.108
	all	**0.243**	0.303	0.260	0.334	0.490	0.066	0.070	0.084	0.045	0.142
avg	10	0.038	0.333	0.300	0.233	0.363	**0.148**	0.053	0.110	0.037	0.065
	50	0.037	0.381	0.300	0.313	0.528	0.133	0.096	**0.116**	**0.055**	0.128
	all	0.032	**0.433**	0.300	0.296	**0.557**	0.121	**0.120**	**0.116**	0.049	**0.174**
sum	10	0.139	0.203	0.305	0.287	0.195	0.054	0.021	0.047	0.038	0.028
	50	0.091	0.166	0.303	0.211	0.169	0.046	0.023	0.046	0.041	0.022
	all	0.098	0.127	0.303	0.214	0.219	0.024	0.014	0.046	0.046	0.021
max	10	0.023	0.246	0.351	0.147	0.355	0.040	0.034	0.074	0.035	0.069
	50	0.058	0.431	0.421	0.287	0.473	0.039	0.032	0.083	0.036	0.142
	all	0.087	0.350	**0.430**	**0.312**	0.479	0.059	0.017	0.087	0.039	0.079

The Covid19 WG analysis shows good agreement between sequence and embedding-based trees, especially for protein embeddings. The best model is again the ESM2 model, with *sum* aggregation and no PCA. Figure 3 shows that agreement is large especially at low and medium resolutions, i.e. for low and medium number of clusters, indicating the embeddings maintain the overall structure of large evolutionary families and some close relations. When zooming in further SeqVec appears to provide slightly better results.

All in all, the ESM2 and DNABert models emerge as best options in terms of preservation of evolutionary information. Performing PCA on the final embeddings before computing distances does not appear to provide important improvements. In terms of aggregation of token/subsequence embeddings, it seems that the *average* and *max* options produce best results, with the exception of the whole genome analysis with protein embeddings where the *sum* appeared to be superior. This could be due to the fact that the sum incorporates information about the length of the sequences studied, which could be important in the case of whole proteome analysis.

Table 5. Covid19 WG dataset: ARI 90th percentile when comparing the evolutionary and embedding-based trees.

aggregator	dimensional PCA	dnabert	seqvec	prose	alphafold	esm2
pca	10	0.171	0.447	0.529	0.422	0.508
	50	0.208	0.482	0.529	0.479	0.501
	all	0.205	0.477	0.529	0.479	0.499
average	10	0.170	0.444	0.468	0.404	0.480
	50	0.203	0.494	0.470	0.474	0.499
	all	0.256	0.490	0.468	0.474	0.502
sum	10	0.269	**0.741**	0.543	**0.627**	0.731
	50	0.246	0.648	**0.575**	0.574	**0.806**
	all	0.272	0.615	**0.575**	0.574	**0.806**
max	10	0.201	0.662	0.530	0.604	0.645
	50	0.370	0.649	0.534	0.585	0.663
	all	**0.382**	0.645	0.536	0.585	0.663

3.2 Biological Annotations

Similar to the previous section, Table 6 shows, for different models and aggregation methods, the 90th percentile of the ARI values obtained when comparing the embedding-based with the annotation-based hierarchical clustering tree. Figure 4 shows the entire ARI series for the best models for each embedding type (in bold font in the corresponding table columns).

The overall results for the Mouse dataset show that there is a good overlap for all models between clusters created with annotations and those created from embeddings, with ARI values around 0.5 in some cases. The best performing model is again the ESM2, with *average* aggregation. While for the evolutionary evaluation the model without PCA was best, here it appears that using only 50 principal components is beneficial. However, the ARI values with and without PCA are similar, both here and in Table 4. Figure 4 shows that for this dataset most protein models maintain a steady ARI level with a different number of clusters, with a slight increase towards the higher resolution (multiple clusters). This indicates that the models are able to preserve both high level biological clusters and also place close together very similar entities.

Instead, most models show modest ARI values for the N. Meningitidis dataset, indicating that the embedding space does not match well with the annotation space. The best result is obtained by the DNABert model, however for a different configuration in terms of aggregation method and PCA compared to the evolutionary analysis. The ARI series in Fig. 4 shows that some agreement exists for a low number of clusters, for DNABert, Alphafold and (partially) ESM2, indicating that high level annotation groups are preserved. However, when zooming in to smaller clusters, the annotation structure is lost by all models. This may indicate that bacterial data could be more problematic for this type of models.

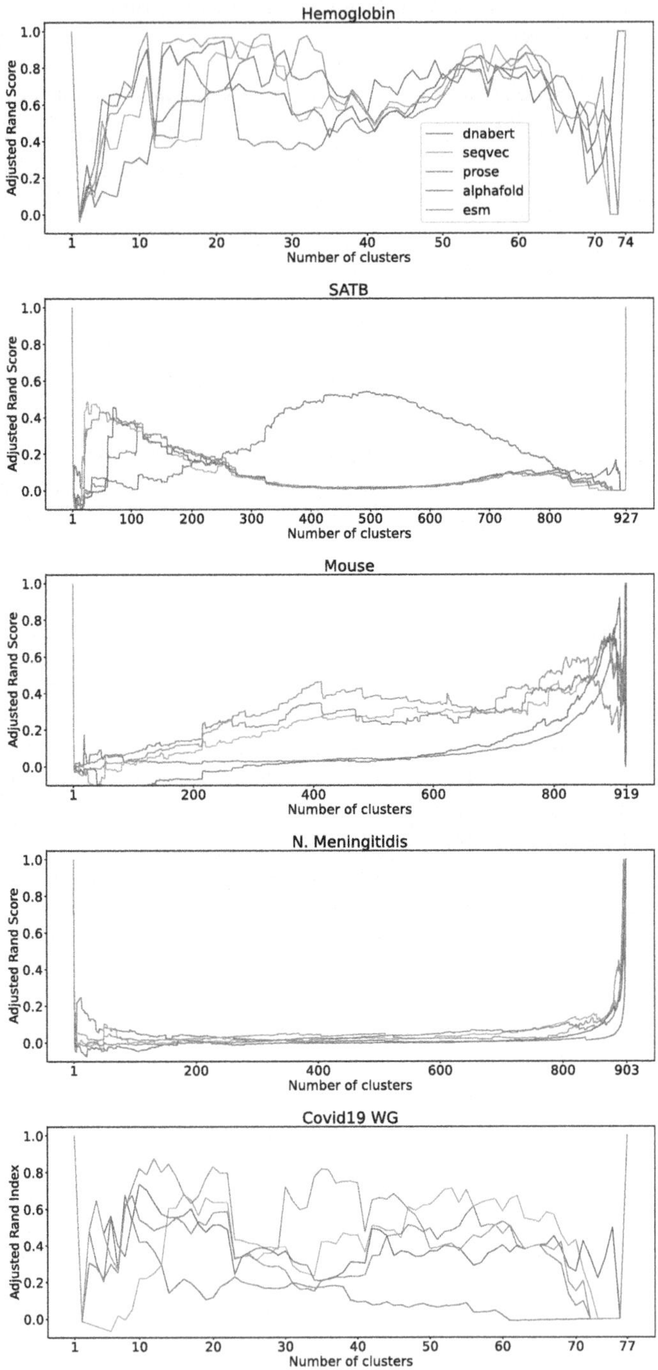

Fig. 3. Evolutionary analysis: ARI increasing the number of clusters (cut level) related to the best result for each embedder.

Table 6. Mouse & N. Meningitidis dataset: ARI 90th percentile when comparing the annotation-based and embedding-based trees.

aggre-gator	dimen-sional PCA	Mouse					N.Meningitidis				
		dna-bert	seqvec	prose	alpha-fold	esm2	dna-bert	seqvec	prose	alpha-fold	esm2
pca	10	0.168	0.192	0.204	0.150	0.300	0.014	0.066	0.071	0.114	0.013
	50	0.201	0.214	0.207	0.249	0.336	0.042	0.100	0.072	0.203	0.054
	all	**0.202**	0.311	0.207	**0.255**	0.441	0.037	**0.110**	0.072	0.197	0.054
average	10	0.033	0.239	0.352	0.178	0.285	0.066	0.022	0.076	0.124	0.022
	50	0.034	0.283	**0.368**	0.251	0.410	0.196	0.046	0.077	0.194	**0.121**
	all	0.030	0.305	0.368	0.251	**0.451**	0.196	0.031	0.077	0.188	0.119
sum	10	0.121	0.169	0.334	0.221	0.170	0.004	0.003	0.052	0.008	0.008
	50	0.081	0.147	0.342	0.206	0.140	0.007	0.005	0.052	0.008	0.009
	all	0.086	0.111	0.342	0.209	0.164	0.006	0.002	0.052	0.009	0.005
max	10	0.021	0.178	0.327	0.137	0.273	0.076	0.008	0.078	0.093	0.014
	50	0.055	**0.314**	0.336	0.208	0.385	0.215	0.011	**0.082**	0.163	0.032
	all	0.062	0.228	0.334	0.210	0.387	**0.276**	0.008	0.081	**0.213**	0.049

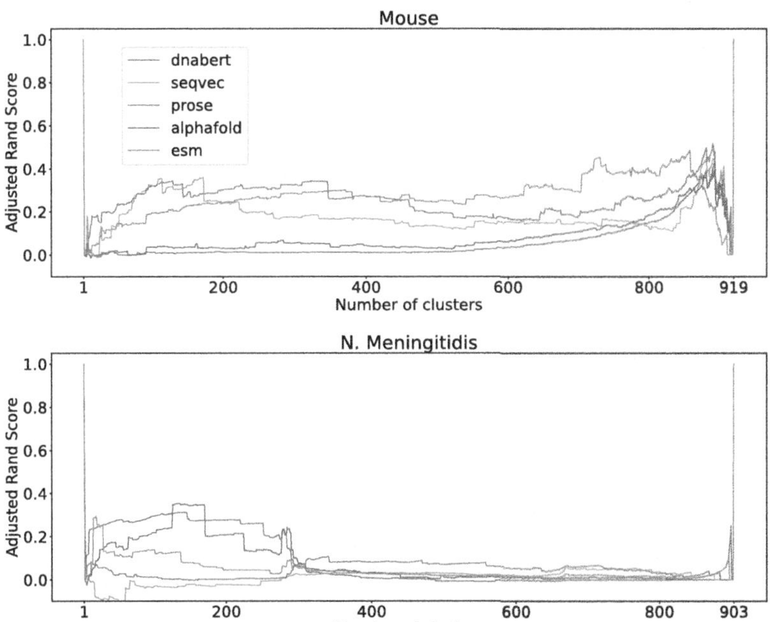

Fig. 4. Annotation analysis: ARI increasing the number of clusters (cut level) related to the best result for each embedder.

4 Related Works

This work is grounded in the evaluation of biological sequence embeddings, which is an active area of research; however, it features some characteristics which distinguish it from previous studies.

Firstly, most works usually evaluate the embeddings quality indirectly through their downstream task performance, e.g., to demonstrate the superiority of newly introduced models compared to existing ones [6,10], or to compare various existing models [41,43,44]. Typical downstream tasks that leverage sequence embeddings include biological probing tasks such as motif mining [38], structure prediction for complex proteins or multiple-chain complexes [11,14,40], single-sequence protein structure prediction [19], ontology-enriched [12] and structure-enriched [4] protein function prediction, protein-protein interactions [45], and also generative tasks [13,16,26,27,30]. Notably, some work couples downstream task performance with other relevant indices such as scalability and energy consumption [44]. With respect to these approaches, our study proposes a novel indicator based on the ARI metric that directly evaluates the structure of the embedding space from the evolutionary and functional point of view. Thus, the evaluation framework proposed in this study is independent on the downstream task, and can be applied regardless of the quality and volume of labelled training data.

Secondly, this work evaluates a seldom overlooked detail of constructing sequence embeddings – how the individual tokens are aggregated to form the final sequence embedding. While most works use averaging [32], here we take an agnostic approach and test different aggregation choices (*mean*, *max*, *average*, and PCA), showing that each dataset requires different aggregations, which by and large aligns to similar findings in related studies [7].

Thirdly, this work brings together and provides a unified analysis tool for both protein and DNA embeddings, while existing work typically concentrates on comparing protein embeddings only [44].

5 Conclusions

In this work we compared five models that create deep embeddings of gene/protein sequences, and four means of aggregating the embeddings. The comparison was based on two criteria: preservation of evolutionary information and of biological knowledge, in the sense of relative positioning of closely related sequences in the embedding space. We introduced a novel methodology which relied on hierarchical clustering of genes/proteins using different distance matrices and comparing the embedding-based hierarchical clustering trees with those based on evolutionary and annotation information. This methodology, when compared with probing downstream tasks present in the literature, has the advantage of not depending on the model performance and data availability for the downstream task, which can influence results. Furthermore, this methodology provides more detailed information than simple comparison of distance matrices, as it allows to explore the similarities at different resolutions. We employed five different datasets covering three analysis scenarios: analysis of orthologues across different species, identification of gene/protein families by studying genes/proteins from one species, whole genome analysis to compare individuals from different species.

Our results showed that in general models that generate embeddings are able to preserve both evolutionary and annotation information, however, the amount depends on the dataset. The top performing methods appear to be ESM2 and DNABert, while the best aggregation functions are *average* and *max*. Only in the whole genome analysis we observed a positive effect of the *sum* aggregation for protein embeddings. On the N. meningitidis data, however, we obtained poor results.

The models tested here are not a complete list of protein and DNA language models, fast progress is being made in introducing new improved models, e.g. [3,6,10], or models that combine existing approaches [19]. Future work will concentrate on integrating these models in our analysis, and expanding the datasets analysed, for an even more comprehensive comparison.

Acknowledgements. This work was partially financed by the co-funding European Union - Next Generation EU, in the context of The National Recovery and Resilience Plan, Investment 1.5 Ecosystems of Innovation, Project Tuscany Health Ecosystem (THE), CUP: B83C22003920001, Spoke 3. We thank Dr. Margherita Bodini and Dr. Alessandro Brozzi from GSK Vaccines, Siena, Italy, for useful discussions on DNA and protein embeddings and their aggregation.

References

1. Abdi, H., et al.: DISTATIS: the analysis of multiple distance matrices. In: 2005 IEEE Computer Society Conference on Computer Vision and Pattern Recognition (CVPR'05)-Workshops. IEEE (2005)
2. Ashburner, M., et al.: Gene ontology: tool for the unification of biology. Nat. Genet. **25**(1), 25–29 (2000)
3. Benegas, G., Batra, S.S., Song, Y.S.: DNA language models are powerful predictors of genome-wide variant effects. Proc. Natl. Acad. Sci. **120**(44), e2311219120 (2023)
4. Bepler, T., Berger, B.: Learning protein sequence embeddings using information from structure (2019). arXiv preprint arXiv:1902.08661
5. Bepler, T., Berger, B.: Learning the protein language: evolution, structure, and function. Cell Syst. **12**(6), 654–669 (2021)
6. Brandes, N., et al.: ProteinBERT: a universal deep-learning model of protein sequence and function. Bioinformatics **38**(8), 2102–2110 (2022)
7. Detlefsen, N.S., Hauberg, S., Boomsma, W.: Learning meaningful representations of protein sequences. Nat. Commun. **13**(1), 1914 (2022)
8. Devlin, J. et al.: BERT: Pre-training of deep bidirectional transformers for language understanding (2018). arXiv preprint arXiv:1810.04805
9. Durbin, R., Eddy, S.R., Krogh, A., Mitchison, G.: Biological sequence analysis: probabilistic models of proteins and nucleic acids. Cambridge university press, Cambridge (1998)
10. Elnaggar, A., et al.: ProtTrans: toward understanding the language of life through self-supervised learning. IEEE Trans. Pattern Anal. Mach. Intell. **44**(10), 7112–7127 (2021)
11. Evans, R., et al.: Protein complex prediction with alphafold-multimer. bioRxiv (2022). https://doi.org/10.1101/2021.10.04.463034

12. Fenoy, E., Edera, A.A., Stegmayer, G.: Transfer learning in proteins: evaluating novel protein learned representations for bioinformatics tasks. Briefings Bioinform. **23**(4), bbac232 (2022)
13. Ferruz, N., Schmidt, S., Höcker, B.: ProtGPT2 is a deep unsupervised language model for protein design. Nat. Commun. **13**(1), 4348 (2022)
14. Gao, M., et al.: AF2Complex predicts direct physical interactions in multimeric proteins with deep learning. Nat. Commun. **13**(1), 1744 (2022)
15. Heinzinger, M., et al.: Modeling aspects of the language of life through transfer-learning protein sequences. BMC Bioinform. **20**(1), 723 (2019)
16. Hie, B.L., et al.: Efficient evolution of human antibodies from general protein language models. Nat. Biotechnol. **42**(2), 275–283 (2024)
17. Hubert, L., Arabie, P.: Comparing partitions. J. Classif. **2**, 193–218 (1985)
18. Ji, Y., et al.: DNABERT: pre-trained bidirectional encoder representations from transformers model for DNA-language in genome. Bioinformatics **37**(15), 2112–2120 (2021)
19. Jing, X., Wu, F., Luo, X., Xu, J.: Single-sequence protein structure prediction by integrating protein language models. Proc. Natl. Acad. Sci. **121**(13), e2308788121 (2024)
20. Jumper, J., et al.: Highly accurate protein structure prediction with AlphaFold. Nature **596**(7873), 583–589 (2021)
21. Khurana, D., et al.: Natural language processing: state of the art, current trends and challenges. Multimedia Tools Appl. **82**(3), 3713–3744 (2023)
22. Kosloff, M., Kolodny, R.: Sequence-similar, structure-dissimilar protein pairs in the PDB. Proteins: Structure Funct. Bioinform. **71**(2), 891–902 (2008)
23. Kusner, M., et al.: From word embeddings to document distances. In: International Conference on Machine Learning. PMLR (2015)
24. Lin, Z., et al.: Language models of protein sequences at the scale of evolution enable accurate structure prediction. bioRxiv (2022)
25. Lin, Z., et al.: Evolutionary-scale prediction of atomic-level protein structure with a language model. Science **379**(6637), 1123–1130 (2023)
26. Madani, A., et al.: Large language models generate functional protein sequences across diverse families. Nat. Biotechnol. **41**(8), 1099–1106 (2023)
27. Mardikoraem, M., et al.: Generative models for protein sequence modeling: recent advances and future directions. Briefings Bioinform. **24**(6), bbad358 (2023)
28. Mikolov, T., et al.: Efficient estimation of word representations in vector space (2013). arXiv preprint arXiv:1301.3781
29. Milligan, G.W., Cooper, M.C.: A study of the comparability of external criteria for hierarchical cluster analysis. Multivar. Behav. Res. **21**(4), 441–458 (1986)
30. Nijkamp, E., Ruffolo, J.A., Weinstein, E.N., Naik, N., Madani, A.: ProGen2: exploring the boundaries of protein language models. Cell Syst. **14**(11), 968–978 (2023)
31. Peters, M.E., et al.: Deep contextualized word representations. In: Proceedings of the 2018 Conference of the North American Chapter of the Association for Computational Linguistics: Human Language Technologies, Volume 1. Association for Computational Linguistics, New Orleans, Louisiana (2018)
32. Rives, A., et al.: Biological structure and function emerge from scaling unsupervised learning to 250 million protein sequences. Proc. Natl. Acad. Sci. **118**(15), e2016239118 (2021)
33. Salant, S., Berant, J.: Contextualized word representations for reading comprehension (2017). arXiv preprint arXiv:1712.03609
34. Sievers, F., et al.: Fast, scalable generation of high-quality protein multiple sequence alignments using Clustal omega. Mol. Syst. Biol. **7**(1), 539 (2011)

35. Sofi, M.Y., Shafi, A., Masoodi, K.Z.: Chapter 6 - multiple sequence alignment. In: Bioinformatics for Everyone, pp. 47–53. Academic Press (2022)
36. Su, J. et al.: RoFormer: Enhanced transformer with rotary position embedding (2021). arXiv preprint arXiv:2104.09864
37. Suzek, B.E., et al.: UniRef clusters: a comprehensive and scalable alternative for improving sequence similarity searches. Bioinformatics **31**(6), 926–932 (2015)
38. Tan, X., Yuan, C., Wu, H., Zhao, X.: Comprehensive evaluation of BERT model for DNA-language for prediction of DNA sequence binding specificities in fine-tuning phase. In: Huang, D.S., Jo, K.H., Jing, J., Premaratne, P., Bevilacqua, V., Hussain, A. (eds.) Intelligent Computing Theories and Application. ICIC 2022. LNCS, vol. 13394. Springer, Cham (2022). https://doi.org/10.1007/978-3-031-13829-4_8
39. The Gene Ontology Consortium, et al.: The gene ontology knowledgebase in 2023. Genetics **224**(1), iyad031 (2023)
40. Tsaban, T., et al.: Harnessing protein folding neural networks for peptide–protein docking. Nat. Commun. **13**(1), 176 (2022)
41. Unsal, S., Atas, H., Albayrak, M., Turhan, K., Acar, A.C., Doğan, T.: Learning functional properties of proteins with language models. Nat. Mach. Intell. **4**(3), 227–245 (2022)
42. Vijaymeena, M., Kavitha, K.: A survey on similarity measures in text mining. Mach. Learn. Appl. Int. J. **3**(2), 19–28 (2016)
43. Villegas-Morcillo, A., Gomez, A.M., Sanchez, V.: An analysis of protein language model embeddings for fold prediction. Briefings Bioinform. **23**(3), bbac142 (2022)
44. Väth, P., et al.: PROVAL: a framework for comparison of protein sequence embeddings. J. Comput. Math. Data Sci. **3**, 100044 (2022)
45. Yao, Y., et al.: An integration of deep learning with feature embedding for protein-protein interaction prediction. PeerJ **7**, e7126 (2019)

Uncovering Dynamic Structures Within Cyclic Attractors of Asynchronous Boolean Networks with Spectral Clustering

Maryam Yousefian[1(✉)], Elisa Tonello[2], Anna-Simone Frank[1], Heike Siebert[3], and Susanna Röblitz[1]

[1] Computational Biology Unit, Department of Informatics, University of Bergen, 5008 Bergen, Norway
maryam.yousefian@uib.no
[2] Department of Mathematics and Computer Science, Freie Universität Berlin, Arnimallee 7, 14195 Berlin, Germany
[3] Department of Mathematics, Kiel University, Heinrich-Hecht-Platz 6, 24118 Kiel, Germany

Abstract. Boolean models provide an intuitive framework for the investigation of complex biological networks. Dynamics that implement asynchronous update rules, in particular, can help embody the complexity arising from non-deterministic behavior. These transition systems allow for the emergence of complex attractors, cyclic subgraphs that capture oscillating asymptotic behavior. Techniques that explore and attempt to describe the structures of these attractors have received limited attention. In this context, the incorporation of process rate information may yield additional insights into dynamical patterns. Here, we propose to use a spectral clustering algorithm on the kinetic rate matrix of time-continuous Boolean networks to uncover dynamic structures within cyclic attractors. The Robust Perron Cluster Analysis (PCCA+) can be used to unravel metastable sets in Markov jump processes, i.e. sets in which a system remains for a long time before it switches to another metastable set. As a proof-of-concept, we apply this method to Boolean models of the mammalian cell cycle. By considering the categorization of transitions as either slow or fast, we investigate the impact of time information on the emergence of significant sub-structures.

Keywords: Time-continuous Boolean networks · Markov chains · Robust Perron cluster analysis (PCCA+) · Mammalian cell cycle

1 Introduction

A deeper understanding of the architecture and dynamics of complex biological networks can be gained through the employment of mathematical modeling. Logic-based models bridge qualitative biological knowledge and quantitative

R. Gori et al. (Eds.): CMSB 2024, LNBI 14971, pp. 226–246, 2024.
https://doi.org/10.1007/978-3-031-71671-3_16

analysis, offering a framework for studying the interactions within complex networks. By affording a more abstract view of the system, they allow comprehensive analysis exploitable for hypothesis generation and experimental design [26]. The agents in Boolean networks (e.g., molecules, genes, proteins) are assigned to nodes with binary states, and logical expressions specify the processing of regulatory effects defining update rules for each node [1,25]. Dynamics in the form of state transition graphs can then be derived using different update strategies. Most common are the so-called synchronous update leading to deterministic dynamics [15] and the asynchronous update where only one node is updated per step, but all possible transitions are considered leading to non-deterministic dynamics [27].

The long-time behavior of a Boolean network is characterized by its attractors, which can often be linked to biological phenotypes. Many efforts have been devoted to the problem of identification of attractors and their reachability (see for instance [2,28] for recent advancements). However, further investigations into the structure of large attractors capturing oscillatory behaviors, called cyclic attractors, are limited. Existing approaches focus, e.g., on complexity reduction by considering meaningful partitions of the state set representing an attractor [7], or on identifying different regions of the original attractor through analysis of sub-modules of the system [5].

Analyses can naturally be further refined if more information about the molecular processes is integrated into the model. If available, state transitions graphs can be endowed with transition rates (e.g., [19,25]) and the resulting stochastic models studied as Markov processes. The assignment of kinetic parameters can reflect a more refined understanding of the interplay of different regulations and time scales, and allow for some investigations of more quantitative flavor. In [25], a simulation algorithm was introduced and run, among others, on a cell cycle model with rates derived from a qualitative classification of processes as fast or slow as proposed in [10]. This classification of rates is not used to simplify the systems, as in the complexity reduction under a quasi steady state assumption applied to dynamical systems with processes categorized as fast and slow. Instead, it still explicitly considers all state changes, allowing to investigate, e.g., probable transition paths and stability properties. Even without further rate information, simulations based on uniform transition rates have been explored, e.g. to identify states with high visitation probability [5].

Previous work, however, did not address the identification of possible substructures within cyclic attractors. When analysing biological systems, this could help uncover different functional phases in cyclic behavior as well as system redundancies and variability. In this work, we, therefore, want to explore the modular structure of cyclic attractors that can emerge from the dynamics, investigating the influence of transition rates on such structures. For this purpose, we present a new approach using spectral clustering to conduct an in-depth analysis of complex attractors and reveal their internal sub-structure, focusing on probability density changes. Our approach applies the complex-valued Robust Perron Cluster Analysis (cPCCA+) [11] to two mammalian cell cycle attractors [7,10]. The paper is organized in the following way. In Sect. 2, we first present a brief background on Boolean networks, as well as the connection between continuous time Boolean networks and their transition matrix and probability density. We then describe the cPCCA+ clustering approach and use a simple toy model from [25] to illustrate our approach. In Sect. 3, we analyze the attractors of two models, presented in [7,10] using our spectral clustering approach. Finally, we discuss some prospects for further investigation opened by our work.

2 Methods

2.1 Boolean Interaction Networks: Definitions and Background

Boolean networks offer a form of qualitative modeling that simplifies the information about all available species (representing genes, proteins, etc.) to two possible levels, 0 for "off" and 1 for "on". The space $\{0,1\}^n$ is, therefore, the set of possible configurations, or state space, under this choice. The following are standard definitions in the analysis of Boolean networks. The influence structure of the network is captured in the *interaction* or *regulatory graph*, a signed directed graph having the species of interest as nodes, and edges representing inhibitions or activations between species. A full description of a Boolean model then requires the specification of how multiple regulators come together to define the update of a given component (logical rules). A Boolean network in n variables can, therefore, be identified with a map

$$f = (f_1, \dots, f_n) \colon \{0,1\}^n \to \{0,1\}^n.$$

Dynamics associated with a Boolean network f are directed graphs with a set of nodes, the state space $\{0,1\}^n$, and edges, also called *transitions*, that depend on the choice of the update rule. The *synchronous* state transition graph is obtained when each state x has a unique successor $f(x)$, and is a sensible choice for scenarios characterized by fully deterministic behaviors. A more common alternative in biological contexts is the *asynchronous* update scheme. This allows for the possibility of multiple out-coming transitions from the same source node, each transition implementing the update of exactly one component.

Definition 1. *The* asynchronous state transition graph *of the Boolean network* $f \colon \{0,1\}^n \to \{0,1\}^n$ *is the directed graph with nodes the set* $\{0,1\}^n$ *and set of edges*

$$\{(x,y) \in \{0,1\}^n \times \{0,1\}^n | \exists i \text{ s.t. } y_i = f_i(x) \neq x_i, y_j = x_j \forall j \neq i\}.$$

Examples of asynchronous state transition graphs for networks with 3 components are shown in Fig. 1. Other definitions of non-deterministic state transitions graphs are possible; while not addressed here, they can also be analyzed with the techniques presented in this work. State transition graphs capture possible dynamic behaviors of the modeled system at a qualitative level. Recall that, in a directed graph, a strongly connected component is a maximal subgraph in which every pair of vertices is mutually reachable. In addition, a strongly connected component is terminal if there are no edges from its vertices to vertices outside the component.

Definition 2. *An* attractor *in a state transition graph is a terminal strongly connected component (SCC).*

Each fixed point of the Boolean network defines a singleton attractor, where all components are stable. Attractors that consist of more than one state, called *cyclic* attractors, capture oscillating asymptotic behaviors and, under the asynchronous update scheme, can involve a complex structure of intersecting cycles. In the example in Fig. 1 on the left, the state 000 is a fixed point and the unique attractor. The state transition graph on the right in the same figure also admits only one attractor, composed of 5 states. For compactness, in the following, we will denote sets of states using the symbol "$\star$" to represent free components, so that an element S of $\{0,1,\star\}^n$ will denote the set of states $\{x \in \{0,1\}^n | x_i = S_i \text{ for all } i \text{ s.t. } S_i \in \{0,1\}\}$.

2.2 Continuous-Time Boolean Networks

In this paper, we use the continuous-time generalization of the asynchronous updating policy as introduced in [25]. There can only be a transition between two states if they differ by the value of only one node. This transition gets assigned a transition rate, which defines the timescale of either activating or deactivating that node (the physical unit of transition rates is 1/time). Transition rates between any two states that differ by the value of more than one node are zero. In other words, any edge in the state transition graph gets assigned a kind of weight, namely the transition rate. The resulting dynamics of this network is a continuous-time Markov process, which can be simulated with the stochastic simulation algorithm (SSA), also known as Gillespie algorithm [12], as implemented for example in the MaBoSS software [24].

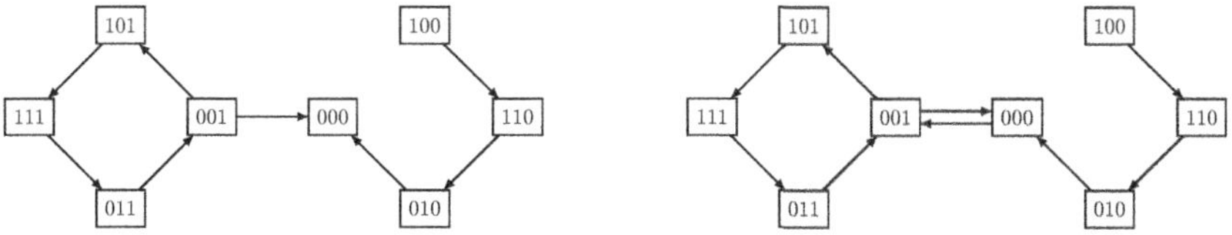

(a) without reactivation of C (b) with reactivation of C

Fig. 1. (a) Asynchronous state transition graph of a toy model involving three species, defined by $f(x_A, x_B, x_C) = ((1-x_B)(x_A+x_C-x_Ax_C), x_A, x_C(x_A+x_B-x_Ax_B))$. The fixed point 000 is the unique attractor. Note that if the transition rate from 001 to 000 is small, the dynamics might cycle a while through the states $\{001, 101, 111, 011\}$ before reaching 000. (b) State transition graph for a different network, including a transition from 000 to 001 (the update function for the third component is $f_C(x_A, x_B, x_C) = (2x_C-1)(x_A+x_B-x_Ax_B)+1-x_C$).

An alternative to simulating individual trajectories is to directly compute the time evolution of probability densities in the state space. For this purpose, the transition rates are gathered into a matrix K of size $2^n \times 2^n$, called transition rate matrix, where n is the number of species in the network, and the matrix entry $K(i,j)$ contains the rate for transitioning from state i to state j. The diagonal entries of this matrix are set to the negative sums of the off-diagonal entries in the corresponding row,

$$K(i,i) = -\sum_{j \neq i} K(i,j).$$

The matrix K is also known as the generator of the continuous-time Markov chain (see, e.g., [22] for the underlying theory).

As an example (taken from [25]), we consider the state transition graph for a toy model (Fig. 1a) involving three species (A, B, C) with different rates for activation of A (rate Au for A going up), deactivation of A (rate Ad for A going down), activation of B (rate Bu for B going up), and deactivation of B (rate Bd for B going down), as well as a small transition rate ε for deactivation of C. If we order the states $\{000, 001, 010, \ldots, 111\}$ from 1 to 8, then the transition rate matrix of this continuous time Boolean network reads

$$K = \begin{pmatrix} 0 & 0 & 0 & 0 & 0 & 0 & 0 & 0 \\ \varepsilon & -(Au+\varepsilon) & 0 & 0 & 0 & Au & 0 & 0 \\ Bd & 0 & -Bd & 0 & 0 & 0 & 0 & 0 \\ 0 & Bd & 0 & -Bd & 0 & 0 & 0 & 0 \\ 0 & 0 & 0 & 0 & -Bu & 0 & Bu & 0 \\ 0 & 0 & 0 & 0 & 0 & -Bu & 0 & Bu \\ 0 & 0 & Ad & 0 & 0 & 0 & -Ad & 0 \\ 0 & 0 & 0 & Ad & 0 & 0 & 0 & -Ad \end{pmatrix}. \tag{1}$$

If reactivation of node C is included (Fig. 1b), then $K(1,2) = \varepsilon$ as well.

The matrix K can be used to compute the time evolution of a probability density vector $p(t)$ of length 2^n via the master equation of a continuous-time Markov chain,

$$\frac{dp(t)}{dt} = K^T \cdot p(t), \quad p(0) = p_0, \tag{2}$$

where the ith entry of $p(t)$ is the probability that the Boolean network occupies the ith binary state at time t. In principle, the exact solution to this system of ordinary differential equations can be calculated by the matrix exponential,

$$p(t) = T^T(t) \cdot p(0) \quad \text{with} \quad T(t) = \exp(tK),$$

where $T(t)$ is a row-stochastic matrix containing the probabilities for transitioning between any two states within a time interval of length t.

Remark 1. Note that the transition probability matrix T obtained via the matrix exponential from the generator K is different from the transition probability matrix P for a random walk on a weighted directed graph. This matrix is typically defined as

$$P(i,j) = \begin{cases} w(i,j)/d_i & \text{if there is an edge from } i \text{ to } j \text{ with weight } w(i,j), \\ 0 & \text{otherwise,} \end{cases}$$

with $d_i = \sum_j w(i,j)$. The difference results from the fact that P, due to normalization, does not differentiate between states that have high-weighted outgoing edges and those that have low-weighted outgoing edges. In other words, the holding probability in the individual states is modeled differently between T and P, resulting in possibly different transient and stationary behavior.

Note that T, in contrast to K, is not a sparse matrix. Though there exist multiple methods for computing the matrix exponential [18], this method, therefore, only works for matrices whose size is less than a few thousands, so that all the elements can be stored in the main memory of a computer. Hence, these calculations quickly become infeasible for Boolean networks whose state space size grows exponentially with the number of species involved.

However, since K is sparse, it is still possible to compute the stationary solution even for larger Boolean networks[1]. The stationary density π satisfies $K^T\pi = 0$ or $\pi = T^T(t)\pi$, respectively. This relationship was exploited by the authors in [16] to efficiently compute the effect of parameter changes, i.e. changes in transition rates, on the stationary solutions. For the toy model matrix in Eq. (1), the stationary density is $\pi = (1, 0, 0, 0, 0, 0, 0, 0)^T$, independent of the values for the rates, since 000 is the only attractor state.

The computation of π as the left eigenvector of K is badly conditioned if K has confluent eigenvalues at zero. This is the case for the toy model as long as the escape rate ε is much smaller than the other rates. In fact, the toy model is an example for a nearly decoupled Markov chain, for which the stationary density is known to be badly conditioned [17]. In the toy model, the stationary density is particularly sensitive to a possible re-activation rate of node C, i.e. when introducing an edge from 000 to 001 with the same transition rate ε as for the deactivation of C (Fig. 1b). In that case, the five states $\{000, 001, 101, 111, 011\}$ form an attractor, which consists of two weakly coupled subsets, namely the single state 000 and the cyclic subset $\{001, 101, 111, 011\}$. We say these subsets are *metastable* because the Boolean network dynamics would spend a long time in one of these two subsets and rarely switch between them. This substructure could not be identified by simply searching for the strongly connected components of the state transition graph. However, as we will demonstrate in the next section, it can be identified by a spectral analysis of the transition rate matrix.

2.3 Spectral Clustering via PCCA+

Clustering algorithms vary because the concept of a "cluster" is imprecise [8]. A popular and computationally efficient method is spectral clustering [29], which uses the eigenvalues of the data's similarity matrix for dimensionality reduction before clustering. The Robust Perron Cluster Analysis (PCCA+) is specifically designed for spectral clustering of Markov chains, using transition probabilities or rates for similarity. Unlike common spectral clustering, which uses the Fiedler

[1] The authors in [16] considered networks with up to 23 nodes. The maximum array size in Matlab is $2^{48} - 1 \approx 2.8 \cdot 10^{14}$. However, the array must fit into memory.

eigenvector, PCCA+ considers more eigenvectors and is linked to invariants and model reduction, thus bringing to light essential but otherwise elusive structures in Markov processes (see [30] for examples and comparisons).

The aim of PCCA+ is to identify *metastable sets*, i.e. subsets of the state space in which the Markov chain spends a long time before transitioning to another metastable set. The algorithm groups all N states of the state space into metastable sets such that transitions between the metastable sets are slow, and transitions within them are fast [6,23]. A prerequisite to identifying clusters in PCCA+ is the computation of the right invariant subspace X that contains the dominant eigenvectors associated with eigenvalues of the matrix T that are close to one. These eigenvectors represent the long-term trends in the dynamics and clustering them allows PCCA+ to identify clusters of states that exhibit similar long-term transition patterns.

The result of PCCA+ is a coarse-grained system representation, known as the Markov State Model (MSM), that describes the system behavior on a long time-scale [13]. A MSM consists of a finite, discrete state space of size N and a coarse-grained transition probability matrix, $T_c(\tau)$, which contains the conditional probabilities for transitions between the n_c ($n_c \ll N$) metastable subsets of the system within a time interval of length τ. The coarse-grained transition matrix T_c is derived from the original transition probability matrix T of the Markov chain. Mathematically, T_c is defined as

$$T_c = (\chi^T D^2 \chi)^{-1} \chi^T D^2 T \chi \tag{3}$$

where D is a positive diagonal matrix (further explained below), and $\chi \in \mathbb{R}^{N \times n_c}$ is a matrix of *membership vectors*, i.e., they are positive and form a partition of unity ($\sum_{j=1}^{n_c} \chi_j(i) = 1,\ \forall i \in \{1, ..., N\}$). The membership vectors are calculated by

$$\chi = XA, \qquad A \in \mathbb{R}^{n_c \times n_c} \text{ non-singular} \tag{4}$$

where $A \in \mathbb{R}^{n_c \times n_c}$ is a non-singular matrix, which transforms the n_c dominant right eigenvectors X of T corresponding to eigenvalues close to one [23]. As shown in [23], the projection (3) represents a form of exact lumping [31], whereby the macrostates are defined by the membership matrix χ.

The eigenvectors X have to furthermore fulfill an *orthonormality relation* through normalization with respect to a density vector w, i.e., $w > 0$ element-wise and $w^T \mathbf{1} = 1$. The matrix $D = \text{diag}(\sqrt{w})$ is a diagonal matrix containing the element-wise square root of this vector. The density vector w can be chosen arbitrarily [9] but is usually the stationary density vector or an equiprobable vector $w = (1/N, \ldots, 1/N)^T$. The latter is typically used in cases where the computation of the stationary density is numerically unstable or when studying graphs that are not strongly connected.

Finding the transformation matrix A has no unique solution [6]. Therefore, the clustering step in PCCA+ employs some optimization procedure (see [23] for details). The results of the clustering process are the membership vectors in the matrix χ, which assign the N states to the n_C clusters. Basically, each entry in a membership vector indicates the degree to which the corresponding state belongs

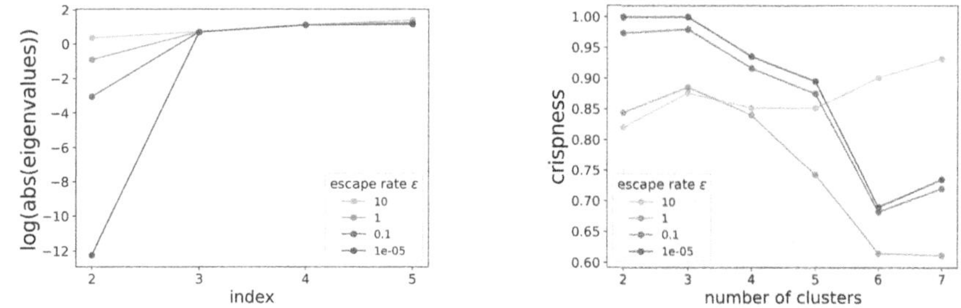

(a) eigenvalues (b) crispness

Fig. 2. Toy model. Eigenvalues (left) and suggested cluster numbers (right) for different values of the escape rate ε.

to a particular cluster. The objective function implemented in PCCA+ ensures that the resulting membership vectors are as close as possible to characteristic vectors with entries of either 0 or 1 [11]. This closeness is measured by an index called *crispness*, which can take values between 0 (worst) and 1 (best).

The above-described original version of PCCA+ [6,23] was developed for real dominant eigenvectors, which made it unable to deal with complex values that can occur for non-reversible Markov chains, including circular chains. This led to an extension of PCCA+ to cPCCA+ [11], which is applicable to Markov chains with complex eigenvalues. In cPCCA+, the possibly complex eigenvectors are first decomposed into their real and imaginary part before the transformation matrix A is computed. This separation is a linear transformation which maintains the invariant subspace condition and the structure of the eigenvectors [11].

Since the clustering algorithm only operates on the corresponding eigenvectors, it can be applied to a rate matrix K in exactly the same way as to the transition probability matrix T, with the only modification being the target eigenvalue. In particular, the transition rate matrix K and the transition probability matrix T have the same eigenvectors, and there is a one-to-one correspondence between the eigenvalue close to 1 for T and eigenvalues close to 0 for K via the exponential transformation. The coarse-grained transition rate matrix,

$$K_c := (\chi^T D^2 \chi)^{-1} \chi^T D^2 K \chi \tag{5}$$

can then be transformed into a coarse-grained transition probability matrix via $T_c(t) = \exp(t \cdot K_c)$. In the following applications, we base our calculations, therefore, on the rate matrix K.

2.4 Spectral Clustering Applied to the Toy Model

The rate matrix K from Eq. (1) with rates $Au = 1$, $Ad = 4$, $Bu = 2$, $Bd = 3$ and escape rate $\varepsilon = 10^{-5}$ (same rates as in [25]) has eigenvalues (in ascending order, starting with the smallest absolute value) $\lambda_0 = 0, \lambda_1 = -4.8 \cdot 10^{-6}, \lambda_2 = -2, \lambda_3 = -3, \lambda_{4/5} = -2.5 \pm 1.94i, \lambda_6 = -4, \lambda_7 = -5$. There is a clear spectral gap after the second eigenvalue (Fig. 2a), i.e. the third eigenvalue is six orders of magnitude larger than the second eigenvalue, suggesting a clustering into two metastable sets. In fact, the crispness for $n_2 = 2$ clusters is close to one (Fig. 2b). Here, we used the equiprobable density vector for normalization because the stationary density is zero on all states except the single attractor state 000. The two resulting membership vectors $\chi_{1,2}$ indicate the assignment of states 2, 4, 6, and 8 into the first cluster and the assignment of states 1, 3, 5, and 7 into the second cluster (Fig. 6a). Projection of the rate matrix K onto these two clusters results in the coarse-grained rate matrix

$$K_c = \begin{pmatrix} -4.8 \cdot 10^{-6} & 4.8 \cdot 10^{-6} \\ 0 & 0 \end{pmatrix}.$$

It has the same two eigenvalues $\lambda_{0,1}$ as the original matrix and indicates that the second cluster is absorbing. The matrix exponential of K_c contains the transition probabilities between the two clusters over a time span of length $\tau = 1$,

$$T_c(1) = \text{expm}(1 \cdot K_c) = \begin{pmatrix} 1 - 4.8 \cdot 10^{-6} & 4.8 \cdot 10^{-6} \\ 0 & 1 \end{pmatrix}$$

indicating that the probability of staying in cluster 2 is one and that staying in cluster 1 is close to one. When the escape rate ε increases, the second eigenvalue gets larger in magnitude (Fig. 2a), and the clustering gets less perfect, i.e. more fuzzy (Fig. 2b). Still, state 6 remains the core state of cluster 1, whereas states 2, 4, and 8 are partially assigned to cluster 2 with increasing membership (Fig. 6).

If we introduce a reactivation rate $\varepsilon = 10^{-5}$ for node C as in Fig. 1b, the second eigenvalue changes slightly ($\lambda_0 = 0, \lambda_1 = -1.48 \cdot 10^{-5}, \lambda_2 = -2, \lambda_3 = -3, \lambda_{4/5} = -2.5 \pm 1.94i, \lambda_6 = -4, \lambda_7 = -5$) but the spectral gap still indicates two clusters. In fact, cPCCA+ delivers the same clustering as before (Fig. 6a), but with projected matrices

$$K_c = \begin{pmatrix} -4.8 \cdot 10^{-6} & 4.8 \cdot 10^{-6} \\ 10^{-5} & -10^{-5} \end{pmatrix}, \quad T_c(1) = \begin{pmatrix} 1 - 4.8 \cdot 10^{-6} & 4.8 \cdot 10^{-6} \\ 10^{-5} & 1 - 10^{-5} \end{pmatrix},$$

indicating that it is now possible to transit from cluster 2 back to cluster 1.

Note that while the stationary density π is very sensitive to changes in the reactivation rate ε (it changes from $\pi = (1, 0, 0, 0, 0, 0, 0, 0)^T$ in the case without reactivation of C to $\pi = (0.32, 0.32, 0, 0.11, 0, 0.16, 0, 0.08)^T$ in the case with reactivation rate $\varepsilon = 10^{-5}$), the clustering and the coarse-grained transition matrices are not sensitive to changes in this rate. The reason is that the two eigenvalues close to 0 remain well-separated from the remaining eigenvalues, preserving the spectral gap, thus making the corresponding invariant subspace spanned by the first two eigenvectors well-conditioned. Since the result of the clustering only depends on this subspace, it is insensitive as well. As demonstrated with this toy example, the cyclic subset $\{001, 101, 111, 011\}$ can be revealed both in the case with and without re-activation of C. As such, our proposed method is able to find metastable sets both within SCCs as well as across multiple SCCs.

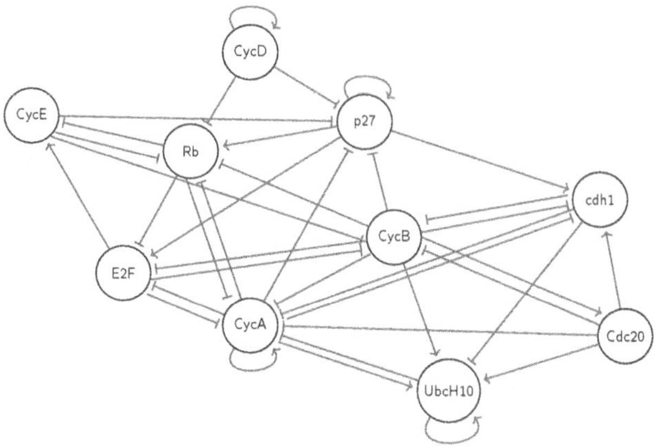

(a) interaction graph

	E2F	CycE	CycA	CycB
	0	1	0	0
G1	1	0	0	0
	1	1	0	0
	0	1	1	0
G1S	1	0	1	0
	1	1	1	0
SG2	0	0	1	0
G2M	0	0	1	1
M	0	0	0	1
MG1	0	0	0	0
	0	1	0	1
∅	0	1	1	1
	1	⋆	⋆	1

(b) phases

Fig. 3. Model B. a) Interaction graph of the mammalian cell cycle network of [7]. Edges terminating in blunt arrows (red) signify inhibition, whereas edges with standard arrowheads (green) symbolize activation. b) Definition of the six cell cycle phases for the four master regulators (E2F, CycE, CycA, CycB) according to Diop et al. (Tab. III in [7]). (Color figure online)

3 Analysis of Mammalian Cell Cycle Models

3.1 Cell Cycle Models

The cell cycle comprises a series of molecular activities that culminate in the duplication of a cell's genome during the Synthesis phase (S phase) and its subsequent division into two daughter cells in the Mitosis phase (M phase) [3]. Computational models have been of great help in better understanding this fundamental process and its regulation. Following the seminal work using differential equations [21], different Boolean models have been developed to allow for a more comprehensive analysis [4,10,14]. In [10], a Boolean model of the mammalian cell cycle captures the cell cycle process as a complex cyclic attractor. It was shown, considering the sequence of value changes of certain components, that in the asynchronous update, the different phases of the cell cycle can be recovered qualitatively; however, since it consists of many different intertwined trajectories, analysis is challenging. Employing some qualitative knowledge on biochemical rates, the authors group updating processes into faster and slower process classes and thus obtain a clearer representation of the cell cycle progression. In [7], they choose a different approach to analyzing the complex attractor of this model, summarizing states according to a meaningful partition of the state set. Based on biological information, they choose four so-called master regulators of the cell cycle that are included in the model, namely E2F, CycE, CycA, and CycB, and use their values to assign states in the attractor to phases in the cell cycle (Fig. 3b). To obtain a more refined partitioning, the phases they consider differ slightly from the usual cell cycle phases G1, G2, M, and S. They consider instead the phases G1, G1S, SG2, G2M, M, and MG1. Based on their analysis, the authors then present a slightly adapted version of the model, which produces a simplified cyclic attractor with a clearer correspondence of states to phases and the cell cycle progression through the phases (see Fig. 3). In the following, we will refer to the model proposed in [10] as *Model A* and to the model presented in [7] as *Model B*.

We start with an analysis of *Model B*, whose cyclic attractor representing the cell cycle contains 53 states as opposed to the 112 states of *Model A*. Comparison with the full partition of the *Model B* attractor into the six phases gives us a straightforward starting point for evaluating the results of our cluster analysis. For our approach, we will include rate information based on the settings discussed in [10] described above and investigate its impact on the clustering results. Finally, we will also test our approach on *Model A* with its more complex attractor.

3.2 Spectral Clustering of the Mammalian Cell Cycle Model

We followed the classification into slow and fast processes by using two different rates, *slow* and *fast*. Throughout our computations, we fixed the *slow* rate to one and varied the *fast* rate between the values $\{1, 10, 100, 1000\}$, thus changing the *fast/slow* ratio. Note that using different value combinations for $fast$ and *slow* while maintaining the *fast/slow* ratio only corresponds to a different scaling of the transition rate matrix K, i.e. the clustering results would be the same as the eigenvectors are invariant to scaling of the matrix. A ratio of one corresponds to the case where no rate information is included in the model. As we will demonstrate below, we observed a convergence of results for increasing ratios, with the results for *fast/slow* = 100 and *fast/slow* = 1000 already being very similar. We did, therefore, not examine higher ratios.

In the following, the ordering of states is lexicographical (CDC20, CycA, CycB, CycE, E2F, UbcH10, cdh1). We fixed CycD = 1 as this is a prerequisite for observing the cyclic attractor. If CycD = 1, then Rb = 0 and p27 = 0 by the definition of their logical rules, hence we omit these nodes in our network analysis.

Starting with *Model B*, for a *fast/slow* ratio of 10 or higher, the transition rate matrix K has a clear cluster of nine eigenvalues, followed by a spectral gap to the 10th eigenvalue (Fig. 4a). This spectral gap is not present if the *fast* and *slow* rates are equal. The spectral gap indicates a well-conditioned invariant subspace. In fact, for a *fast/slow* ratio of 100 or 1000, the *crispness* as an indicator for the quality of the clustering gets close to one, meaning that the membership vectors are close to characteristic vectors (Fig. 4b). Here, we used the stationary density for normalizing the eigenvectors. Using the equiprobable density for normalization leads to slightly different values for the *crispness* (Fig. 7), favoring two over nine clusters, but if a clear spectral gap exists, the final cluster number should always be determined by this gap.

The clusters identified by cPCCA+ represent metastable sets, i.e. a dynamic trajectory spends a long time within these clusters and only rarely switches between them. This dynamic behavior is nicely illustrated by the coarse-grained transition probability matrix for a time interval of length *1/fast* = 0.001, which is diagonally dominant, i.e. the holding probabilities within the clusters are close to one (see Appendix A.3).

This high-quality clustering deteriorated for decreasing *fast/slow* ratio. For a ratio of 10, there is still a small spectral gap, and the *crispness* still peaks at 9 clusters, but only to a value of around 0.5, indicating that the membership vectors are no longer close to $\{0, 1\}$-vectors, making it difficult to uniquely assign states to clusters. Interestingly, this behavior can also be observed with other measures for the quality of a clustering, e.g. modularity (see Appendix A.4 Figs. 8 and 9).

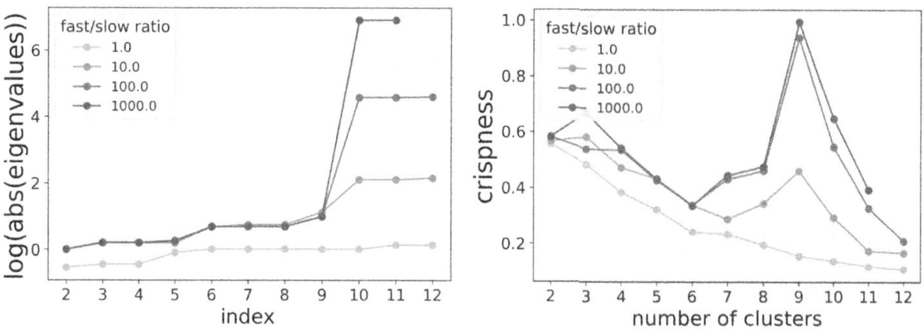

(a) eigenvalues

(b) crispness

Fig. 4. Model B. (a) The logarithm of the absolute value of the eigenvalues and (b) the cluster numbers suggested by cPCCA+ for different *fast/slow* ratios. Note that the first eigenvalue is zero, so it is not plotted here because its logarithm is not defined. For increasing *fast/slow* ratio, the spectral gap after 9 eigenvalues becomes larger; hence, the corresponding eigenspace is better conditioned.

Table 1 shows how the attractor states are assigned to the nine different clusters for *fast/slow* ratio equal to 100 or 1000. A threshold of $\chi_j(i) > 0.7$ was used to assign state i to cluster j. Only four states are identified as transition states (111000⋆, 111001⋆). They were assigned to two different clusters with probability values around 0.5 each. Figure 5 illustrates the results. The colors of the nodes indicate the node memberships to the phases defined in [7, Table III], whereas the clusters found by cPCCA+ are displayed by the dotted boxes numbered from 0 to 8. Edges in gray and black have slow and fast rates, respectively.

We can observe that states belonging to G1 are grouped into two clusters (clusters 3 and 4). These differ only in the fourth species (CycE) and have fast internal connections and slow transitions to external nodes. The slow update of UbcH10 and fast update of CycE and E2F are probably responsible for the

Table 1. Model B. Assignment of states to clusters for *fast/slow* ratio equal to 100 or 1000. Bold states are clustered equally in *Model A*. The numbers in parentheses next to the transition states are the indices of the clusters to which theses states belong.

cluster	G1	G1S	SG2	G2M	M	MG1
0					**0010000**	
1					**0010010**	
2				**0110000**		
3	**⋆0001⋆⋆**					
4	**⋆0011⋆⋆**					
5		**010010⋆** 0101⋆0⋆	**010000⋆** 110000⋆			
6					**10100⋆⋆**	**⋆0000⋆⋆**
7		**0100110** **0101⋆10**	**0100010** 0100011 110001⋆			
8				**0110010**		
trans				111000⋆ (5,6) 111001⋆ (6,7)		

genesis of clusters 5 and 7, which cover G1S and SG2, but contain states of both phases. Cluster 6 contains all of the states labeled as MG1, as well as four fast-connected states in M. The remaining four clusters consist of 4 isolated nodes, which are in fact connected to the remainder of the attractor only by slow transitions. Finally, the four transitions states are states in G2M. They are connected to other nodes by fast transitions, but cannot be assigned to a single cluster with sufficiently high probability.

The division of G1 into two clusters shows some interesting properties. All transitions from the prior phase MG1 reach only cluster 3. The dynamics can then progress in a few steps through G1 only traversing cluster 3 or additionally transitioning to cluster 4, possibly executing further secondary processes. In clusters 5 and 7 we see a grouping together of G1S and SG2 states that are connected with fast transitions. As in the case for clusters 3 and 4, we see here alternative scenarios of passing through the two phases.

3.3 Comparison Between Model A and Model B

We now turn our attention to *Model A*. We run the clustering algorithm on the state transition graph restricted to the attractor, which in this case includes 112 nodes, again using the stationary density for normalization of the eigenvectors. We use the same scenarios of *fast/slow* ratios considered for the previous model. Notably, we observe the same trends obtained with the previous model (Fig. 4b). In particular, the largest spectral gap and highest levels of crispness are obtained with 9 clusters for high *fast/slow* ratio (Appendix A.5, Figs. 10a and 10b). The

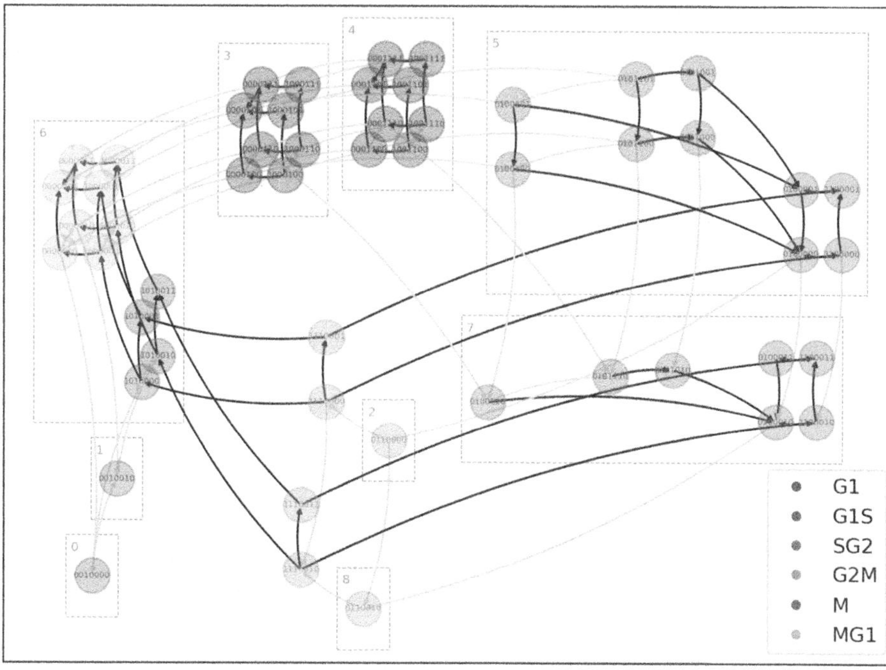

Fig. 5. Model B. Cyclic attractor. The 9 clusters obtained with a ratio equal to 100 or 1000 are represented by the dashed gray boxes.

composition of these 9 clusters is shown in Table 2, while the transition states, which are significantly more numerous than for *Model B*, are listed in Table 3. We can observe a certain degree of similarity between the 9 clusters of the two models. The main difference is that clusters 0, 1, 2, 6, and 8 in *Model A* contain additional states that are not attractor states in *Model B*, with cluster 6 in *Model A* being the most affected since it contains states that fall outside of the defined phases, as well as states from multiple phases. In addition, there are a few states which are transition states in one model whereas they are assigned to a cluster in the other model (for a detailed list of differences, see Appendix A.6). This cluster analysis offers another perspective for the comparison of the two models. It suggests that the simplified attractor proposed by *Model B* might capture the general structure of the 9 dynamical modules of the original attractor. It also indicates a high degree of uncertainty around the dynamical role of cluster 6 in *Model A* since it spans all the phases proposed in [7].

4 Conclusion

Extracting information from mathematical models is a crucial process, necessary for their validation, as well as for the exploration of their predictive utility. Logical models are created to summarize intricate networks of interactions, and, despite their apparent simplicity, can lead to dynamics that are quite complex and difficult to describe. Here we proposed an investigation of modular structures in asynchronous dynamics of Boolean networks, in particular in complex attractors. The objective is to identify regions of the state space that constitute "steps" in the dynamics that are somewhat separable from the rest of the process, not in an exact way—like strongly connected components—but with a certain margin for exceptions. Spectral approaches to clustering, and in particular for clustering of dynamics, are very popular because of their consolidated interpretation and ease of implementation. The algorithm we choose, cPCCA+, is particularly suited for Markov processes defined on directed graphs. We choose to analyze models of the cell cycle since these can naturally be studied for their ability to implement detectable steps that lead the cell to successful duplication. cPCCA+ proved useful in detecting structures in the asynchronous attractors we considered. These modules could then be compared to the division in phases proposed by modelers. In the examples we considered, we observed that the definition of transition rates is crucial for the emergence of modular structures. A preliminary exploration of other Boolean models, not discussed here, suggests that this might not be an isolated case: setting transition rates to uniform values often leads to dynamics that do not exhibit modular behavior. The incorporation of biological knowledge becomes, therefore, decisive. On the other hand, we saw in the examples discussed here that even coarse-grained information, such as the classification of rates into only two classes ("fast" and "slow"), might induce some detectable structures. In addition to the investigation of biological hypotheses, we think a systematic analysis of *fast/slow* rate assignments that can lead to the emergence of meaningful structures might be of interest, for the models we considered as well as in other scenarios exhibiting sustained cyclic behavior. Such analysis could, in turn, yield testable predictions about the fundamental steps in complex cyclic behavior. It should also be noted that the approach we presented is not restricted to the analysis of terminal strongly connected components, but can be applied without any change to investigate transient behavior. A comparison of results from different clustering methods for directed networks would, of course, also be beneficial, in particular with a view to establishing the robustness of the observed insights.

Acknowledgements. ET was funded by the Deutsche Forschungsgemeinschaft (DFG) under Germany's Excellence Strategy - The Berlin Mathematics Research Center MATH+ (EXC-2046/1, project ID 390685689). The work of MY and SR was funded by the Norwegian Research Council (NFR), project "Markov State Models for Cellular Phenotype Switching" (grant no. 324080).

Data Availability Statement. All data and figures presented in this paper can be reproduced with the Python code provided at Zenodo (10.5281/zenodo.12608344).

Disclosure of Interests. The authors have no competing interests to declare that are relevant to the content of this article.

A Appendix

A.1 Clustering Result for the Toy Model

Figure 6 shows the assignment of the eight states to the two clusters for different values of ε. As ε increases, the two clusters become less separated.

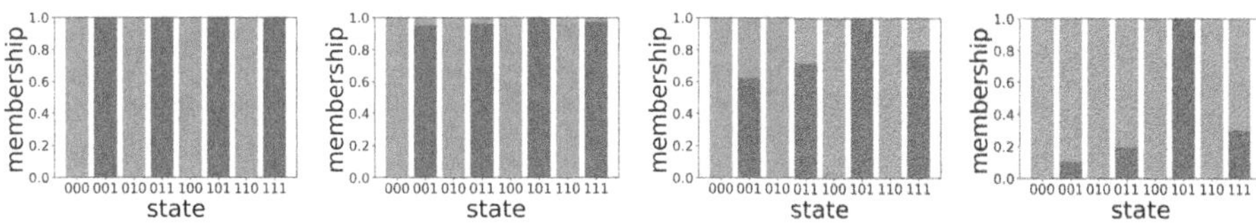

(a) $\varepsilon = 10^{-5}$ (b) $\varepsilon = 0.1$ (c) $\varepsilon = 1$ (d) $\varepsilon = 10$

Fig. 6. Toy model. Membership values for 2 clusters, for different values of ε.

A.2 Clustering for Model A and B with Equiprobable Density

Using the equiprobable density $w = (1, \ldots, 1)$ instead of the stationary density for normalizing the eigenvectors, the *crispness* indicates a possible partitioning into two clusters for both Model A and Model B (Fig. 7. However, this clustering has a low modularity (Fig. 8). In fact, for a *fast/slow* ratio of 1000, this clustering simply assigns the two states 0010000 and 0010010 into a separate cluster. This clustering, however, is not robust because of the lack of a spectral gap after the second eigenvalue.

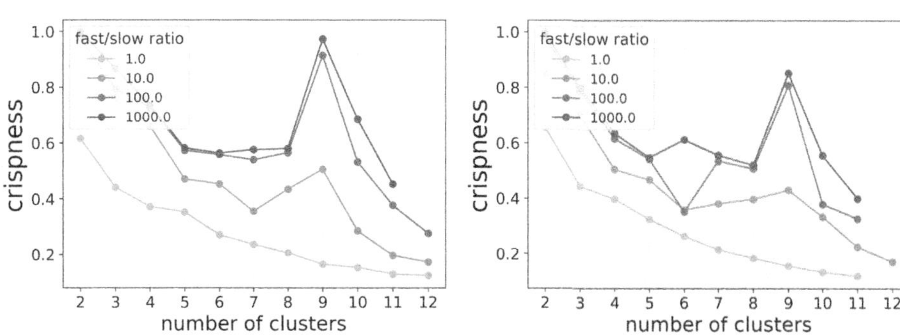

(a) Model B (b) Model A

Fig. 7. Crispness of the optimal membership vectors for different cluster numbers using the equiprobable density for the normalization of eigenvectors.

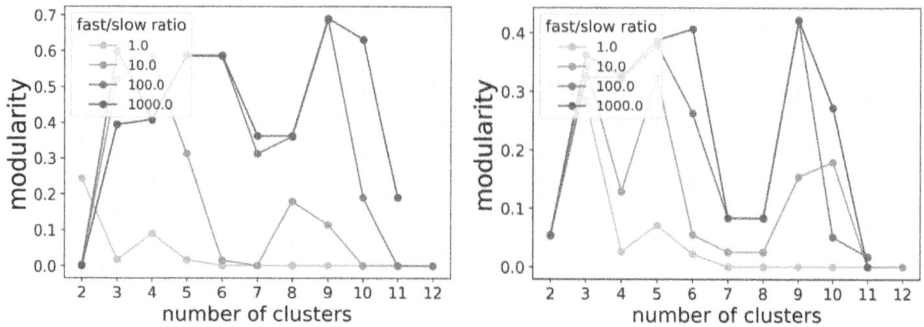

(a) Model B (b) Model A

Fig. 8. Modularity of the clusters obtained by cPCCA+ using the equiprobable density for the normalization of eigenvectors.

A.3 Markov State Model for Model B

The coarse-grained transition probability matrix for the model by Diop et al. [7] reads:

$$T_c(0.001) = \begin{pmatrix} 0.9980 & 0.0010 & 0.0000 & 0.0000 & 0.0000 & 0.0000 & 0.0010 & 0.0000 & 0.0000 \\ 0.0000 & 0.9990 & 0.0000 & 0.0000 & 0.0000 & 0.0000 & 0.0010 & 0.0000 & 0.0000 \\ 0.0000 & 0.0000 & 0.9980 & 0.0000 & 0.0000 & 0.0005 & 0.0005 & 0.0000 & 0.0010 \\ 0.0000 & 0.0000 & 0.0000 & 0.9980 & 0.0010 & 0.0010 & 0.0000 & 0.0000 & 0.0000 \\ 0.0000 & 0.0000 & 0.0000 & 0.0000 & 0.9990 & 0.0010 & 0.0000 & 0.0000 & 0.0000 \\ 0.0000 & 0.0000 & 0.0010 & 0.0000 & 0.0000 & 0.9980 & 0.0000 & 0.0010 & 0.0000 \\ 0.0000 & 0.0000 & 0.0000 & 0.0010 & 0.0000 & 0.0000 & 0.9990 & 0.0000 & 0.0000 \\ 0.0000 & 0.0000 & 0.0000 & 0.0000 & 0.0000 & 0.0000 & 0.0000 & 0.9990 & 0.0010 \\ 0.0000 & 0.0000 & 0.0000 & 0.0000 & 0.0000 & 0.0000 & 0.0005 & 0.0005 & 0.9990 \end{pmatrix}$$

A.4 Other Measures of Modularity

In addition to crispness which, in the context of fuzzy clustering, measures the degree to which nodes in a network are assigned exclusively to a single cluster [23], we test the goodness of the partitions identified by assigning each node to a single cluster (the one with the highest probability). In Fig. 9 we show the *modularity* of the identified partitions. This is a measure of the fraction of edges found inside clusters, minus the fraction of edges to be expected if these were distributed at random. It is a popular measure of strength for static network clustering [20]. The values we find confirm that the partition identified when setting the number of clusters to 9 is the best solution in both models. A *fast/slow* ratio of 100 is sufficient to obtain the highest of the calculated modularity values.

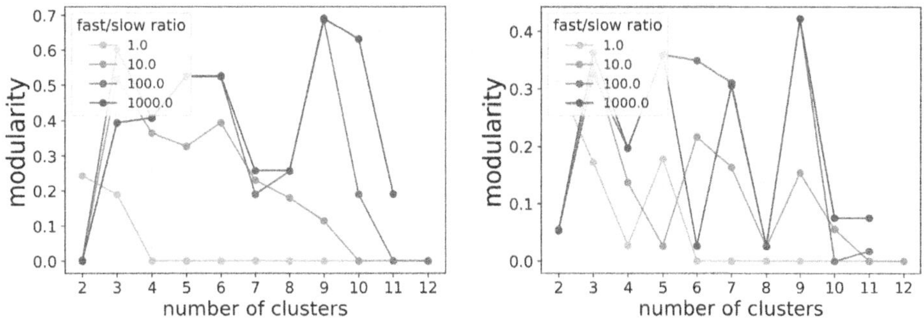

(a) Model B (b) Model A

Fig. 9. Modularity for the clusters obtained via cPCCA+ for the cyclic attractor of [7] (Model B) and of [10] (Model A).

A.5 Clustering of States in Model A

Figure 10a illustrates how the spectral gap after the first nine eigenvalues increases with increasing *fast/slow* ratio. Accordingly, crispness peaks at nine clusters (Fig. 10b). The assignment of states to these nine clusters is contained in Table 2, and transition states are listed in Table 3.

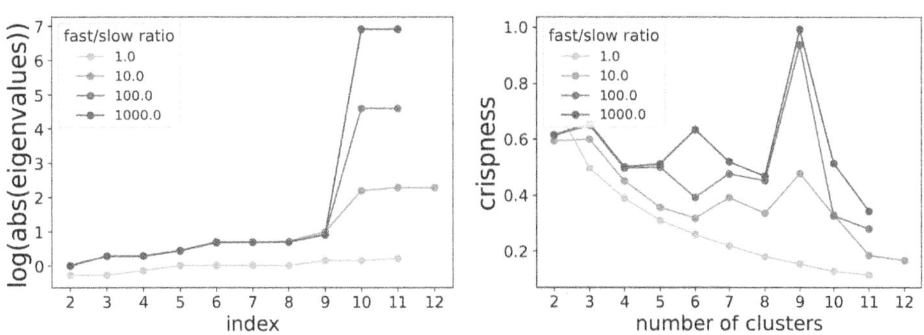

(a) eigenvalues (b) crispness

Fig. 10. Model A. (a) The logarithm of the absolute value of the eigenvalues and (b) the cluster numbers suggested by cPCCA+ for different *fast/slow* ratios. Note that the first eigenvalue is zero, so it is not plotted here because its logarithm is not defined. For increasing *fast/slow* ratio, the spectral gap after 9 eigenvalues becomes larger; hence, the corresponding eigenspace is better conditioned.

Table 2. Model A. Assignment of states to clusters (threshold 0.7). Bold states are clustered equally in *Model B*. Transition states are not listed.

cluster	G1	G1S	SG2	G2M	M	MG1	∅
0					**0010000**		0010100 0011⋆00
1					**0010010**		0010110 0011⋆10
2				**0110000**			0110100 0111⋆00
3	**⋆0001⋆⋆**						
4	**⋆0011⋆⋆**						
5		**010010⋆** **0101⋆0⋆**	**010000⋆**				
6	⋆0010⋆⋆	1101011	1100011	11100⋆⋆	**10100⋆⋆**	**⋆0000⋆⋆**	1⋆110⋆⋆
7		**0100110** **0101⋆10**	**0100010**				
8				**0110010**			0110110 0111⋆10

Table 3. Model A. Transition states (threshold 0.7) and the clusters to which they partially belong.

G1	G1S	SG2	G2M	M	MG1	∅
	0100111 (3,6,7) 0101011 (6,7) 0101111 (4,6,7) 110010⋆ (3,5,6) 110011⋆ (3,6,7) 110100⋆ (5,6) 1101010 (6,7) 110110⋆ (4,5,6) 110111⋆ (4,6,7)	0100011 (6,7) 1100000 (5,6)				10101⋆⋆ (3,6) 10111⋆⋆ (4,6) 111010⋆ (3,5,6) 111011⋆ (3,6,7) 111110⋆ (4,5,6) 111111⋆ (4,6,7)

A.6 Differences in the Clustering Between Model A and B

The following differences can be observed between the clusters in the two models:

- state 1100001 from cluster 5 and state 1100010 from cluster 7 in *Model B* are not contained in the attractor of *Model A*;
- state 1100000 from cluster 5 in *Model B* becomes a transition state between clusters 5 and 6 in *Model A*;
- state 1100011 is assigned to cluster 7 in *Model B*, and to cluster 6 in *Model A*;
- state 0100011 from cluster 7 in *Model B* becomes a transition state between clusters 6 and 7 in *Model A*;

– the four transition states 1110000 (5,6), 1110001 (5,6), 1110010 (6,7), and 1110011 (6,5) from *Model B* are reassigned to cluster 6 in *Model A*;

References

1. Albert, R., Robeva, R.: Chapter 4 – Signaling networks: asynchronous Boolean models. In: Robeva, R.S. (ed.) Algebraic and Discrete Mathematical Methods for Modern Biology, pp. 65–91. Academic Press, Boston (2015). https://doi.org/10.1016/B978-0-12-801213-0.00004-6
2. Beneš, N., Brim, L., Huvar, O., Pastva, S., Šafránek, D., Šmijáková, E.: AEON.py: python library for attractor analysis in asynchronous Boolean networks. Bioinformatics **38**(21), 4978–4980 (2022). https://doi.org/10.1093/bioinformatics/btac624
3. Cooper, G.M.: The Cell: A molecular approach, 2nd edn. Sinauer Associates, Sunderland (2000)
4. Davidich, M.I., Bornholdt, S.: Boolean network model predicts cell cycle sequence of fission yeast. PLoS ONE **3**(2), e1672 (2008). https://doi.org/10.1371/journal.pone.0001672
5. Deritei, D., Rozum, J., Ravasz Regan, E., Albert, R.: A feedback loop of conditionally stable circuits drives the cell cycle from checkpoint to checkpoint. Sci. Rep. **9**(1), 16430 (2019). https://doi.org/10.1038/s41598-019-52725-1
6. Deuflhard, P., Weber, M.: Robust Perron cluster analysis in conformation dynamics. Linear Algebra Appl. **398**, 161–184 (2005). https://doi.org/10.1016/j.laa.2004.10.026
7. Diop, O., Tourniel, L., Fromion, V.: Summarizing complex asynchronous Boolean attractors, application to the analysis of a mammalian cell cycle model. In: 2019 18th European Control Conference (ECC), pp. 1677–1682. IEEE (2019). https://doi.org/10.23919/ECC.2019.8795712
8. Estivill-Castro, V.: Why so many clustering algorithms: a position paper. SIGKDD Explor. **4**(1), 65–75 (2002). https://doi.org/10.1145/568574.568575
9. Fackeldey, K., Sikorski, A., Weber, M.: Spectral clustering for non-reversible Markov chains. Comput. Appl. Math. **37**(5), 6376–6391 (2018). https://doi.org/10.1007/s40314-018-0697-0
10. Fauré, A., Naldi, A., Chaouiya, C., Thieffry, D.: Dynamical analysis of a generic Boolean model for the control of the mammalian cell cycle. Bioinformatics **22**(14), e124–e131 (2006). https://doi.org/10.1093/bioinformatics/btl210
11. Frank, A.S., Sikorski, A., Röblitz, S.: Spectral clustering of Markov chain transition matrices with complex eigenvalues. J. Comput. Appl. Math. **444**, 115791 (2024). https://doi.org/10.1016/j.cam.2024.115791
12. Gillespie, D.T.: Exact stochastic simulation of coupled chemical reactions. J. Phys. Chem. **81**(25), 2340–2361 (1977). https://doi.org/10.1021/j100540a008
13. Husic, B.E., Pande, V.S.: Markov state models: from an art to a science. J. Am. Chem. Soc. **140**(7), 2386–2396 (2018). https://doi.org/10.1021/jacs.7b12191
14. Irons, D.J.: Logical analysis of the budding yeast cell cycle. J. Theor. Biol. **257**(4), 543–559 (2009). https://doi.org/10.1016/j.jtbi.2008.12.028
15. Kauffman, S.: Homeostasis and differentiation in random genetic control networks. Nature **224**(5215), 177–178 (1969). https://doi.org/10.1038/224177a0
16. Koltai, M., Noel, V., Zinovyev, A., Calzone, L., Barillot, E.: Exact solving and sensitivity analysis of stochastic continuous time Boolean models. BMC Bioinform. **21**, 1–22 (2020). https://doi.org/10.1186/s12859-020-03548-9

17. Meyer, C.D.: Sensitivity of the stationary distribution of a Markov chain. SIAM J. Matrix Anal. Appl. **15**(3), 715–728 (1994). https://doi.org/10.1137/S0895479892228900
18. Moler, C., Van Loan, C.: Nineteen dubious ways to compute the exponential of a matrix, twenty-five years later. SIAM Rev. **45**(1), 3–49 (2003). https://doi.org/10.1137/S00361445024180
19. Müssel, C., Hopfensitz, M., Kestler, H.A.: BoolNet-an R package for generation, reconstruction and analysis of Boolean networks. Bioinformatics **26**(10), 1378–1380 (2010). https://doi.org/10.1093/bioinformatics/btq124
20. Newman, M.E., Girvan, M.: Finding and evaluating community structure in networks. Phys. Rev. E **69**(2), 026113 (2004). https://doi.org/10.1103/PhysRevE.69.026113
21. Novák, B., Tyson, J.J.: A model for restriction point control of the mammalian cell cycle. J. Theor. Biol. **230**(4), 563–579 (2004). https://doi.org/10.1016/j.jtbi.2004.04.039
22. Privault, N.: Understanding Markov Chains: Examples and Applications. SUMS, 2nd edn, p. 358. Springer, Singapore (2018). https://doi.org/10.1007/978-981-13-0659-4
23. Röblitz, S., Weber, M.: Fuzzy spectral clustering by PCCA+: application to Markov state models and data classification. Adv. Data Anal. Classif. **7**, 147–179 (2013). https://doi.org/10.1007/s11634-013-0134-6
24. Stoll, G., et al.: MaBoSS 2.0: an environment for stochastic Boolean modeling. Bioinformatics **33**(14), 2226–2228 (2017). https://doi.org/10.1093/bioinformatics/btx123
25. Stoll, G., Viara, E., Barillot, E., Calzone, L.: Continuous time Boolean modeling for biological signaling: application of Gillespie algorithm. BMC Syst. Biol. **6**(1), 1–18 (2012). https://doi.org/10.1186/1752-0509-6-116
26. Thakar, J.: Pillars of biology: Boolean modeling of gene-regulatory networks. J. Theor. Biol. **578**, 111682 (2023). https://doi.org/10.1016/j.jtbi.2023.111682
27. Thomas, R.: Regulatory networks seen as asynchronous automata: a logical description. J. Theor. Biol. **153**(1), 1–23 (1991). https://doi.org/10.1016/S0022-5193(05)80350-9
28. Trinh, V.G., Hiraishi, K., Benhamou, B.: Computing attractors of large-scale asynchronous Boolean networks using minimal trap spaces. In: BCB 2022: Proceedings of the 13th ACM International Conference on Bioinformatics, Computational Biology and Health Informatics, pp. 1–10 (2022). https://doi.org/10.1145/3535508.3545520
29. Von Luxburg, U.: A tutorial on spectral clustering. Stat. Comput. **17**, 395–416 (2007). https://doi.org/10.1007/s11222-007-9033-z
30. Weber, M., Rungsarityotin, W., Schliep, A.: Perron cluster analysis and its connection to graph partitioning for noisy data. Technical Report 04-39, ZIB (2004). https://webdoc.sub.gwdg.de/ebook/serien/ah/reports/zib/zib2004/paperweb/reports/ZR-04-39.pdf. Accessed 24 June 2024
31. Wei, J., Kuo, J.C.: Lumping analysis in monomolecular reaction systems. Analysis of the exactly lumpable system. Ind. Eng. Chem. Fundam. **8**(1), 114–123 (1969). https://doi.org/10.1021/i160029a019

Author Index

R. Gori et al. (Eds.): CMSB 2024, LNBI 14971, pp. 247–248, 2024.
https://doi.org/10.1007/978-3-031-71671-3

The manufacturer's authorised representative in the EU is Springer Nature Customer Service Centre GmbH, Europaplatz 3, 69115 Heidelberg, Germany. If you have any concerns regarding our products, please contact ProductSafety@springernature.com

Printed and bound by CPI Group (UK) Ltd, Croydon, CR0 4YY
28/09/2026
02224082-0002